ENGINEERING SIMULATION

USING

SMALL SCIENTIFIC COMPUTERS

Prentice-Hall
Series in Automatic Computation

MARTIN, *Security, Accuracy, and Privacy in Computer Systems*
MARTIN, *Systems Analysis for Data Transmission*
MARTIN, *Telecommunications and the Computer*
MARTIN, *Teleprocessing Network Organization*
MARTIN AND NORMAN, *The Computerized Society*
MCKEEMAN, et al., *A Compiler Generator*
MEYERS, *Time-Sharing Computation in the Social Sciences*
MINSKY, *Computation: Finite and Infinite Machines*
NIEVERGELT, et al., *Computer Approaches to Mathematical Problems*
PLANE AND MCMILLAN, *Discrete Optimization:*
    Integer Programming and Network Analysis for Management Decisions
POLIVKA AND PAKIN, *APL: The Language and Its Usage*
PRITSKER AND KIVIAT, *Simulation with GASP II:*
    A FORTRAN-based Simulation Language
PYLYSHYN, ed., *Perspectives on the Computer Revolution*
RICH, *Internal Sorting Methods Illustrated with PL/1 Programs*
RUDD, *Assembly Language Programming and the IBM 360 and 370 Computers*
SACKMAN AND CITRENBAUM, eds., *On-Line Planning:*
    Towards Creative Problem-Solving
SALTON, ed., *The SMART Retrieval System:*
    Experiments in Automatic Document Processing
SAMMET, *Programming Languages: History and Fundamentals*
SCHAEFER, *A Mathematical Theory of Global Program Optimization*
SCHULTZ, *Spline Analysis*
SCHWARZ, et al., *Numerical Analysis of Symmetric Matrices*
SHAH, *Engineering Simulation Using Small Scientific Computers*
SHAW, *The Logical Design of Operating Systems*
SHERMAN, *Techniques in Computer Programming*
SIMON AND SIKLOSSY, eds., *Representation and Meaning:*
    Experiments with Information Processing Systems
STERBENZ, *Floating-Point Computation*
STOUTEMYER, *PL/1 Programming for Engineering and Science*
STRANG AND FIX, *An Analysis of the Finite Element Method*
STROUD, *Approximate Calculation of Multiple Integrals*
TANENBAUM, *Structured Computer Organization*
TAVISS, ed., *The Computer Impact*
UHR, *Pattern Recognition, Learning, and Thought:*
    Computer-Programmed Models of Higher Mental Processes
VAN TASSEL, *Computer Security Management*
VARGA, *Matrix Iterative Analysis*
WAITE, *Implementing Software for Non-Numeric Application*
WILKINSON, *Rounding Errors in Algebraic Processes*
WIRTH, *Algorithms + Data Structures = Programs*
WIRTH, *Systematic Programming: An Introduction*
YEH, ed., *Applied Computation Theory: Analysis, Design, Modeling*

To my wife Margarita and my daughters Yasmin and Sheila.

ENGINEERING SIMULATION
USING
SMALL SCIENTIFIC COMPUTERS

MANESH J. SHAH

Senior Industry Analyst
International Business Machines Corporation
General Systems Division
Menlo Park, California

PRENTICE-HALL, INC.

ENGLEWOOD CLIFFS, NEW JERSEY

Library of Congress Cataloging in Publication Data

SHAH, MANESH J. date
    Engineering simulation using small scientific
computers.

    (Prentice-Hall series in automatic computation)
    Bibliography
    Includes index.
    1. Engineering–Data processing. 2. Engineering
–Mathematical models. I. Title.
TA345.S5   620'.001'84   75-28088
ISBN   0-13-279422-5

10   9   8   7   6   5   4   3   2   1

Printed in the United States of America

PRENTICE-HALL INTERNATIONAL, INC., *London*
PRENTICE-HALL OF AUSTRALIA PTY. LIMITED, *Sydney*
PRENTICE-HALL OF CANADA, LTD., *Toronto*
PRENTICE-HALL OF INDIA PRIVATE LIMITED, *New Delhi*
PRENTICE-HALL OF JAPAN, INC., *Tokyo*
PRENTICE-HALL OF SOUTHEAST ASIA PTE. LTD., *Singapore*

CONTENTS

4

DSL LANGUAGE RULES   44

5

APPLICATION OF SIMULATION   68

6

PROBLEMS IN THE FIELD OF CHEMISTRY AND CHEMICAL ENGINEERING   125

7 SETUP AND EXECUTION OF THE DSL PROGRAM   172

8 DSL ADVANCED LANGUAGE FEATURES   184

12

DSL PROGRAM DESCRIPTION   230

13

AIDS TO PROGRAM MODIFICATION   248

PREFACE

This book is a result of several years of simulation experience the author has gathered during his career at IBM in the fields of engineering, science, and, particularly, process control. We have seen the computing capabilities of the available machines grow at a very rapid pace in the past decade, while the so-called price/performance ratio has plummeted to unbelievably low figures, compared with those figures of only fifteen years ago. No doubt, all this phenomenal success is a result of the innovative technical revolution in electronics, with smaller and smaller circuits being packed to compact large memories which perform arithmetic manipulations. The basic functions of all digital computers are to move and manipulate data, perform arithmetic operations, and make logical decisions to move from one operation to another based on the result of an operation. However, the solution of an engineering or scientific problem requires a great number of these steps repeatedly. Only fifteen years ago this solution required a substantial insight into numerical techniques as well as programming methodology for handling even simple problems with the computers of those days.

We can compare the convenience offered by the present day small and large scientific computers with the convenience offered by slide rules in the early days when they were introduced; however the digital computers provide a true problem solution on a much larger scale, with visual simulation capability when graphics are employed on the computer system. The complexity of the engineering problems solvable with the 'electronic sliderule' (if I may call it) is several orders of magnitude greater than that of the problems solved manually using slide rules.

I remember the days in 1957 when I had to learn to use an IBM 650 electronic computer to solve the problem of a nonNewtonian liquid flow past a flat plate described by a single third order nonlinear differential equation. It took me over two months to formulate, flowchart, write the program in machine language commands with the help of an experienced programmer, and solve the equation for about four or five case studies. Each case study took more than an hour's computer time to complete, which was an expensive proposition in those days.

A more generalized and more difficult case of the same problem is illustrated in this book, which the reader can write, program, and prepare for computer solution in less than one hour and complete execution on an IBM 1130 or an equivalent machine in less than five minutes for multiple cases. This dramatic productivity change is the result of two basic improvements in computers. One is the widely heralded price/performance computer hardware improvement. Another, more important improvement from the application engineer's standpoint, is the ability provided to program in his own language, using the automatic built-in capability of the computer to search and provide the various standard engineering mathematical functions he needs to solve his problem. At best, a large amount of work has been done to build a library of programs, which provide this function. It is this capability of problem solution that the book is intended to address in the hope of attracting and converting those engineers and scientists in industry and universities who remain to be convinced that indeed, digital computers, small and large, provide an excellent tool for rapid problem solution in engineering and science to improve their productivity.

The book is intended for use by practical engineers in industry who wish to use the engineering simulation language via self-study, as well as students in senior and graduate years who are taking simulation and computer science courses, and need hands-on training in engineering problem solution with small and large scientific computers. The faculty members in engineering may find this book as a useful text to supplement their class notes in teaching simulation. Of course, as the book clearly implies, the programs DSL and CSMP are available only for the IBM 1130, IBM 1800, and IBM Systems 360/370. The exercise described in Chapter 11 in converting to IBM System/3 can be followed to make the program operational on any scientific computer with the required attributes described. I have no doubt that university faculty members who have access to computers can find an enterprising graduate student willing to convert and make the system operational on their available system in a summer three month period.

The book is organized to provide as many examples in the various fields as possible. Many more examples were left out since they provided only redundancy in the use of the language even though they dealt with systems in totally different fields. After introductory Chapter 1, Chapter 2 provides a short outline of the IBM 1130/1800 systems and IBM System/3 and System/7. The main purpose of this chapter is to familiarize the reader with the many variations in input (card reader, keyboards) and output (printer, console typewriter, card punch, and disk) devices each machine has. These devices must be recognized in the building of the program for the system to be utilized for simulation. The reader should consult the appropriate references for a more comprehensive description of the various hardware systems.

Chapters 3 and 4 provide some basic information on the simulation language to be used with the programs described in the book. The rules to be followed when using the language statements together with examples are shown in these chapters.

Chapters 5 and 6 provide the heart of the book. These chapters show the simulation examples in the various fields. Sufficient variations are suggested in some of the more complex problems in which the student may develop a better feel for the language and the program. The references quoted in Chapters 5 and 6 may be consulted for more examples which will be useful in industry as well as universities.

The remaining chapters deal with program operation, error diagnostics, and program preparation on the various machines. These chapters will be especially of interest to those responsible for setting up the simulation system. A listing of the simulation program DSL is given for those who for some reason are unable to obtain the program from the IBM program information library.

Many people have been responsible for sustaining my effort in producing this book over a span of three years. Among those responsible in providing moral support are Mr. Sterling Weaver, Dr. Donald R. Kuehn, and Mr. Don Sallan. I am indebted to Don Wyman for his assistance in program conversion to System/3 and Wai Mun Syn for providing some examples. These two gentlemen were largely responsible for authorship of the DSL program. Examples in Chapter 5, Sections 5.1.4 and 5.1.5 are reprinted by permission of the November 1967 edition of *Simulation,* pp. 237-247 (copyright Simulation Councils Inc., 1967). Examples in Chapter 6, Sections 6.3, 6.4, 6.5, and 6.6 are reprinted by permission of *Digital Computations for Chemical Engineers* by Leon Lapidus, pp. 119-127, McGraw Hill, 1962.

The International Business Machines Corporation not only permitted me to use some of the program material but also gave me manuscript preparation assistance for which I am deeply indebted. Finally, without the patience of my wife Margarita and my children Yasmin and Sheila, I could not have produced this book.

MANESH J. SHAH

1 COMPUTER SIMULATION

1.0 Introduction

In the past twenty years, digital and analog computers have brought about a major revolution in the field of science and engineering. The computers have provided a tool for the solution of scientific problems that can be formulated mathematically. The formulation of complex engineering and scientific problems employs, in most cases, differential or integral equations. Generally, these equations, except for rather simple cases, are nonlinear and therefore create difficulties in the use of analytical methods for their solution. In many cases, the boundary conditions for the equations are not available at the starting point of solution due to the physical nature of the system under investigation. A large number of the physical systems under investigation, when formulated mathematically, yield nonlinear partial differential equations, which are even more difficult to solve than ordinary differential equations. The approach for the analytical solution of these differential equations then has been to make simplifying assumptions to obtain manageable equations. Asymptotic solutions to examine the system behavior under one or more extreme conditions is one such example. The other answer is to use a form of series expansion to obtain the solution of the differential equation, a procedure which is laborious and time consuming. The multipoint boundary conditions require repeated solution of the differential equation until the boundary conditions at multiple points are met. The analytical solution of these equations then becomes an even more tedious task.

1.1 Analog and Digital Computers

The introduction of analog computers in the early 1950s led to their use in simulating electrical networks, servomechanisms, and physical systems which could be described in terms of resistance-capacitance (RC) networks. As their application expanded in examination of control systems and space/missile applications, the analog computers were expanded to large systems with multiple amplifiers and circuits. In the mid-1950s, digital computers began to appear and offered an alternative method of solution for scientific and engineering problems. In the present days of credit card society, the commercial use of digital computers far exceeds computer use in the scientific area, but as the digital computer hardware has become more and more attractively priced with continuously increased performance and increasingly large central and bulk storage, even the most complex large scientific problems seem within reach of solution with these computers. The digital computers, in addition, provide speed, precision, and relatively fast access to information from mass, bulk storage, a tremendous tool for problem solution in science and engineering.

1.2 Steps Required for Computer Usage

Unfortunately, the use of both the analog and digital computers requires a considerable amount of learning and setup time for a scientist before he can attempt the solution of his problem. In the case of an analog computer he must set up a network of resistors, integrators, differentiators, adders, etc., to simulate either his physical process or the differential equations which describe his system under study. For the digital computer he must program a sequence of commands to the computer to perform the solution of his problem. As the number and size of the computers have proliferated even in the case of one manufacturer, it has become increasingly difficult for a scientist not only to choose a particular computer for problem solution but also to learn the language and idiosyncrasies of each machine and its usage. While the introduction of high-level languages such as FORTRAN, ALGOL, and PL/1 has eased the problem to some extent, there still remains the need for a computer language which would enable a digital computer to perform like an analog machine, thereby capitalizing on the latter's attractive block-oriented approach to problem solution and at the same time providing powerful algebraic capabilities for those who want to handle complex systems of differential equations directly, bypassing the block diagram setup. In addition, the language needs to relieve the engineer from the myriad programming details which usually accompany digital computers. This can be achieved only by providing features such as a library of basic functions for modeling the components of a physical system, comprehensive prepackaged input/output facilities for simplified data handling, and a choice of integration methods for problems of varying complexity. The two-point boundary-value problems keep haunting the engineer, who has to play a trial-and-error game on the computer and try to guess his missing initial conditions to satisfy the known boundary conditions at a finite point along the solution.

1.3 Tools to Run Computers

What the engineer or the scientist needs then is a tool with which he can communicate his engineering problem in an engineering or scientific language and can observe the processes of his solution not only in terms of printed results but also, whenever desired, in terms of one or more graphic outputs, with the ability to plot any one variable against another. He can thus interpret his results directly in terms of the physical system under investigation. He also does not need to be bogged down with details of programming, searching the various program libraries to find an adequate program which will perform the task of integration or defining a functional block. He also would prefer not to be tied down by a procedural language that will force him to define each variable to be initialized before being used in the following statement. Errors in this procedure alone can cause many hours of debugging a complex program.

A further requirement for a practical engineer, scientist, or student is that he should be able to communicate with the digital computer during the progress of the problem solution. Thus, he can observe the results and can abort the computer run, choose another parameter or boundary condition, and restart. This type of man-machine interaction may be obtained on a computer used in a time-shared environment or on a small scientific computer with multiple data switches.

1.4 Computers and Tools
Described in this Book

We intend to address this book to the group of engineers and scientists who are not expert programmers or computer scientists but are looking for a simplified, nonprocedural, engineering-oriented language which provides a great deal of user interaction capability and a large library of programs built into the system. As the problem solver progresses in his use of the language, he can then expand the built-in library by adding his own programs to it.

We shall briefly describe the historic evolution of simulation languages presently available and then quickly explore a specific engineering simulation language on a small scientific computer. We shall show the use of the digital simulation language (DSL) on the IBM 1130, IBM 1800, IBM System/7, and IBM System/3 with examples in the various fields of science and engineering, emphasizing particularly the capabilities of man-machine interaction in solving the engineering problems. For those interested in using this language on small computers other than the IBM systems discussed, Appendices A, B, and C are included to show where the differences in the source language form ASA FORTRAN are and what the user may do to modify the simulator for his system.

1.5   Evolution of Simulation Languages

One of the most important groups of programs for the simulation of continuous systems is that of the digital-analog simulators, i.e., digital programs which simulate the elements and organization of the analog computer. The pioneering effort in this area has been credited to R. G. Selfridge in 1955 (29).* Among the groups of digital analog simulators, PACTOLAS, MIDAS, MIMIC, CSMP, and DSL/90 have had the largest literature exposure.

The languages employing analog computer simulation started with the conventional analog block diagram of a system evolved by breakdown of the mathematical system model into its component parts and functional blocks. The blocks have one-to-one correspondence with analog computing elements such as integrators, summers, and limiters and usually appear as subroutines or subprograms within the simulation program. To use the simulation language, the programming of the engineering problem sometimes involves interconnecting the functional blocks by a sequence of connection statements according to some format of rules laid down by the language. Thus, the interconnection of the blocks is akin to the wiring of the patchboard on an analog computer. The use of the digital computer provides the features of the analog computer with these languages and the superior precision and reliability of the digital computer.

PACTOLAS was the first digital-analog simulator implemented on a small digital computer, viz., IBM 1620, that provided some degree of man-machine interaction during the solution of the problem. Subsequently, the program CSMP-1130 (16) evolved to further improve this interaction capability and take advantage of the superior performance of the IBM 1130.

*The references (their numbers appearing in parentheses) are given in Appendix D.

1.6   With and Without Analog
Block Diagrams

While there are many problems in the area of electrical networks and other similar areas of investigation where the preferable route may be to solve the problem after setting up an analog block diagram to describe the system, there are many engineers and scientists who would choose to proceed with the solution directly from the differential equations describing the physical system under investigation. It is clear that in these cases many of the interconnections of the system are provided by simple mathematical formulas which can be described by a language such as FORTRAN.

1.6.1   An Example Without a Block Diagram

To illustrate this point, let us consider the problem of boundary-layer flow of a non-Newtonian liquid past a wedge, which reduces to the nonlinear differential equation

$$F''' + FF'' = B(F'^2 - 1) \tag{1.1}$$

with the boundary conditions

$$\begin{aligned} F' = F = 0 & \quad \text{at } X = 0 \\ F'' = 0 & \quad \text{at } X \to \infty \end{aligned} \tag{1.2}$$

It is clear that the following four statements describe the various relationships needed for computer simulation:

$$F''' = -FF'' + B(F'^2 - 1) \tag{1.3}$$

$$F'' = \int_0^L F''' \, dx \tag{1.4}$$

$$F' = \int_0^L F'' \, dx \tag{1.5}$$

$$F = \int_0^L F' \, dx \tag{1.6}$$

A language which takes the four statement equations (1.3)-(1.6) with the boundary conditions (1.2) is much more preferable in this case than the route of constructing an analog block diagram to solve the fluid flow problem.

Another example where both the analog block diagram and the differential equation may be of equal interest is the examination of a mass-spring damper system, shown in Fig. 1.1. If the mass is displaced from its rest position and then released, it will oscillate until the energy is dissipated by the damper. The purpose of the simulation may be to analyze the effects of different spring constants on the motion of the mass.

The equation describing the motion of the mass is given by

$$MX'' + BX' + KX = 0 \tag{1.7}$$

with the boundary conditions $X' = 0$ and $X = A$ at $t = 0$, where t is the independent variable.

The statements describing the various relationships for this system are

$$X'' = \frac{-1}{M}(BX' + KX) \tag{1.8}$$

$$X' = \int_0^T X'' \, dt \tag{1.9}$$

$$X = \int_0^T X' \, dt \tag{1.10}$$

For those familiar with the analog computer concepts, Fig. 1.2 shows the equivalent analog block diagram representing the mass-spring damper system. Depending on the user's background and choice, he may wish to pursue his

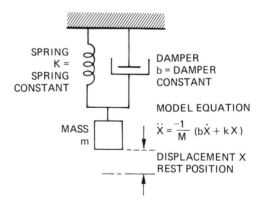

Figure 1.1   Mass spring damper system.

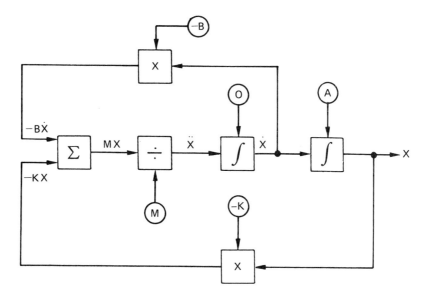

Figure 1.2   Analog block diagram for the mass spring damper system.

problem solution directly from differential equations, shown above, or to proceed via the analog diagram of Fig. 1.2.

Another simulation language, DSL 1130/1800 (17) to which this book is primarily devoted, is, on the other hand, suited to problems requiring both direct solution of differential equations and solution through the use of analog block diagrams.

In Chapter 5, we shall show how a more general case of Eq. (1.1) can be solved with only seven statements of equations using DSL without constructing the analog block diagram.

1.7   Simulation Languages on
IBM Computers

In 1966, a simulation language which provided the capabilities of analog computer simulation with a library of functional blocks as well as the facility of direct differential equation solution was developed for a large computer such as an IBM 7090 and offered as DSL/90. In addition, the graphic capabilities for plots of variables enhanced the features of this simulation language. This program was available on the IBM 7090/7094 and, subsequently, on an IBM 7040. It was later modified for the IBM System/360 and was made available under the name of Continuous System Model Program 360 (CSMP/360) (15). With some further modification for real-time communication, it was converted to run on the IBM

360/44 with special graphic display on oscilloscopes. The latest version of this program on the IBM 360/370 is CSMP III (15), available as a program product for a license fee. It has further enhancements over CSMP II, especially in providing interactive graphics using the IBM 2260 video display system.

This book is devoted to the use of this language, and we shall discuss in great depth its use on two small scientific computers, the IBM 1130 and the IBM 1800; a commercial/scientific computer, the IBM System/3; and a small sensor-based computer, the IBM System/7. The DSL 1800/1130 language has been available through the IBM Program Library since 1969. Although we shall describe in some detail the general usage of the language, the primary aim of the author is to illustrate the various language features with as many examples as possible so that the problem solver can immediately get a feel for the use of the language. For the sophisticated programmer, we shall provide ideas for extension, additions, and modifications to the program to suit his applications.

Although the user who has an access to a large system such as an IBM 360/370 may not be interested in the descriptions and use of the small computers, he will still find Chapters 3 and 4 useful. He will not have the burden of learning about installation on the small computer; however, he will not have the interactive capability of the small system.

1.8 Text Organization

Chapter 2 is devoted to an introduction to the IBM 1130, 1800, System/7, and System/3, for those who are unfamiliar with these computers. A minimum of essential information required for the operation of these systems using their data switch feature and the logical unit number information required to build the DSL system is given in Chapter 2. The chapter may be skipped by readers already familiar with these systems or those who will use the simulation only on large IBM systems using CSMP on the 360/370 (15).

Chapter 3 is devoted to the features of the simulation language required for small scientific computers such as an IBM 1130/1800 or System/3, following which a brief description of the two languages is given. The first language, CSMP 1130, is especially suited to simulation of physical systems, which are easier to understand from an analog block diagram. On the other hand, readers wishing a greater depth of information on these systems are directed to other publications for their description (16, 17).

In Chapter 4 we shall describe the rules of DSL language which the user must follow to be able to arrive at problem solution. These rules, though somewhat dull to read, are illustrated with sufficient examples to clarify their usage. In Chapters 5 and 6, the use of the language is illustrated with solutions of problems in the various engineering fields.

In Chapter 7 we shall describe the procedures for the setting up and execution of the translation and simulation steps required for the IBM 1130 and

1800, including the user interaction with the data switches. Error messages printed by the DSL program are also explained in this chapter. The users of the 360/370 systems (CSMP II and CSMP III) must refer to their appropriate reference manuals for this information. In Chapter 8 we shall discuss the advanced language features of DSL 1130/1800 and provide suggestions for extensions to the reader who wishes to modify the DSL program by himself.

In Chapter 9 we shall give a brief overview of the mathematical basis for some of the important subroutines and functions in the DSL library. Chapters 10, 11, 12, and 13 and the appendices are primarily meant for the reader who will have the responsibility of installing the program on the IBM 1130, 1800, System/3, and System/7. The last chapters (10–13) and the appendices will also help the enterprising user who wishes to convert the DSL program to run on any other small scientific computer which meets the specific criteria described in these chapters.

It is pointed out that while the program used in this book is for specific small IBM computers, the principles used here in simulation using a small scientific computer are valid for other small scientific systems, provided that a bulk storage is available for mass memory, provided that a graphic device as well as a printer are attached to the main memory, and, most important of all, provided that the system has programmable data switches available to the user. In addition, the user must be able to compile the DSL programs which are in FORTRAN and store them on the system hardware which is chosen. The system hardware discussion in the next chapter gives the reader an idea of the various input/output units available and what units are necessary for him to build the DSL system for his particular system configuration.

2 COMPUTER SYSTEMS

2.0 A Brief Note on the Computer Systems

In this chapter we shall briefly discuss some of the small IBM computer systems on which the two programs discussed in this book operate. As indicated in Chapter 1, two other programs—the Continuous System Modeling Program, versions II and III (CSMP II and CSMP III)—are also available for simulation on the IBM System 360/370, and although they do not provide the user interactive capability of the small system via system control by the user, the larger systems provide larger problem solution capability, because of larger storage size availability. The larger systems (IBM 360/370, IBM 7090) are not discussed here simply because the user need not necessarily know the system operation details, although he does require knowledge of the job control language (JCL) for the program use. The JCL may be found in the referenced program materials when using the respective programs.

The use of the simulation language on the smaller systems such as the IBM 1130, 1800, System/3, and System/7 requires some familiarity not only with the hardware but also with peripherals (input and output devices) for their operation. In addition, the logical unit numbers (device addresses) for these devices on each system are different, and their proper referral is required in the program preparation. The user should at least become familiar with the small system he is to operate the simulation language on, by reading the brief exposé here and should consult the referenced IBM manuals for further details.

Those enterpreneurs wishing to convert the program to operate on other systems should consult the chapters on program preparation and become familiar with the operating systems and FORTRAN of the system under consideration prior to embarking on the project.

2.1 Description of the IBM 1130 and 1800

Figures 2.1 and 2.2 show the IBM 1130 and IBM 1800 Systems, respectively. While the 1130 is primarily a computer for *batch* processing of programs, read into the system memory by means of a card reader or from the bulk disk memory, the 1800 performs batch processing as well as execution of real-time programs. The real-time program calculations may be a result of an interrupt of the computer processing because of time-scheduled tasks (computer core clocks are used to provide software timers) or may result because an interrupt has occurred from an external hardware, for example, as a result of a thermocouple or a pressure transducer output value exceeding its set limit. The latter function assures an immediate action to serve the plant control function. Due to the time-shared facility on the 1800, the batch-processed programs will execute concurrently with the plant control programs and may therefore take a longer time for execution of simulation and translation on the 1800 than on an 1130 with an equivalent machine cycle time.

In this book we shall not deal with the real-time function of the 1800 since simulation will operate utilizing its time-sharing capability.

2.1.1 *Peripheral Devices*

In Figs. 2.3 and 2.4, we show a schematic of the 1130 and 1800 computer systems. These systems have a central processing unit (CPU) with arithmetic capability and central storage of anywhere from 8192 to 32,768 words (65,536 words for the 1800). The 1800 CPU has, in addition, an external timer clock which is used for periodic interrupt calculations. The CPU also has access to the large bulk memory on one or more removable disks, depending on the number attached to the hardware. A card reader is used to enter program and data via punched cards, whereas the output of the computer results may be obtained from a line printer, a typewriter (console typewriter on the 1130), an X-Y plotter (IBM 1627 Plotter), or a storage oscilloscope. In addition, extremely useful features of the machines are the input keyboards (console keyboard on the 1130 and one or more typewriter keyboards such as 1816 on the 1800 system) and a set of 16 data switches on the two machines. As we shall see later, these data switches permit interruption of a simulation run to perform abortion, restart, or continuation of a simulation run after a new parameter specification with the use of the typewriter keyboard.

The card reader is also used for output of programs in the form of punched cards. As we shall discuss in subsequent chapters, the digital simulation

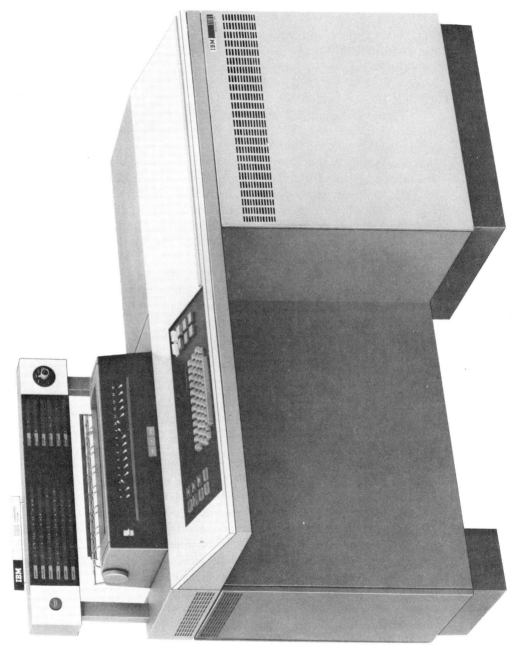

Figure 2.1   The 1130 system. (Robert K. Louden & George Ledin, Jr., *Programming the IBM 1130*, 2nd ed., © 1972. By permission of Prentice-Hall, Inc., Englewood Cliffs, New Jersey.)

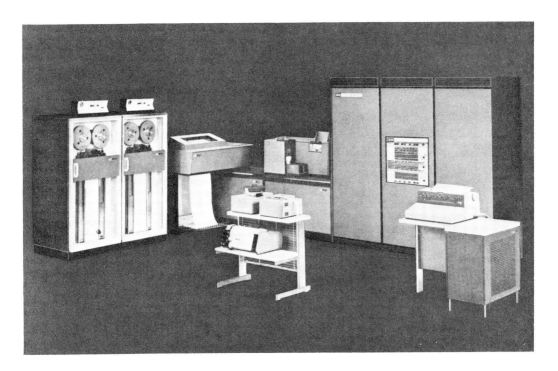

Figure 2.2   The 1800.

language (DSL/1800) is a two-pass computer program. In the first pass, the language translator produces output in the form of punched cards, which will then become the input to the FORTRAN compiler of the systems. Although this inconvenience has to be imposed on the user because of the limited storage of the CPU, it has the advantage that if the user wishes to make alterations of parameters, constants, and boundary conditions of a run after language translation, these alterations can be made in the translated deck, thus eliminating one pass of language translation for this purpose.

On the 1130 system, two types of line printers are available, namely, the IBM 1132 printer and the IBM 1403 printer (300 lines per minute and 600 lines per minute). In addition, the system may have an IBM 1442 card read and punch device, or it may have an IBM 2501 card reader and 1442 model 5 card punch device.

If the 1130 system has more than one disk, housed in enclosures of the CPU, the additional disks (up to a total of five) are in a separate housing. The expanded capability of the 1130 for the 1403 printer, the 2501 reader, and multiple disks requires a multiplex enclosure 1133. The 1130 system may also have a 2311 disk system as a bulk storage unit rather than 2305 multiple disk units.

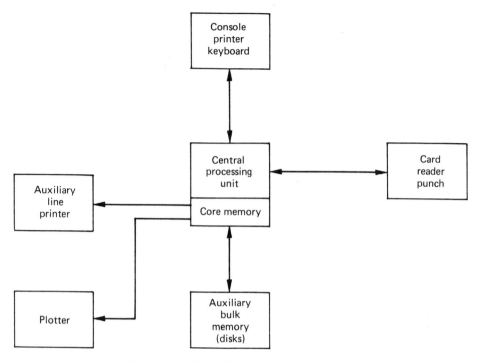

Figure 2.3. Schematic of the 1130 component units.

In Fig. 2.5 additional peripherals for the 1130 are shown. The IBM 1800, on the other hand, has only one printer 1443 and one model card reader. It may have multiple typewriters. The 1800 can have up to three 1810 disks, or it may have multiple 2311 disk drives.

The discussion about peripherals is presented here to point out the availability of various combinations of input and output units, which create the problem of specifying logical unit numbers in input/output operation for the digital simulation language as well as for the user. The user must determine what input/output units are available for his system so that their logical unit numbers may be specified to the DSL translator. These numbers are fixed for the 1130 system and are shown in Table 2.1. For the 1800 system these numbers are chosen by the system programmer at the time of the system executive program generation and hence may vary from user to user even for identical 1800 systems. In Table 2.2, the conventional logical unit numbers for the 1800 systems are given. The information in this table is used for specifying parameters in a subroutine prior to program installation on the system, discussed in Chapters 10 and 11.

It is in the interest of the user to become familiar with the hardware system before embarking upon the use of the language discussed in this book, especially if he has to generate the program from a source obtained from the

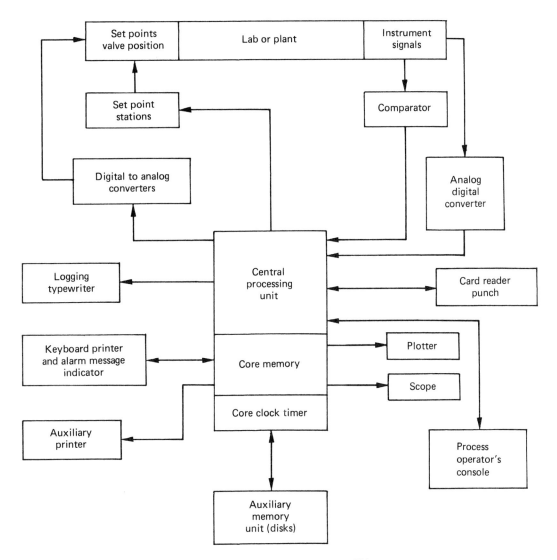

Figure 2.4   Showing the various units of the 1800 computer.

IBM Program Information Department. For further details on these systems, the reader is referred to the publications (11), (13), and (17) on the IBM 1130 and 1800. Since almost all 1130 and 1800 systems are operated on the basis that the user/programmer will also be an operator of the system, the reader should become familiar with the procedures of starting up the computer and loading a program on the two systems. He should also become familiar with the proper procedures for terminating the use of the system without affecting the bulk memory of the system disk and without affecting the real-time function of the 1800 system.

Figure 2.5   The 1130 peripherals: 1403 printer, 2501 reader, 2310 disk
with two drives, 1627 plotter.

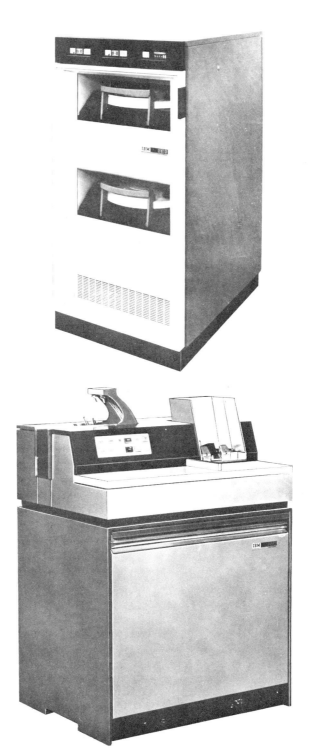

TABLE 2.1

FORTRAN I/O Logical Unit Designations and Record Sizes on the 1130

Logical unit no.	Device	Kind of transmission	Record size allowed
1	Console printer	Output only	120
2	1442 card read punch	Input/output	80
3	1132 printer	Output only	1 carriage control + 120
4	1134/1055 paper tape reader punch	Input/output	80
5	1403 printer	Output only	120
6	Keyboard	Input only	80
7	1627 plotter	Output only	120
8	2501 card reader	Input only	80
9	1442 card punch	Output only	80
10	UDISK	Unformatted input/output without data conversion	320

TABLE 2.2

FORTRAN I/O Normal Unit Designations on the 1800

Logical unit no.	Device	Kind of transmission
1	1816 printer keyboard	Input/output
2	1442 card read punch	Input/output
3	1443 printer	Output only
4	1810 disk	Input/output
5	1627 plotter	Output only
6	1443 printer*	Output only
8	1810 disk no. 2	Input/output
9	1810 disk no. 3	Input/output

*Alternative assignment.

2.2 System/3

In Fig. 2.6 the IBM System/3 is shown with its card reader and printers. The System/3 Models 15, 10, 8, and 6 are a combination of commercial and scientific machines and have card readers with characteristics different from the normal 80-column card readers, which are most popular. The scientific use of the System/3 is with FORTRAN and BASIC. The DSL program can be used on System/3 with the FORTRAN facility. Table 2.3 gives the various input and output devices available on System/3. The corresponding logical unit numbers for these devices are also shown in Table 2.3. This information is to be used, as shown in Chapter 11, for generation of the proper DSL system by the user, depending on the hardware components of System/3 which are available to him.

System/3 can have four disks, two fixed and two removable. The DSL system should be built and stored on a removable disk pack which the user should mount on drive R1. The disk unit is somewhat different from the 1130/1800. The reader should refer to the publications on System/3 hardware

Figure 2.6   System/3 with dial face.

TABLE 2.3

Logical Unit Assignment for the IBM System/3

Device	Function	Logical unit no.
	Model 10 Installation	
5424 MFCU primary hopper (MFCU1)	Read cards	1
5424 MFCU secondary hopper (MFCU2)	Read/punch/print	2
5203 or 1403 printer	Print	3
5471 printer/keyboard	Read	5
5471 printer/keyboard	Print	6
1442 reader/punch	Read/punch	9
Tapes	Read/write sequential formatted and unformatted records	10–13
5444 Disk	Read/write sequential and direct access formatted and unformatted records	$n*$
5445 Disk	Read/write sequential and direct access formatted and unformatted records	$n*$
	Model 6 Installation	
5406 console keyboard	Read card images	1
5496 data recorder	Read/punch	2
5213 or 2222 printer	Print	3
5444 disk	Read/write sequential and direct access formatted and unformatted records	$n*$

*May be any number from 1 to 32,767 which has not been used in the source program to define another device.

and its operation for further details, as well as the FORTRAN publications for System/3, before attempting to generate the source for compilation of the DSL program on System/3. The DSL system generation for System/3 is described in detail in Chapter 11.

An important difference between System/3 and the 1130/1800 systems is in the data switches. The System/3 has rotary dials which are mapped into or interpreted as data switches. The user must become familiar with the use of these rotary dials. The correspondence between the dial position and data switches is discussed in Chapter 11, where System/3 simulation program generation and operation are described.

2.3   System/7

The IBM System/7 is another small scientific computer used in a process control environment, except that it is a smaller system in some respects than the IBM 1800. It has a batch operating system called DSS/7 under which a simulator can operate. However, in the batch environment the system cannot operate the real-time functions concurrently unless the user generates his own special supervisor nucleus. The reader who wishes to operate the simulation program discussed here on System/7 should first read some introductory reference material (21) on the system and then follow Chapter 11 for program preparation.

Figure 2.7 shows a typical System/7 used in a batch environment.

The System/7 has only two card readers and two printers (7431 and 5024) available on it. It also has a console keyboard/typewriter (5028) which is used, as in

Figure 2.7.   System/7 with 5028 printer.

the case of the 1130, for data entry and error messages if desired. In general, the logical unit numbers for these three devices are fixed by the systems programmer when generating his System/7 DSS/7 system. They may, however, be changed by the user prior to execution of the simulation program.

There are no data switches on the System/7. However, the user may, using a 10- to 12-volt power supply and a few toggle switches, connect these switches to a digital input group of the System/7 and perform the same function as the 1130/1800 data switches. In Chapter 11 we shall show the programming method to make use of the digital input feature of the System/7 to incorporate data switch usage.

It must be pointed out that neither System/3 nor System/7 has any program-supported graphics available on the system. In System/3 this constraint is possibly due to hardware. In System/7 it is possible to connect a plotter or an oscilloscope. However, since no system support for graphics is available, we shall not delve into the graphic aspects of the system preparation on System/7 either.

It is not within the scope of this book to describe the large variety of other small scientific computers available on the market which use the standard FORTRAN with a batch system and which could operate the simulation program discussed in the book. It would take but small effort on the part of the enterprising reader familiar with his particular hardware/software system to examine and convert the program discussed in the book to his system. Chapters 8–12 will be very valuable for this reader.

Having discussed the system hardware requirements for simulation, the next most logical subject is programming requirements in terms of the features of a simulation language from a user's standpoint, which follow in Chapter 3.

3 LANGUAGE FEATURES

3.0 Features of the Simulation Language on the 1130/1800

In Chapter 1 we briefly touched on some of the requirements of a simulation language which would provide tools for solution of engineering problems. We shall summarize these requirements here before proceeding with how they are met.

1. Easy analog computer simulation capability. Facility to bypass the analog block diagram requirement when the problem can be directly formulated in the form of a set of differential equations.

2. Simplified structural statements, with provision for sorting of statements by the language.

3. Compatability with FORTRAN so that when required the simulation language statements and FORTRAN statements can be intermixed.

4. Availability of multiple integration methods.

5. Provision for adding user-specified functions and programs to the built-in library.

6. Support for graphic output, oscilloscope as well as hard copy plotter, whenever possible.

7. Start/stop capabilities during the simulation run, to inspect intermediate results and to change parameter values and continue or restart the run with altered parameters.

8. Capability of error recognition and correction without the necessity of a restart.

We shall devote this chapter to the language description with these requirements in mind and shall illustrate the use of the language features as we describe two specific simulation languages on the 1130. We must caution the reader that the discussion of the first language will be, of necessity, brief, because it serves a special purpose on the IBM 1130. It is different from the second language, which is more universal, in that it can be used on the 1800, System/3, and System/7 and at least from the user input standpoint provides the same interface on the System 360/370.

The first of these two is the language CSMP-1130 for those who are used to the path of the analog simulation. A detailed description of this language may be found in Reference (16). The most useful features of this program are its facility to completely define the problem through the use of the 1130 data switches and console keyboard and its ability to provide step-by-step solution of the engineering problem by the use of these two devices with recovery from error at every step. It must be pointed out that the CSMP-1130 is a true analog simulator; i.e., to solve an engineering problem the user must first construct an analog block diagram of his system to simulate on the computer.

3.1 Continuous System Modeling
Program for the 1130
(CSMP-1130)

This program was originally available from IBM as a type II program (developed and maintained by IBM) supplied from the IBM Program Library for 1130 users. An expanded version of the program with additional features (increase in number of blocks and functions) is available as a program product (license fee charged) from IBM. It was developed for simulation of continuous processes which can be described in terms of a variety of functional blocks. These blocks are interconnected to describe the total system of the problem to be solved. A brief description of the CSMP program with an illustration is given here to show the highlights. Readers requiring more detailed information on the program should refer to the CSMP-1130 manuals (11, 16).

This program can be used once the block diagram of an equivalent analog system for the problem has been developed. The diagram is then translated into a corresponding set of CSMP language statements which are read in as data cards through the card reader or through keyboard entries on the 1130 console.

Each operation in CSMP is with a block, and each block is identified by type. There are three basic blocks, namely, integrator, summer, and limiter, and a number of additional element blocks which provide a large variety of functions. Table 3.2 illustrates the various functional blocks together with the symbol for each block. For example, a function generator block can be used to define a function $f(X)$ if $f(X)$ can be broken into 10 line segments in the range of X of interest as shown in

Fig. 3.1. In addition a special function may be defined with a user-written FORTRAN routine.

Each of the blocks has a gain of 1, and this gain is modified by specifying associated parameters. For example, in Fig. 3.2 block number 10 is an integrator with $e_0 = \int (e_1 + e_2 + e_3) dt$ unless parameters P_1, P_2, and P_3 are specified in which case $e_0 = \int (P_1 e_1 + P_2 e_2 + P_3 e_3) dt$.

The interconnections between the various blocks may be specified in any order, since the sorting of the program is done by CSMP language. Up to 75 (98 for the revised CSMP program) blocks can be specified in a problem. Block number 76 (99 for the revised program) is set as the time variable. Up to 25 (50 for the revised program) integrators may be specified for a problem, and a maximum of three parameters can be associated with each block.

The language CSMP, although limited to some extent in scope, is extremely useful from the point of interactive mode of operation between the user and the 1130 computer with the data switches and keyboard input. The entire operation of problem solution proceeds in several phases whereby the user can completely define his problem with the 1130 console keyboard entry, provide the necessary parameters, correct his errors, and follow the solution of the problem as the value of the independent variable, time, increases. The user has output as either printed results on the console typewriter or as graphs on the 1627 plotter. Alternatively, a print plot capability is available if the user has no 1627 attached to the 1130.

Output of any block at any value of time is available with interruption by data switches, and the value of the outputs may be saved on the disk for starting at a subsequent date with the interrupted values of time and the dependent variables as starting values for the solution.

TABLE 3.1

Showing Use of the 1130 Console Switches to Perform Various Phases of CSMP

Option	Switch
Configuration	1
Initial conditions or parameters	2
Function generator intercepts	3
Integration specifications	4
Print interval	5
Print variables	6
New plot frame	7
Plotter X-axis	8
Plotter Y-axis	9
Suppress typing of card input data	11
Punch updated data deck	12
Interrogate block outputs	13
Save status at interrupt time	14
Restart at previous interrupt point	15

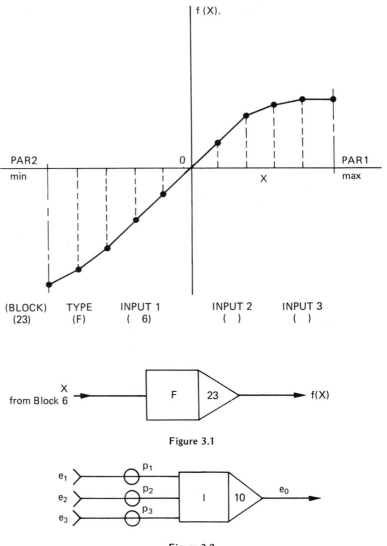

(BLOCK)	TYPE	INPUT 1	INPUT 2	INPUT 3
(23)	(F)	(6)	( )	( )

Figure 3.1

Figure 3.2

The control phase of the program requires that the integration interval and the total time of integration be specified. The initial condition for the integrator is the first parameter.

In the case of nonspecified data or incorrectly specified data the CSMP translator prints out the appropriate diagnostic messages on the console typewriter and requests that a corrected value be entered through the keyboard. Thus, in case of programming errors, program recovery is instantaneous.

3.1.1 An Example for Use of CSMP-1130

The user interested in the use of this language is referred to the publication on CSMP (11, 16) for further detail.

We shall illustrate the use of CSMP-1130 with a sample problem.

Let us use the solution of Eq. (1.1) in Chapter 1 to show the use of the language. The equation is

$$F''' + FF'' = B(F'^2 - 1) \tag{3.1}$$

We first note that the various integrations of F''', F'', and F' yield F'', F', and F. See Fig. 3.3.

Now we must construct Eq. (1.1) from these blocks. First we construct the blocks for $B(F^2 - 1)$; then we add FF'' and sum it, which is the input to block 1. Figure 3.4 illustrates the arrangement in a block diagram.

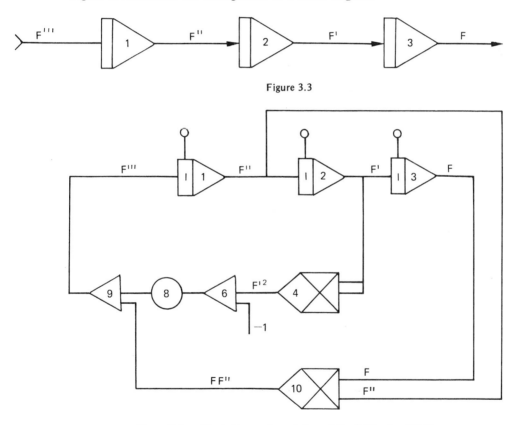

Figure 3.3

Figure 3.4.   Block diagram for solution of Eq. (3.1) using CSMP.

TABLE 3.2

Functional Blocks Available in CSMP

ELEMENT TYPE	LANGUAGE SYMBOL	DIAGRAMMATIC SYMBOL	DESCRIPTION
BANG-BANG	B	e_i —[B | n]— e_0	
DEAD SPACE	D	e_i —[D | n]— e_0	
FUNCTION GENERATOR	F	e_i —[F | n]— e_0	
GAIN	G	e_i —(n) P_1 — e_0	$e_0 = P_1 e_i$
HALF POWER	H	e_i —[H | n]— e_0	$e_0 = \sqrt{e_i}$ Square root
INTEGRATOR	I	e_1, e_2 (P_2), e_3 (P_3), P_1 | n — e_0	$e_0 = P_1 + \int(e_1 + e_2 P_2 + e_3 P_3)\,\mathrm{d}t$
JITTER	J	[J | n]— e_0	Random number generator between ± 1
CONSTANT	K	(n) P_1 — e_0	$e_0 = P_1$
LIMITER	L	e_i —[L | n]— e_0	

n REPRESENTS THE BLOCK NUMBER

TABLE 3.2 (continued)

ELEMENT TYPE	LANGUAGE SYMBOL	DIAGRAMMATIC SYMBOL	DESCRIPTION
MAGNITUDE	M		$e_0 =$ absolute value (e_i)
NEGATIVE CLIPPER	N		
OFFSET	O		$e_0 = e_i + P_1$
POSITIVE CLIPPER	P		
QUIT	Q		Quit (terminate run) if $e_1 > e_2$
RELAY	R		$e_0 = \begin{array}{l} e_2 \\ e_3 \end{array}$ for $e_1 \begin{array}{l} \geqslant 0 \\ < 0 \end{array}$
TIME PULSE GENERATOR	T		Generates pulse train with period equal to P_1 starts when $e_i \geqslant 0$
UNIT DELAY	U		$e_0 = e_i \, (t - \triangle t/2)$
VACUOUS	V		$e_0 \neq f(e_i)$ used in conjunction with element wye

n REPRESENTS THE BLOCK NUMBER

TABLE 3.2 (continued)

ELEMENT TYPE	LANGUAGE SYMBOL	DIAGRAMMATIC SYMBOL	DESCRIPTION
WEIGHTED SUMMER	W		$e_0 = P_1 e_1 + P_2 e_2 + P_3 e_3$
MULTIPLIER	X		$e_0 = e_1 e_2$
WYE	Y		Logical branch element used for implicit operations
ZERO ORDER HOLD	Z		$e_0 = e_1$ for $e_2 > 0$ e_0 unchanged for $e_2 \leqslant 0$
SUMMER	+		$e_0 = \pm e_1 \pm e_2 \pm e_3$ Only element where negative sign is permissible in configuration specification
DIVIDER	/		$e_0 = e_1 / e_2$
SIGN INVERTER	—		$e_0 = -e_i$
SPECIAL	1-5		Subroutines supplied by user

n REPRESENTS THE BLOCK NUMBER IN THE DIAGRAMMATIC SYMBOL COLUMN

The console switches to be used for the various phases are shown in Table 3.1 An overview of the functional blocks available in CSMP is shown in Table 3.2, which is spread over three pages. For the input to CSMP for this example we shall use Tables 3.1 and 3.2 and assume that we are in phase I, the configuration phase. That is, the 1130 system has already been cold-started, and execution of the CSMP program has already been initiated. Table 3.1 shows the input to the CSMP program for the configuration phase, with data switch 1 on and all other data switches off.

3.1.1.1 Parameter Phase

Table 3.3 shows the input to the console typewriter in this phase—to enter parameter values and/or initial conditions to the translator.

TABLE 3.3
Configuration Specification

OUTPUTNAME	BLOCK	TYPE	INPUT1	INPUT2	INPUT3
FDPRO	(1)	(I)	(9)	( )	( )
FPR	(2)	(I)	(1)	( )	( )
F	(3)	(I)	(2)	( )	( )
FPRSQ	(4)	(X)	(2)	(2)	( )
FPS1	(6)	(O)	(5)	(4)	( )
FPSB	(8)	(G)	(6)	(6)	( )
FIPR	(9)	(+)	(8)	(−10)	( )
FFPR	(10)	(X)	(1)	(3)	( )

3.1.1.2 Control Phase

In the control phase we must specify certain information for the plotter as well as information for the independent variable, time. See Table 3.4. Table 3.5 shows the input on the typewriter for the variable time.

The complete sequence of how a problem solution run is made with CSMP using the 1130 console typewriter is thus shown by Table 3.6. The values typed in parentheses are supplied by the program printed text.

TABLE 3.4
Initial Conditions and Parameters

IC/PAR NAME	BLOCK	IC/PAR1	PAR2	PAR3
FDPRO0	(1)	(+1.2275)	( )	( )
FPR0	(2)	(0.)	( )	( )
FO	(3)	(0.)	( )	( )
ONE	(6)	(−1.)	( )	( )
BETA	(8)	(1.)	( )	( )

TABLE 3.5

(0.001)	INTEGRATION INTERVAL				
(5.)	TOTAL TIME				
(0.5)	PRINT INTERVAL				
(2)	BLOCK FOR Y-AXIS	(0.) MINIMUM VALUE	(2.0) MAXIMUM VALUE		
(99)	BLOCK FOR X-AXIS	(0.) MINIMUM VALUE	(5.) MAXIMUM VALUE		

TABLE 3.6

CSMP Printout and User Input on the
1130 Console Printer

```
1                    1130 CONTINUOUS SYSTEM MODELING PROGRAM II
                     A DIGITAL ANALOG SIMULATOR PROGRAM FOR THE IBM 1130

TURN ON SWITCH 1 TO RUN IN NORMAL INTERPRETIVE MODE
TURN OFF SWITCH 1 TO RUN IN COMPILER MODE
PRESS PROGRAM START AFTER SELECTION


INSTRUCTIONAL COMMENTS MAY BE SUPPRESSED AT ANY TIME BY TURNING ON SWITCH 10
TURN ON SWITCH 1 TO ENTER OR MODIFY CONFIGURATION STATEMENTS VIA THE KEYBOARD
TURN ON SWITCH 2 TO ENTER OR MODIFY INITIAL CONDITIONS OR ELEMENT PARAMETERS VIA THE KEYBOARD
TURN ON SWITCH 3 TO ENTER OR MODIFY FUNCTION GENERATOR INTERCEPTS VIA THE KEYBOARD


                         CONFIGURATION SPECIFICATION

OUTPUT NAME    BLOCK      TYPE    INPUT 1    INPUT 2    INPUT 3

ENTER SPECIFICATIONS WITHIN THE PARENTHESES
BLOCK NUMBERS MUST BE RIGHT JUSTIFIED
CAREFULLY CONFIRM TYPING -- IF OK, PRESS EOF KEY
IF ERROR, PRESS ERASE FIELD KEY AND RE-ENTER

FDPRO         ( 1)      (I)      ( 9)       (   )      (   )
FPR           ( 2)      (I)      ( 1)       (   )      (   )
F             ( 3)      (I)      ( 2)       (   )      (   )
FPRSQ         ( 4)      (X)      ( 2)       ( 2)       (   )
FPSI          ( 6)      (O)      ( 4)       (   )      (   )
FPSB          ( 8)      (G)      ( 6)       (   )      (   )
FIPR          ( 9)      (+)      ( 8)       (-10)      (   )
FFPR          (10)      (X)      ( 1)       ( 3)       (   )

                         INITIAL CONDITIONS AND PARAMETERS

IC/PAR NAME    BLOCK      IC/PAR1          PAR2           PAR3

FDPRO0        ( 1)      (  1.2275)    (        )    (        )
FPRO          ( 2)      (  0.    )    (        )    (        )
FO            ( 3)      (   0.   )    (        )    (        )
ONE           ( 6)      ( -1.    )    (        )    (        )
BETA          ( 8)      (  1.    )    (        )    (        )

(.001    ) INTEGRATION INTERVAL

( 5.     ) TOTAL TIME

( 2) BLOCK FOR Y-AXIS    (    0.   ) MINIMUM VALUE     (    2.   ) MAXIMUM VALUE

(99) BLOCK FOR X-AXIS    (    0.   ) MINIMUM VALUE     (    5.   ) MAXIMUM VALUE
PREPARE PLOTTER AND PRESS PROGRAM START
SET PEN ABOUT ONE INCH FROM RIGHT MARGIN

( 0.5    ) PRINT INTERVAL
     TIME      OUTPUT( 1)   OUTPUT( 2)   OUTPUT(  )   OUTPUT(  )   OUTPUT(  )   OUTPUT  2   OUTPUT 99
     0.000       1.2275       0.0000       0.0000       0.0000       0.0000       0.0000      0.0000

     0.500       0.7529       0.4921       0.0000       0.0000       0.0000       0.4921      0.5000
     1.000       0.3912       0.7723       0.0000       0.0000       0.0000       0.7723      1.0000
     1.500       0.1670       0.9064       0.0000       0.0000       0.0000       0.9064      1.5000
     2.000       0.0513       0.9574       0.0000       0.0000       0.0000       0.9574      2.0000
     2.500       0.0003       0.9684       0.0000       0.0000       0.0000       0.9684      2.5000
     3.000      -0.0204       0.9626       0.0000       0.0000       0.0000       0.9626      3.0000
     3.500      -0.0300       0.9497       0.0000       0.0000       0.0000       0.9497      3.5000
     4.000      -0.0362       0.9331       0.0000       0.0000       0.0000       0.9331      4.0000
     4.500      -0.0416       0.9136       0.0000       0.0000       0.0000       0.9136      4.5000
     5.000      -0.0467       0.8915       0.0000       0.0000       0.0000       0.8915      5.0000
```

TABLE 3.7
Console Input for Parameter Changes After
Run End on CSMP

AFTER SELECTING DESIRED OPTION PRESS PROGRAM START

SWITCHES SET ON WERE   2   5

INITIAL CONDITIONS AND PARAMETERS

IC/PAR NAME	BLOCK	IC/PAR1	PAR2	PAR3
FDPRO0	(1)	(1.233)	( )	( )

(1.   ) PRINT INTERVAL

TIME	OUTPUT 1	OUTPUT 2	OUTPUT 0	OUTPUT 0	OUTPUT 0	OUTPUT 2	OUTPUT 99
0.000	1.2330	0.0000	0.0000	0.0000	0.0000	0.0000	0.0000
1.000	0.3985	0.7782	0.0000	0.0000	0.0000	0.7782	1.0000
2.000	0.0667	0.9742	0.0000	0.0000	0.0000	0.9742	2.0000
3.000	0.0066	1.0005	0.0000	0.0000	0.0000	1.0005	3.0000
5.000	0.0026	1.0060	0.0000	0.0000	0.0000	1.0060	5.0000

The results from the simulation are shown in Table 3.6 as obtained from the 1130 console printer. We notice that the solution is for an assumed value of $F''(0)$, since this is an unknown condition. To solve the problem discussed in Chapter 1, a trial and error procedure must be used to guess at $F''(0)$ and find the value which will satisfy F' when the independent variable, time, has a large value.

The user at the console may repeat the solution procedure by returning to the parameter phase and entering another value of FDPRO0. He does this as often as necessary to converge upon an acceptable solution to the problem. Table 3.7 shows how a second guess of $F''(0)$ was supplied for a run. Figure 3.5 shows plot output for the two runs from CSMP program on the 1130. Table 3.8 shows a complete printout of the problem on the 1130 output printer.

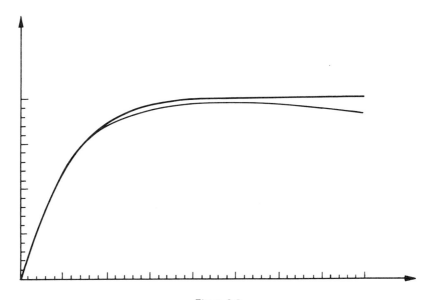

Figure 3.5

TABLE 3.8

CSMP Simulation Results on the 1132 Printer

```
                1130 CONTINUOUS SYSTEM MODELING PROGRAM II
                A DIGITAL ANALOG SIMULATOR PROGRAM FOR THE IBM 1130


                        INTERPRETIVE MODE
```

CONFIGURATION SPECIFICATION

OUTPUT NAME	BLOCK	TYPE	INPUT 1	INPUT 2	INPUT 3
FDPRO	1	I	9	0	0
FPR	2	I	1	0	0
F	3	I	2	0	0
FPRSQ	4	X	2	2	0
FPS1	6	0	4	0	0
FPSB	8	G	6	0	0
FIPR	9	+	8	-10	0
FFPR	10	X	1	3	0

INITIAL CONDITIONS AND PARAMETERS

IC/PAR NAME	BLOCK	IC/PAR1	PAR2	PAR3
FDPROO	1	1.2275	0.0000	0.0000
FPRO	2	0.0000	0.0000	0.0000
FO	3	0.0000	0.0000	0.0000
ONE	6	-1.0000	0.0000	0.0000
BETA	8	1.0000	0.0000	0.0000

(0.0010) INTEGRATION INTERVAL

(5.0000) TOTAL TIME

TIME	OUTPUT 1	OUTPUT 2	OUTPUT 0	OUTPUT 0	OUTPUT 0	OUTPUT 2	OUTPUT 99
0.000	1.2275	0.0000	0.0000	0.0000	0.0000	0.0000	0.0000
0.500	0.7529	0.4921	0.0000	0.0000	0.0000	0.4921	0.5000
1.000	0.3912	0.7723	0.0000	0.0000	0.0000	0.7723	1.0000
1.500	0.1670	0.9064	0.0000	0.0000	0.0000	0.9064	1.5000
2.000	0.0513	0.9574	0.0000	0.0000	0.0000	0.9574	2.0000
2.500	0.0003	0.9684	0.0000	0.0000	0.0000	0.9684	2.5000
3.000	-0.0204	0.9626	0.0000	0.0000	0.0000	0.9626	3.0000
3.500	-0.0300	0.9497	0.0000	0.0000	0.0000	0.9497	3.5000
4.000	-0.0362	0.9331	0.0000	0.0000	0.0000	0.9331	4.0000
4.500	-0.0416	0.9136	0.0000	0.0000	0.0000	0.9136	4.5000
5.000	-0.0467	0.8915	0.0000	0.0000	0.0000	0.8915	5.0000

SWITCHES SET ON WERE   2   5

INITIAL CONDITIONS AND PARAMETERS

IC/PAR NAME	BLOCK	IC/PAR1	PAR2	PAR3
FDPROO	1	1.2330	0.0000	0.0000

TIME	OUTPUT 1	OUTPUT 2	OUTPUT 0	OUTPUT 0	OUTPUT 0	OUTPUT 2	OUTPUT 99
0.000	1.2330	0.0000	0.0000	0.0000	0.0000	0.0000	0.0000
1.000	0.3985	0.7782	0.0000	0.0000	0.0000	0.7782	1.0000
2.000	0.0667	0.9742	0.0000	0.0000	0.0000	0.9742	2.0000
3.000	0.0066	1.0005	0.0000	0.0000	0.0000	1.0005	3.0000
4.000	0.0022	1.0037	0.0000	0.0000	0.0000	1.0037	4.0000
5.000	0.0026	1.0060	0.0000	0.0000	0.0000	1.0060	5.0000

Although we have not discussed in this section uses of function generation and special user functions, these facilities exist in CSMP-1130, and the interested reader is referred to the manuals on CSMP-1130 for details on the use of these features.

In a later chapter, we shall show how the same problem, rather a more general case of it, can be solved with the use of DSL (digital simulation language) without the use of the analog block diagram shown in the figures above.

The CSMP-1130 program is useful for the 1130 users. In the next section, we shall present the discussion on the second program, namely, DSL, which is available for the IBM 1130 and IBM 1800 and, as shown in subsequent chapters, can be converted to run on IBM System/3, IBM System/7, and other small scientific computers. The program in a similar form for the user is also available on IBM 7090 and Systems 360 and 370.

3.2 Digital Simulation Language for 1130, 1800, System/3, System/7, etc.

This program is also distributed as a type III program (supplied but not maintained by IBM) by the IBM program Information Department. The storage requirement for this program is a 10K variable core in 1800, either TSX or MPX. The program will also function in an 8K minimum core 1130 system. If larger storage size is available on the 1130, either the dimension of the program can be changed to provide more structural statements and a larger number of integrators or some of the *LOCAL cards can be removed to improve the speed of the problem solution. These points are covered in some detail in Chapter 10, where system generation is discussed.

The language DSL is a programming system to simulate continuous system dynamics or, in the case of differential equations describing a system with an independent variable other than time, to simulate the behavior of an engineering system under steady-state conditions along the value of its independent variable (such as length or depth). The input to this language is applications-oriented, non-procedural, and free format. An intermix of DSL and FORTRAN statements, together with a library of functions which model analog components, will describe the engineering system under investigation. Alternatively, the language can also accept input directly in differential equation format. Extensive user interaction is provided at program execution time.

The language will have applications in the following areas:

1. Where analog computer applications exist.

2. Where solution of a set of ordinary differential equations describing a physical system is needed.

3. Where simulation of continuous process dynamics is required.

The DSL program converts the user input to a FORTRAN subprogram, re-ordered and sorted according to proper information flow. The FORTRAN-translated subprogram which represents the model is then used to compute values of derivatives and other variables of interest. Where integration is required, choice of one of four integration methods is available to the user.

3.2.1 DSL Program Highlights

The program highlights and the features are given below:

1. A library of DSL system blocks which mathematically model analog components or provide specific mathematical or control functions. A description of these functions will be found in Table 4.2.

2. The input language is simple, nonprocedural, and applications-oriented. The rules for connecting the library blocks are specified.

3. The model description can be at the block diagram or differential equation level so that both analog block notation and algebraic notation are permitted.

4. FORTRAN language statements and DSL language statements can be inter-mixed. The FORTRAN arithmetic and function capabilities within DSL statements are provided.

5. The input language statements are automatically sorted (sequenced) for any or all sections of the model description. It permits model segmentation and pre-simulation as well as postsimulation calculations.

6. Parameters, constants, and other data are entered through an input routine which permit data entry by variable name in free format.

7. Results of simulation are provided in print output as well as plot output on the 1627 plotter. In addition, SCOPE (cathode ray tube) output is provided on the 1800.

8. Facility to add to the DSL library any user-defined blocks, *at execution time* or before, in the form of subroutines. These subroutines may be in FORTRAN, assembler, or object decks. By storing these programs on the disk in the user area, they become a permanent part of the user's DSL library.

9. Choice of numerical integration techniques: The user may specify not only one of four integration methods but also the integration step size, which may be fixed or variable, in which case automatic step size selection will be made to satisfy the error bounds. The integration scheme may be centralized (simultaneous) or noncentralized.

10. Diagnostics capability (debugging and error messages) at language translation time, program compile time, and problem simulation time (solution time).

11. Variables and blocks are labeled symbolically. For example, in the solution of Eq.(1.1) in Chapters 1 and 2, instead of blocks 1, 2, 3, etc., the symbols F3PRM, F2PRM, FPRM, and F can be used for the blocks/variables.

12. Multiple nesting of functions to any level is provided. An example is

$$Y = \text{INTGRL (SIN (LOG (1 /X**2.1)),}$$
$$\text{ERFC (ATANB (LOG (1-}$$
$$\text{EXP (3.-X**2.3) + COS (LOG (X)))*X**3)))}$$

This expression is valid, provided the function ERFC is part of the user's library of programs.

13. Unspecified algebraic loops are identified by warning messages. If an implicit function is to be evaluated iteratively, an implicit function subroutine is provided.

14. The data required at execution time are stored dynamically in the machine core storage, providing all available space to the user. Only those DSL library routines used by the problem are loaded from bulk storage to core storage.

15. Use of logical control statements of FORTRAN are allowed provided certain sorting rules are followed.

16. Complete user interaction with the simulation model by the use of data switches on the computer console. A run may be interrupted at any time; any data statement may be altered by card and/or typewriter input. The run may then be continued from the point of interruption, or the problem solution may be restarted from the initial point (independent variable = 0). Scope and plotter output may be redrawn, continued, or superimposed.

Before delving into the details of the DSL program and its structure, it will be beneficial for the reader to get a feel for the language by examining the method of solution with DSL of two simple examples, one of which was already shown earlier in the case of CSMP.

3.2.2 A DSL Example Using an Analog Block Diagram

The mass-spring damper system shown in Fig. 1.1 in Chapter 1 is of interest for analysis via simulation. If the mass is displaced from its rest position and then released, the mass will oscillate until the energy of the system is dissipated by the damper. We want to observe the effect of different spring constants on the motion of the mass and determine how long it takes to reduce displacements due to oscil-lations below certain values. The equation for the motion of the spring-mass system is given by

$$MX'' + BX' + KX = 0 \qquad (3.2)$$

with the boundary condition $X' = 0$ and $X = A$ at time $= 0$.

Although the DSL formulation from Eq. (3.2) is straightforward, one can also describe the system using an analog block diagram as shown in Fig. 1.2 in Chapter 1. The function of each block is described by the symbol contained within the block. A constant or initial condition (in the case of an integrator) is supplied to the block by a parameter in a circle.

The input/output relationships for variables in each block can thus be written as follows:

1. MX2DT   = M1 + M2
2. X2DT   = MX2DT/M
3. XDT   = INTGR (0., X2DT)
4. X   = INTGR (A, XDT)
5. M1   = –B*XDT
6. M2   = –K*X

If, on the other hand, the block diagram is bypassed, Eq. (3.2) can be structured in DSL as follows:

$$\text{XDT} = \text{INTGRL } (0., \; (\text{–B*XDT–K*X})/M)$$

$$\text{X} = \text{INTGRL } (A, \text{XDT})$$

To complete specifications for solution of this problem in DSL, we need only supply the values of parameters, the initial conditions, the maximum value of the independent variable, the method of integration, and the starting step size for integration. The user can provide his title, and if he wishes to print and plot the results, he can specify the variables for output and the interval of the independent variable at which output is desired. Multiple runs can be made as shown in Table 3.9. The program terminates after encountering the STOP statement.

An examination of Table 3.9 shows that the CONTRL statement has provided the integration parameters, PARAM has provided values of parameter constants, and the INCON statement has provided the initial conditions. The PRINT statement specifies the printing interval of TIME and other variables. To obtain a plot of the variables the plot interval, the X- and Y-axis dimensions, and the respective variables to be printed are required in the GRAPH statement. Furthermore, a SCALE statement specifies the scaling factors for the variables to be plotted so that the curves remain within the boundaries of the specified dimensions of the X- and Y-axes on the GRAPH statement.

Figure 3.6 shows the family of the curves generated by the translator after problem solution, whereas a portion of the printed output is shown in Table 3.10.

TABLE 3.9
Input for the DSL Translator for the
Mass-Spring Damper Problem

```
TITLE MASS SPRING DAMPER SYSTEM SIMULATION
        XDT=INTGRL (0.,  (–B*XDT–K*X)/M))
        X = INTGRL (A, XDT)
INCON   A=10.
PARAM   M=9.2, B=1.5, K=3.
CONTRL  DELT=0.05, FINTI=8
INTEG   MILNE
PRINT   .1, X, XDT
GRAPH   .01, 8, 6, TIME, XDT
SCALE   1.5, .05
LABEL CURVES FOR MASS-SPRING-DAMPER SYSTEM
END
PARAM   K = 8.6
END
PARAM   K = 14.
END
STOP
```

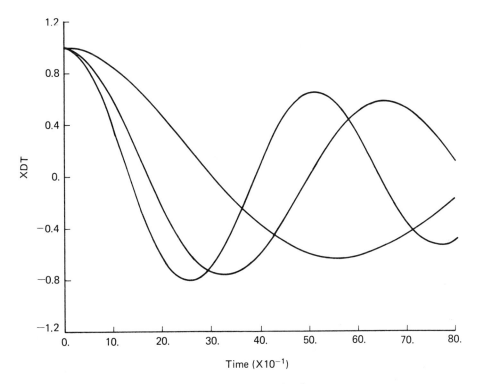

Figure 3.6.   Plot output for mass spring damper system.

TABLE 3.10

Print Output for the Mass-Spring Damper System Using DSL

MASS, SPRING, DAMPER SYSTEM IN DSL

TIME	X	XDOT
0.0	1.0000E 01	0.0
2.0000E-01	9.9356E 00	-6.4026E-01
4.0000E-01	9.7458E 00	-1.2518E 00
6.0000E-01	9.4372E 00	-1.9276E 00
8.0000E-01	9.0176E 00	-2.3616E 00
1.0000E 00	8.4958E 00	-2.8481E 00
. . . .	. . . .	. . . .
. . . .	. . . .	. . . .
7.2000E 00	-3.9755E 00	2.5673E 00
7.4000E 00	-3.4458E 00	2.7231E 00
7.6000E 00	-2.8889E 00	2.8390E 00
7.8000E 00	-2.3128E 00	2.9148E 00
8.0000E 00	-1.7256E 00	2.9508E 00

3.2.2 A DSL Example Without a Block Diagram

In the second example we would like to show the solution of the flow of a Newtonian fluid past a wedge, for which the boundary-layer equations in steady state reduce to the nonlinear differential equation given in Eq. (3.1). This problem was solved in Section 3.1.1 using CSMP-1130.

In this particular example, we shall bypass the analog block diagram and write the statement for the DSL translator input directly:

$$F \quad = \text{INTGRL}(0., \text{FDT})$$

$$\text{FDT} \quad = \text{INTGRL}(0., \text{F2DT})$$

$$\text{F2DT} \quad = \text{INTGRL}(X, (B*(\text{FDT}**2-1.)-F*\text{F2DT})$$

Table 3.11 shows the complete set of DSL input for three different parametric runs for the solution of this problem. Figure 3.7 shows the plots of F', versus time (the independent variable). If the user wishes to name the independent variable as other than TIME, he may change the name by a RENAME statement in DSL. This is especially useful in the plots of variables.

In Chapter 5 we shall show the solution of a more generalized case of the problem above with boundary conditions at two different points which require a trial and error solution and therefore interaction between the user and the 1130/1800 system during the solution.

From the above two examples, which are elementary from the standpoint of DSL usage, the reader will recognize the power of the language facility in the

TABLE 3.11
DSL Input and Simulation Output for the
Newtonian Fluid Problem

TITLE   DSL SIMULATION OF NEWTONIAN FLOW PAST A WEDGE
         F3DT=B*(FDT*FDT−1.)−F*F2DT
         F2DT=INTGR (A, F3DT)
         FDT=INTGR (0., F2DT)
         F=INTGR (0., FDT)
PARAM   B=1., A=1.233
CONTRL   DELT=.005, FINTI = 5.
ABSERR   FDT=.0001
INTEG MILNE
GRAPH .05, 8, 8, TIME, FDT
LABEL PLOT OF VELOCITY VS DIST
SCALE 1., 5.
PRINT .2, F, FDT, F2DT
END
PARAM   A = 1.235
END
STOP

OUTPUT VARIABLE SEQUENCE
    1   2   3   4

STORAGE USED/MAXIMUM
INTGR   3/50, IN VARS   8/300, OUT VARS   4/100, PARAMS   8/60, SYMBOLS 28/200

DSL SIMULATION OF NEWTONIAN FLOW PAST A

TIME	F	FDT	F2DT
0.000E 00	0.0000E 00	0.0000E 00	0.1233E 01
0.199E 00	0.2333E−01	0.2266E 00	0.1034E 01
0.399E 00	0.8808E−01	0.4146E 00	0.8467E 00
0.599E 00	0.1867E 00	0.5665E 00	0.6756E 00
0.799E 00	0.3125E 00	0.6862E 00	0.5256E 00
0.999E 00	0.4594E 00	0.7783E 00	0.3985E 00
0.119E 01	0.6223E 00	0.8472E 00	0.2944E 00
0.139E 01	0.7970E 00	0.8975E 00	0.2118E 00
0.159E 01	0.9801E 00	0.9333E 00	0.1484E 00
0.459E 01	0.3965E 01	0.1007E 01	0.7676E−02
0.479E 01	0.4161E 01	0.1008E 01	−0.4650E−02
0.499E 01	0.4364E 01	0.1008E 01	0.3200E−02
0.499E 01	0.4364E 01	0.1008E 01	0.3200E−02

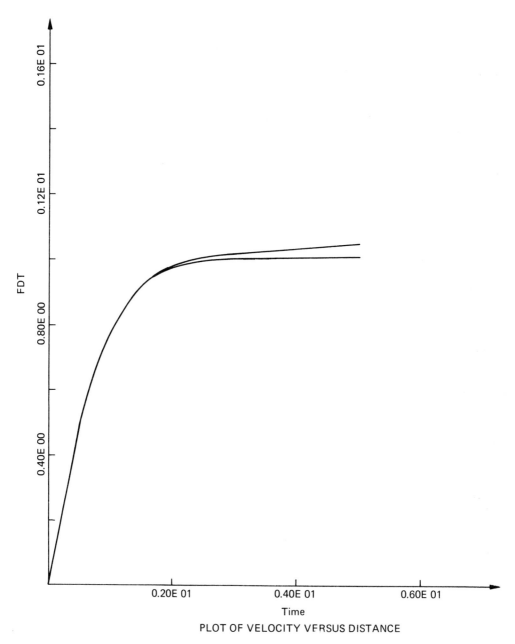

PLOT OF VELOCITY VERSUS DISTANCE

Figure 3.7. Simulation plot output from DSL for Newtonian fluid flow past a wedge.

solution of nonlinear differential equations, which arises in two different engineering fields. Once the user becomes somewhat familiar with the use of the facilities of engineering simulation described here, problems of the nature shown above should take no more than a few minutes to formulate on the small scientific computer, and he should be able to obtain the solution in less than an hour, the actual computer execution time being dependent on the problem complexity.

4 DSL LANGUAGE RULES

4.0 DSL Language Usage Description

In Chapter 3, we showed that input to the language CSMP followed a certain description format, conventions, and rules. The CSMP usage also required execution in several phases. Although the use of DSL does not follow phases in that manner, the statements in the language can be categorized into three types:

1. Structure statements, which provide definition of the simulation model structure as input to the translator.

2. Data statements, which specify model parameters, constants, initial conditions, etc., for problem solution.

3. Translator commands, which specify, for example, output control, method of integration, and problem rerun with different parameter values.

The structure statements not only provide model statements but include functions or blocks and their connections and the inclusion of a subblock of FORTRAN logical type statements. The structural statements are made up of named variables, constants, function references, expressions, and operators. As we mentioned earlier, DSL translates the user input into a FORTRAN program, so that some of the rules of FORTRAN in the definition of symbols, operators, and functions are followed in DSL. For details on the description of FORTRAN, the user is referred to the manual entitled *IBM 1130/1800 FORTRAN IV Language* for 1130/1800 users or other FORTRAN manuals for the particular system the reader is using for his simulation.

4.1 Variables, Constants, Expressions, and Function Blocks

4.1.1 Variable Names

A variable name may contain five alphanumeric characters with the first character always alphabetic. L532, XDTSQ, K, and MQ23 are valid examples of variable names. Unlike FORTRAN, all variables in DSL are considered real; i.e., their values are assigned floating-point numbers. If the user wishes to specify the variable as integer, he must so declare them via the INTGER data statement as described in a later section. User-supplied functions (or blocks) also follow the convention of variable names; i.e., the function names are of five characters or less and must begin with an alphabetic character.

Subscripted variables are permitted in DSL to reference a one-dimensional table; i.e., only a single subscript may be used. However, the dimensioned variable when referred to on the left-hand side of an equation (to define its value, for example) must appear within the NOSORT block. If a subscripted variable is used, the variable name must appear in a STORAG data statement with the maximum possible value of its subscript. As in FORTRAN the subscript may be an integer or an integer name enclosed in parentheses.

If a user functional block is referred to in the DSL structure statements, it must refer to a function program supplied by the user. The arguments of the functional blocks follow the rules of FORTRAN.

Example:

$$X(1) = T*2.3 + SIN(T) \tag{4.1}$$

$$X(2) = T**2.3 + COS(T) \tag{4.2}$$

Equations (4.1) and (4.2) can be accepted if the input to DSL is in the format shown in Table 4.1 (Section 4.1.4), within the NOSORT block. Similarly,

$$TX(1) = ERFC (T, XC1) \tag{4.3}$$

is acceptable if the user provides the function in FORTRAN, Assembler, or object form named ERFC in such a way that the value is defined using two arguments.

4.1.2 Constants

A real (floating-point) constant consists of one to seven significant decimal digits written with a decimal point. The constant may also be followed by a decimal exponent, with the use of letter E (as in FORTRAN) followed by a signed or unsigned integer. If the sign is missing on the exponent, the exponent is assumed positive.

Examples of constants are 21., 2135., .20587, −8.0067, 5.23E-3, and −.546E32. The magnitude of the constant is limited by the computer hardware to be between 10^{38} and 10^{-39}. At least seven-digit precision is provided.

An integer constant may be from one to five decimal digits, must be without a decimal point, and can attain a maximum absolute value of 32,767. Examples are 3, −45, and 8500.

The user must note that when integers are used, if the result of an arithmetic operation with two integers is greater than 32,767, the answer will be 32,767. This limitation is due to FORTRAN when one-word integers are used.

4.1.3 Expressions and Operators

From the examples in Chapter 3, the reader must have realized that the structure statements in DSL are in the form of equations, with output variables to the left of the equal sign and an expression to the right. The expression on the right-hand side may be a single variable or constant, an output from a lower-level function or subroutine. It may also consist of a string of variables, constants, and functions connected by arithmetic operation symbols.

Operation symbols +, −, *, /, and ** denote addition, subtraction, multiplication, division, and exponentiation, respectively. The equal sign is used to specify replacement rather than equivalence.

The use of operators to form expressions is subject to the following rules:

1. Two operators may not appear consecutively. A*−B is not a valid expression, but using parentheses the expression A*(−B) is valid.

2. Parentheses may also be used to establish the order of computation of an expression. A+B*C will be evaluated by taking the produce B*C first and adding the result to A, while the parentheses in (A+B)*C cause the sum to be evaluated before the product.

3. All operations must be specified explicitly. A times COS(B) must be written A*COS(B), not ACOS(B). ACOS will be treated as a new function name if the user specifies ACOS(B).

4. The order in which expressions are evaluated is as follows:

   a. Expressions within parentheses.
   b. Function references.
   c. Exponentiation (**).
   d. Multiplication and division (* and /).
   e. Addition and subtraction (+ and −).

Where operators having the same precedence appear in succession, the evaluation proceeds from left to right. A*B/C is equivalent to (A*B)/C. An exception to this is exponentiation (**), in which evaluation proceeds from right to left. A**B**C is equivalent to A**(B**C).

5. Integer division yields an integer result. The fractional part, if any, is truncated.

Examples:
Assume that K1=5, K2=3, and K3=−2.

K1/K2 yields a result of 1.
K2/K1 yields a result of 0.
K1/K3 yields a result of −2.

The last example illustrates how the result of an integer division may be greater than the result of a floating-point (real) division when the result is negative.

The expressions in DSL must in addition conform to the following rules:

1. Any of the elements of an expression may be nested (i.e., contained one within the other) or may appear in combination as terms on the same level.

   Example: AL=B*COS(ALOG(X*SIN(BX/T)) − CX).

2. Any expression may be enclosed in parentheses, and expressions may be connected by arithmetic operators.

   Example: A=(SQRT(AX*BX))/(SIN(A*/BX)).

3. Function arguments in an expression are separated by commas and enclosed by parentheses.

   Example: A=ERFC(AX, BX). where ERFC is a function of two variables.

4. An expression may be continued on as many as six data cards, this limitation being imposed by 1130/1800 FORTRAN.

4.1.4 Function Blocks in DSL

DSL function blocks, like algebraic functions, consist of a statement of the relationships among a number of variables. The basic DSL function block is characterized by one or more output variables which are functionally related to one or more input variables (or constants). The *primitive form* of a function block reference in a DSL problem program is as follows:

[output-names]=block-name ([initial-conditions], [parameters], [inputs])

The output-names field is made up of one or more variable names.

Where there is more than one output name, the names are separated by commas. The block-name field contains the name of the function block. The three fields initial conditions, parameters, and inputs make up the function block argument list. The argument list is enclosed in parentheses; individual arguments in the list are separated by commas. In most cases, any valid expression may be used as an argument to a function block. Exceptions to this rule are stated in the description of the individual blocks.

Where there is more than one output from a function block, the function block reference must take the primitive form, i.e., outputs to the left of an assignment symbol (=) and the function name and argument list to the right of the assignment symbol. Thus,

$$\text{OUT1, OUT2} = \text{VALVE(LEVEL, INHI, IMMED, INLO)} + \text{X2}$$

is an *invalid* statement, because the function VALVE has multiple outputs and its reference statement is not in primitive form.

A reference to a function block with only one output is not restricted to the primitive form. For example,

$$Y = X2 + \text{LIMIT (P1, P2, X)}$$

is a valid statement, but not in primitive form. Here Y is an output variable whose value is the sum of X2 and the output of the LIMIT block.

The DSL system provides a large number of operational elements which may be used to model the components of an analog computer such as integrators, relays, and pulse and function generators. The arithmetic operators +, −, *, /, and ** are used to represent addition, subtraction, multiplication, division, and exponentiation. Named function blocks are not provided for these functions. The mathematical function subprograms available in FORTRAN may also be used as DSL function blocks. Subroutines available from the 1130 scientific subroutine package library or from any other program library may be used, provided the usual requirements regarding storage, order, number and type of arguments, etc., are met.

Following are several examples of the use of DSL function blocks.

Example 1:

Figure 4.1 illustrates the algebraic representation, the block diagram notation, and the use in DSL of the SQRT function block. It happens that this is one of the blocks from the FORTRAN library. In this usage, the input is called TEMP and the output is called OUT and there are no "initial condition" or other "parameters." We should note that each reference to a function block in DSL is made unique by the use of a unique variable name as output. Input names need not be unique. Thus, another unique use of the SQRT block might be

$$X = \text{SQRT(Y)}$$

Since the SQRT block has but one output, the function reference may be used

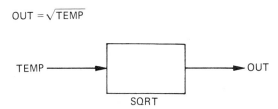

$$OUT = \sqrt{TEMP}$$

SQRT

$$OUT = SQRT\ (TEMP)$$

Figure 4.1.   The SQRT function block.

as an expression, as in the statement

$$XOUT=A+SQRT(TEMP)+SQRT(Y)$$

Example 2:

A block which is supplied by the DSL system and is basic to most DSL problems is the INTGRL block. Figure 4.2 illustrates the algebraic and block representations of the INTGRL block and a DSL statement referencing the INTGRL block. Here, Y is the output, IC2 is the initial condition, and YDOT is the input. Note that there is no "parameter" argument.

The DSL program provides a large number of operational elements that are similar to an analog computer. These elements include items such as integrators, summers, multipliers, relays, and pulse and function generators. Programming in DSL consists of "interconnecting" these elements to meet the requirements of a particular problem. This is analogous to the use of patchcords on an analog computer to interconnect the electronic operational elements. The user can add on to the basic set of operational elements or alter those existing as part of the system. This can be done by writing the user functions in FORTRAN or Assembler, compiling on the 1130 or 1800 system to be used, and adding to the disk library by storing on the user area. The user routine then becomes a

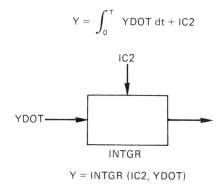

$$Y = \int_0^T YDOT\ dt + IC2$$

IC2

YDOT

INTGR

$$Y = INTGR\ (IC2, YDOT)$$

Figure 4.2.   The INTGR block.

permanent part of the user's DSL library. During a particular problem simulation, the user routine will be loaded from the disk only if required for simulation.

The FORTRAN library function blocks available to the user are shown in Table 4.1. Several additional functional blocks are provided by DSL, shown in Table 4.2, which is spread over three pages. The input/output form and the function performed by each block are shown in the table. Note that System/3, System/7, and 360/370 FORTRAN systems provide additional functions which may also be used by the reader. However, in System/3 and System/7, the reader must utilize only the functions which permit one-word options when using integers as arguments.

4.2 Structure Statements

TABLE 4-1
Some Standard of Functions (Blocks) Available
From the FORTRAN Library in 1130/1800

Function name	Purpose	Definition
ABS (Arg)	Absolute value	Abs (Arg)
ALOG (Arg)	Natural logarithm	Log_e (Arg)
ALOGX (Arg)	Common logarithm*	Log_{10} (Arg)
AMAX1 (Arg)	Choosing largest value*	Max (Arg1, Arg2)
AMIN1 (Arg)	Choosing smallest value*	Min (Arg1, Arg2)
AMOD (Arg)	Forcing a cycle*	Arg – (Arg1/, Arg2)*Arg (Only the integral part of the term within parentheses is used.)
ATAN (Arg)	Arctangent	Arctan (Arg)
COS (Arg)	Trigonometric cosine	Cos (Arg)
EXP (Arg)	Exponential	exp (Arg)
SIGN (Arg)	Transfer of sign from Arg to Arg	
SIN (Arg)	Trigonometric sine	Sin (Arg)
SQRT (Arg)	Square root	Square root of Arg

*These functions are not in 1130/1800 FORTRAN but are added as DSL blocks because they normally exist in a FORTRAN system.

Structure statements are used to describe the network or model to be simulated. They are translated into a FORTRAN subroutine called UPDAT, which is executed during the simulation phase to make derivative calculations as well as other logical and algebraic computations associated with the model, at each iteration cycle. In general, structure statements may be written in any order and intermixed freely with data statements. The system establishes the correct

TABLE 4-2

FUNCTION GENERATORS

GENERAL FORM	FUNCTION	
= AFGEN (FUNCT, X) RBITRARY LINEAR FUNCTION GENERATOR	$Y = FUNCT(X) \quad X_0 \leqslant X \leqslant X_n$ LINEAR INTERPOLATION $Y = FUNCT(X_0) \quad X < X_0$ $Y = FUNCT(X_n) \quad X > X_n$	
Y = NLFGN (FUNCT, X) ION-LINEAR FUNCTION GENERATOR	$Y = FUNCT(X) \quad X_0 \leqslant X \leqslant X_n$ QUADRATIC INTERPOLATION (LA GRANGE) $Y = FUNCT(X_0) \quad X < X_0$ $Y = FUNCT(X_n) \quad X > X_n$	
= LIMIT (P_1, P_2, X) IMITER	$Y = P_1 \quad X < P_1$ $Y = P_2 \quad X > P_2$ $Y = X \quad P_1 \leqslant X \leqslant P_2$	
= QNTZR (P, X) UANTIZER	$Y = kP \quad (k - 1/2)P \leqslant X \leqslant (k + 1/2)P$ $k = 0, \pm 1, \pm 2, \pm 3, \ldots$	
= DEADS (P_1, P_2, X) EAD SPACE	$Y = 0 \quad P_1 \leqslant X \leqslant P_2$ $Y = X - P_2 \quad X > P_2$ $Y = X - P_1 \quad X < P_1$	
= HSTRS (IC, P_1, P_2, X) (0) = IC YSTERESIS LOOP	$Y = X - P_2 \quad (X - X_{n-1}) > 0 \text{ AND}$ $Y_{n-1} \leqslant (X - P_2)$ $Y = X - P_1 \quad (X - X_{n-1}) < 0 \text{ AND}$ $Y_{n-1} \geqslant (X - P_1)$ OTHERWISE $\quad Y = $ LAST OUTPUT	
= STEP (P) TEP FUNCTION	$Y = 0 \quad t < P$ $Y = 1 \quad t \geqslant P$	
= RAMP (P) AMP FUNCTION	$Y = 0 \quad t < P$ $Y = 1 - P \quad t \geqslant P$	
= IMPUL (P_1, P_2) MPULSE GENERATOR	$Y = 0 \quad t < P_1$ $Y = 1 \quad (t - P_1) = kP_2$ $Y = 0 \quad (t - P_1) \neq kP_2$ $k = 0, 1, 2, 3, \ldots$	
= PULSE (P, X) ULSE GENERATOR	$Y = 1 \quad T_k \leqslant 1 = (T_k + X)$ $Y = 0 \quad$ otherwise $k = 1, 2, 3, \ldots$ $T_k = t$ of pulse k, P_k	
= SINE (P_1, P_2, P_3) = FREQUENCY IN RADIANS/SEC = PHASE SHIFT IN RADIANS RIGONOMETRIC SINE WAVE WITH MPLITUDE, PHASE, AND DELAY	$Y = 0 \quad\quad\quad 1 < P_1$ $Y = SIN [P_2 (1 - P_1) + P_3], 1 \geqslant P_1$	
= NORML $(P_1, P_2, P_3,)$ OISE GENERATOR ORMAL DISTRIBUTION)	$Y = $ GAUSSIAN DISTRIBUTION WITH MEAN, P_2, AND STANDARD DEVIATION, P_3 $(P_1 = $ ANY ODD INTEGER)	
= UZRPI (P_1) = UMIPI (P_1) = UATOB (P_1, P_2, P_3) OISE GENERATOR NIFORM DISTRIBUTION)	$Y = $ UNIFORM DISTRIBUTION 0 TO 1 $(P_1 = $ ANY ODD INTEGER) $-$ USE VARIABLE $Y = $ UNIFORM DISTRIBUTION, -1 TO $+1$ $Y = $ UNIFORM DISTRIBUTION, P_2 TO $P_2 + P_3$	

TABLE 4-2 (cont)

GENERAL FORM	FUNCTION
Y = INTGR (IC, X) Y(O) = 1C INTEGRATOR	$Y = \int_0^t X\,dt + 1C$ EQUIVALENT LAPLACE TRANSFORM $\cdot \dfrac{1}{S}$
Y = MODIN (IC, P_1, P_2, X) MODE-CONTROLLED INTEGRATOR	$Y = \int_0^t X\,dt + 1C \qquad P_1 = 1, P_2 = 0$ $Y = 1C \qquad\qquad\quad P_1 = O, P_2 = 1$ $Y = $ LAST OUTPUT $\quad P_1 = O, P_2 = 0$
Y = REALP (IC, P, X) Y(O) = IC 1ST ORDER SYSTEM (REAL POLE)	$P\dot{Y} + Y = X$ EQUIVALENT LAPLACE TRANSFORM, $\dfrac{1}{PS + 1}$
*	A, Y = LEDLG (IC, P_1, P_2, X) CALL LEDLG (IC, P_1, P_2, X, A, Y) A(O) = IC LEAD − LAG	$P_2\dot{A} + A = P_1\dot{X} \qquad Y = \int \dfrac{X - Y}{P_2} \qquad A = Y + \dfrac{(X - Y)P_1}{P_2}$ EQUIVALENT LAPLACE TRANSFORM $\dfrac{P_1 S + 1}{P_2 S + 1}$
*	Y, $\dot{Y}$ = CMPXP (IC_1, IC_2, P_1, P_2, X) CALL CMPXP (IC_1, IC_2, P_1, P_2, X, Y, $\dot{Y}$) Y(O) = IC_2, Y(O) = IC_1 2ND ORDER SYSTEM (COMPLEX POLE)	$\ddot{Y} + 2P_1 P_2 \dot{Y} + P_2^2 Y = X$ EQUIVALENT LAPLACE TRANSFORM, $\dfrac{1}{S^2 + 2P_1 P_2 S + P_2^2}$
**	Y = DERIV (IC, X) Y(O) = IC DERIVATIVE	$Y = \dfrac{dX}{dt}$ QUADRATIC INTERPOLATION EQUIVALENT LAPLACE TRANSFORM, S
***	Y = DELAY (N, P, X) P = TOTAL VAR. DELAY IN TERMS OF INDEP. VAR. N = MAX NO OF POINTS DELAY (INTEGER) DEAD TIME (DELAY)	$Y(t) = X(t − P) \qquad t \geqslant P$ $Y = 0 \qquad\qquad\qquad t < P$ EQUIVALENT LAPLACE TRANSFORM, e^{-PS}
Y = ZHOLD (P, X) Y(O) = 0 ZERO-ORDER HOLD	$Y = X \qquad\qquad\qquad P = 1$ $Y = $ LAST OUTPUT $\quad P = 0$ EQUIVALENT LAPLACE TRANSFORM $: \dfrac{1}{S}(1 − e^{-st})$
Y = IMPL (IC, ERROR, FUNCT) IMPLICIT FUNCTION	$Y = IC \qquad\qquad t = 0$ $Y = $ FUNCT (Y) $\quad t = 0$ $\mid Y − $ FUNCT $(Y) \mid \leqslant $ ERROR $\cdot \mid Y \mid$

SPECIAL FUNCTIONS

K = LOOK (X) LOCATE SYMBOL	K = RELATIVE LOCATION OF SYMBOL X AND VALUE OF X IN COMMON.
Y = DEBUG (N,T) N = INTEGER	DUMP CURRENT VALUES AND SYMBOLS FOR ALL MODEL VARIABLES N INTERATIONS STARTING AT TIME T.

Y = TRNFR (N, M, A, B, X)
    A = NUMER. COEFF. VECTOR
    B = DENOM. COEFF. VECTOR
 	
N, M = INTEGERS
Mth ORDER TRANSFER FUNCTION
	$\dfrac{Y}{X} = \dfrac{\displaystyle\sum_{i=0}^{N} A_j S^{N-i}}{\displaystyle\sum_{I=0}^{M} B_j S^{M-i}} \qquad \begin{array}{l} j = \exp + 1 \\ (N < M) \end{array}$

TABLE 4-2 (cont)

SWITCHING FUNCTIONS

$Y = FCNSW (P, X_1, X_2, X_3)$ FUNCTION SWITCH	$Y = X_1$ $Y = X_2$ $Y = X_3$	$P < 0$ $P = 0$ $P > 0$
$Y = INSW (P, X_1, X_2)$ INPUT SWITCH (RELAY)	$Y = X_1$ $Y = X_2$	$P < 0$ $P \geqslant 0$
*$Y_1, Y_2 = OUTSW (P, X)$ CALL OUTSW (P, X, Y_1, Y_2) OUTPUT SWITCH	$Y_1 = X, Y_2 = 0 \qquad P < 0$ $Y_1 = 0, Y_2 = X \qquad P \geqslant 0$	
$Y = COMPR (X_1, X_2)$ COMPARATOR	$Y = 0$ $Y = 1$	$X_1 < X_2$ $X_1 \geqslant X_2$
$Y = RST (P_1, P_2, P_3)$ RST FLIP-FLOP	$Y = 0$ $Y = 1$ $Y = 0$ $Y = 1$	$P_1 > 0$ $P_2 > 0, (P_1 \leqslant 0)$ $P_3 > 0, Y_{n-1} = 1, (P_2 \leqslant 0, P_1 \leqslant 0)$ $P_3 > 0, Y_{n-11} = 0.$
$Y = AND (X_1, X_2)$ $Y = OR (X_1, X_2)$ $Y = EOR (X_1, X_2)$ BOOLEAN OPERATORS	$Y = X_1 \cdot AND \cdot X_2$ $Y = X_1 \cdot OR \cdot X_2$ $Y = X_1 \cdot EOR \cdot X_2$ AND, OR, EOR = BOOLEAN OPERATORS	

*USE CALL FORM IF USING DSL SIMULATOR ONLY.
**USE WITH CAUTION—MATEMATICALLY UNSTABLE.
***P MAY VARY WITHIN THE SIMULATION BUT INITIAL P MUST BE MAX.

sequence of computation based on inputs and outputs to each element. An element is considered to be in sequence when all the inputs to that element have been processed previously in the current iteration cycle.

Structure and data statements of a DSL program may be written conveniently on a standard FORTRAN coding sheet by observing the conventions described below.

4.2.1. Structure Statement Format

1. Columns 2–5 of the first line of a statement may contain a statement number for cross reference. Blanks and leading zeros are ignored in these columns.

2. Column 6 of the first line in a statement must be either blank or zero.

3. Columns 7–72 contain the actual DSL statement. All blanks are ignored in these columns.

4. A statement may be continued on as many as six cards. Any card concluded with three consecutive decimal points is considered to be followed by a continuation card. Columns 1–72 of the next line will be considered a

continuation of the same statement. The continuation symbol (. . .) may appear any place in columns 7–72.

5. Cards with a D in column 1 are not processed, but the D is removed and the card is transferred as it is to the UPDAT subroutine. This feature may be used to insert a FORTRAN specification statement, such as DIMENSION, TYPE, and DATA, into the problem definition. Continuation cards for statements with a D in column 1 must be in the usual FORTRAN sense (i.e., with a nonzero column 6); these cards must also contain a D in column 1. All D cards appear at the beginning of the UPDAT subroutine. (*Caution*: Two DSL variables cannot be EQUIVALENCED because they will appear in COMMON.)

6. An asterisk (*) in column 1 denotes a comments card; the card is listed with other statements but is otherwise ignored. FORTRAN comment cards with a C in column 1 are *not* acceptable to DSL.

7. Columns 73–80 are not processed by DSL and may be used for identification.

8. Certain variable names are reserved for systems use and cannot appear in a DSL structure statement. These names are NALRM and ZZnnn (where n = any number). The reserved names TIME, DELT, DELMI, FINTI, DELTP, and DELTC may appear only in their intended context as explained under Data Preparation. In addition, DSL subroutine names must be used only as intended. This includes subroutines within the DSL simulator (ERR, FXINT, INPT, INTEG, MILNE, MVORG, ORGSC, PLOTS, PLTSW, PRINT, REMSW, RKS, SCDAC, TYPIN, and UPDAT) and all standard DSL and FORTRAN block names (explained under Use of Operational Elements). A standard block may be replaced by placing a substitute block of the same name on the disk.

4.2.2. Connection Statements

Connection statements are the most commonly used structure statements in a DSL problem. Each connection statement states the computation necessary to evaluate one or more output variables in terms of one or more input expressions. The general form of a connection statement is exactly the same as the form of the FORTRAN arithmetic assignment statement:

$$A=B$$

where A is a single output variable name and B is any valid expression. The equals sign (=) denotes assignment, or replacement, rather than equivalence.

If there is more than one output, the connection statement must be in primitive form, such as

$$A1, A2,..., AN=BLOCK(ARG2, AR2,...,ARGM) \tag{4.4}$$

No terms other than the block name and argument list may appear to the right of the assignment symbol. Some examples of connection statements are

$$\text{MULTI=-D*XDOT} \tag{4.5}$$

$$\text{R=SQRT(X**2+Y**2) + LIMIT (-3.25, 6.4, RATE) + ...} \tag{4.6}$$
$$\text{A2*P1}$$

$$\text{Y2=INTGRL(0.0, INTGRL (0., (2.0*FK1+FK2)/MX))} \tag{4.7}$$

$$\text{YDOT, Y=CMPXP(IC1, IC2, PAR2, PAR2, X2)} \tag{4.8}$$

Statement (4.5) illustrates use of (–) as a negative prefix to a variable name. Statement (4.6) shows the way in which a DSL statement may be continued on a following line. Statement (4.7) illustrates nested or cascaded INTGRL blocks and use of compound expressions in an argument list. Statement (4.8) illustrates the use of a DSL-supplied block, CMPXP, which has two outputs. This block also requires initial conditions, parameters, and an input.

Connection statements (except FORTRAN CALLs) may be written in any order and intermixed freely with DSL data statements. This feature gives DSL its nonprocedural nature. The user does not have to specify the order in which computations are to be performed. The DSL system establishes the correct sequence of computation based on the criterion that a statement should not be executed until the inputs to the statement have been evaluated for the current calculation interval (also called SORTING).

The FORTRAN CALL statement may be used to reference certain function blocks or subroutines from other libraries such as the 1130 scientific subroutine package. This statement must appear in a NOSORT section, since no outputs are specified for the statement.

4.2.3 Branching or Switching Statements

Groups of DSL connection statements may be effectively switched in or out of the model by the use of switching or branching statements. These statements are the FORTRAN "control" statements, used to alter the sequence of computations within a subroutine. These statements must appear within a NOSORT section in the DSL problem. A list of the FORTRAN statements which may be used for switching is given below. The definition and use of these statements is described in the FORTRAN manual.

Unconditional GO TO.

Computed GO TO.

CONTINUE.

4.3 Data Statements

Before discussing data statements, the terms *run, parameter study,* and *job* should be clarified. Simulation begins with the independent variable (TIME) set to zero. The model is then evaluated at successive points (in increments of DELT) until some terminating condition is reached. This constitutes a unit of simulation which will be referred to as a *run.* If a second run follows, TIME is reset to zero, and the simulation procedure is repeated. However, using a CONTIN statement (to be described later), one may continue a simulation without resetting TIME to zero. Such a run is part of the same *parameter study.* In other words, a parameter study consists of one or more runs of simulation extending over some positive range of the independent variable. The terms *job, problem,* and *simulation* will be used to refer to a computer application, which may include several runs or parameter studies.

There are two ways in which data may be entered into a DSL program: by means of DSL data statements discussed in this section, and by using FORTRAN READ statements, in the NOSORT section of the DSL structure statements.

DSL data statements are used to specify problem and system parameters for the simulation phase of a DSL job. Data statements may be classified according to the function they perform under the headings of (1) data inputs, (2) execution (simulator) control, and (3) data output. Data statements, with a few exceptions which are noted as they appear, are nonprocedural and may appear in any order. They may also be intermixed with the structure statements. A helpful point to remember is that data statements are collected by the DSL translator into a separate data set. Later, during the simulation phase, a DSL system routine named INTRAN processes them to generate input data for the simulation, one run at a time. Data statements do *not* appear in the FORTRAN subroutine UPDAT. In the following paragraphs we shall describe the format and function of each data statement. A summary of all DSL data statements appears in Section 4.3.3.

4.3.1 DSL Systems Variables

Variable names which have a special meaning to the system and may be set by the CONTRL data card are shown as follows:

Name	Description
TIME	Current value of simulation independent variable (set to zero by the system unless CONTIN is used)
DELT	Initial integration interval or step size for the independent variable (must be set)

DELMI	Minimum allowable integration interval or step size if a variable step integration is used (zero if not specified)
FINTI	Maximum simulation value for the independent variable (must be set)
DELTP	Print interval
DELTC	Not used by DSL—may be set by the user

4.3.2 Data Statement Format

1. Each data type is identified by a specific label that is punched left-adjusted in the first six columns of the card (or typed).

2. A statement may be continued on any number of cards. Any card concluded with three consecutive decimal points is considered to be followed by a continuation card.

3. Columns 7–72 contain the actual DSL data. All blanks are ignored by these columns. Alphabetic and numeric data may appear anywhere on the data card (or on a continuation card) following the label that specifies the type of data; i.e., data specification is free-form, as are the structure statements.

4. Columns 73–80 are not processed by DSL and may be used for sequencing or identification.

5. Data statements, with few exceptions, contain fields of variable names and real or integer constants which must follow the same specifications as in structure statements. An equals sign, comma, or statement ending is considered to be the end of a field. Since these data statements imply some action on the model structure, all variable names used must also appear in the structure or must be reserved system names. Several cards require an alphabetic field followed by a numeric which is to be converted to a real or integer constant and considered as the current value of the preceding variable, e.g., XDOT=250. Numeric fields may be integers or floating-point numbers; the latter are identified by a decimal point and follow the rules for real constants. A blank numeric field is read as zero. A minus sign precedes a negative number. Examples of numeric fields are

$$1.524, \quad -1.5E\text{-}5, \quad .426E3, \quad 4$$

Exceptions to the above format are card types LABEL and TITLE, where columns 7–46 are considered a single field, and card types SORT, NOSORT, ENDPRO, END, STOP, and CONTIN, which are blank in columns 7–72.

4.3.3 Summary of DSL Data Statement Formats

TABLE 4.3

Problem Data Input

Label, cols. 1–6	Function (by example), cols. 7–72
PARAM	TAU=25., PAR=3.158E3, C4=2.0E–5
CONST	CON1=45.3, PI=3.14159, K=3
INCON	IC1=20., A=50.2, IC3=0
AFGEN	FCN=3., 25., 5.2, 26.4, 6.0, 24., 7.5, 21.3
NLFGEN	FY3=0., 850., 5., 1234., 8., 1.574E3, 12.4, 1.3E03
TABLE	PAR1(8)=4.5, INPUT(1-4)=2., 2*8.6, 3.52E3

TABLE 4.4

Problem Output Control

Label, cols. 1–6	Function (by example), cols. 7–72
PRINT	0.1, X, XDOT, VELOC
TYPE	X, XDOT, Y
TITLE	MASS, SPRING, DAMPER SYSTEM IN DSL
SCOPE	.05, 6, 4, TIME, XDOT
GRAPH	.05, 10, 8, TIME, X
SCALE	2.0, .06
LABEL	MASS, SPRING, DAMPER SYSTEM – 4/1/67
RANGE	X, XDOT, VELOC, DELT
RESET	GRAPH, PRINT, RANGE

TABLE 4.5

Problem Execution Control

Label, cols. 1–6	Function (by example), cols. 7–72
CONTRL	DELT=.002, FINTI=8.0, DELMI=1.OE–10
FINISH	DIST=0., ALT=5000
INTEG	MILNE
RELERR	X=1.E-4, XDOT=5.E-5
ABSERR	X=1.E-3, XDOT=1.E-4
CONTIN	
RESET	RELERR, FINISH

TABLE 4.6
DSL Translator Pseudooperations

Label, cols. 1–6	*Function (by example), cols. 7–72*	
RENAME	TIME=DISPL, DELT=DELTX	
INTGER	K, GO	
MEMORY	INT(4), DELAY(100)	
STORAG	IC(6), PARAM(10)	
INTGRL	Y, YDOT	
SORT		
NOSORT		
PROCED	X=FCN(A, B, PAR5, IC3)	
.		
.		
.		
ENDPRO		
*LOCAL (UPDAT, INTGR, AFGEN, IMPL)		(1800)
*LOCALSIMUL, PRINT, XYPLT, TYPIN, AFGEN, IMPL		(1130)
DUMP		
END		
STOP		

4.3.4 Problem Data Input

These are the "twiddling" statements. Data statements in this group are used to set problem program parameters for successive runs or parameter studies. In the following description, each of the problem data input statements is listed by example, followed by a brief description of the statement function. Section 4.3.3 shows examples of the statements as they appear in card formats.

1. CONST	PI=3.14159, N1=5, STEP=0.
2. INCON	XO=1.2, Y0=0.
3. PARAM	MULT=5.2, RATE=5000.

Columns 7–72 contain assignment fields which state the name of the program variables and the value which is to be assigned to the variable. The three labels perform identical functions and differ only in mnemonic content. The value specified is always treated as a real constant unless the variable has been declared as an integer type (see Translator Commands section).

4. AFGEN	CURV1= 1., 300., 1.5, 400., 2., 250.
NLFGEN	CURV2= -3., 2., -1., 1., .7, 4., 1.5, 6.

This statement stores tabular information for use by an AFGEN (arbitrary function generator) block or a NLFGEN (nonlinear function generator) block. Columns 7–72 contain the unique name (CURV1 in the above example) which identifies the set of ordered pairs of data points that follow. This name is the means by which this table is referenced. The first value and alternating values thereafter are those of the independent variable and must be presented in algebraically increasing order. Increments may be of unequal size. Each of these values must be followed by its corresponding coordinate value. The list of values may extend to continuation cards. An AFGEN (or NLFGEN) data statement which immediately follows another AFGEN (or NLFGEN) data statement with the same variable name will effect a continuation of the preceding table. There may be any number of unique AFGEN (or NLFGEN) functions, each with its own name, in a DSL problem. There is no limit to the number of data points in each function. Data values are stored in COMMON, the size of which can be altered by the user to accommodate more or less points.

5. AFGEN   OVRLY, CURV1= 1., 20., 2., 30., 3.5, 60.
   NLFGEN   OVRLY, CURV2= 3., 2., 0., 1., 1.2, -1.

The OVRLY option is used with the AFGEN and NLFGEN statements to conserve memory by replacing the previous table of the same function name with a new table of values. This feature can be used only if the overlaying table does not contain more entries than it did when defined without overlay. The identifier, OVRLY, following AFGEN or NLFGEN must be separated from the function name by a comma. The TABLE statement may also be used to alter individual points in an AFGEN table.

6. TABLE   STRNG(9)=4.3, GROUP(1-7)=2., 3., 4*0., 8.

This statement is used to generate data values into single dimensioned arrays defined by the STORAG translator command, or in AFGEN (or NLFGEN) tables. More than one variable name may appear on a TABLE statement, and up to six continuation cards may be used. In the example, the ninth element of the array named STRNG will be set equal to 4.3. The appearance of two integer constants separated by a minus sign (–) within parentheses signifies that a range of elements is to be filled. Elements 1–7 in the array named GROUP will be assigned values from the string of constants following the assignment symbol. The asterisk (*) in the string of constants is used to indicate a repetition factor. 4*0.0 is equivalent to writing 0.0, 0.0, 0.0, 0.0.

4.3.5. Execution Control

Statements in this group are used to set DSL system parameters for execution of the simulation. Statements are listed by example, followed by a description of the use and function of each statement.

1. CONTRL   DELT=0.01, FINTI=1.15, DELMI=4.E-5

The format of this statement is identical to the PARAM card. The CONTRL statement is used to initiate values for the following reserved system variables:

TIME: current value of the independent variable of simulation (set to zero by the system unless control statement CONTIN is used).

DELT: Initial integration interval or step size of the independent variable. This value must be set greater than zero; otherwise the program will go into a loop.

DELMI: Minimum allowable integration interval or step size. This variable is used when an integration method such as the Milne or the fourth-order Runge-Kutta scheme is used where the step size is determined by the error criterion specified. If the value for DELMI is not specified, it will be set for zero, and when a very tight error criterion is specified (for example, relative error less than .00006%) the integration time can become extremely large for the computer solution of the problem.

FINTI: Value of the independent variable at which simulation is to be terminated. This value must be set.

DELTP: Print interval. The program produces the print output at intervals of DELTP.

2. FINISH   ALT=0, X=5000., etc.

In addition to simulated and actual run times reaching some prescribed value, a run can be ended when any dependent variable reaches some significant point. Thus, simulation of the current problem terminates if ALT reaches 0 (or X reaches 5000.) before the specified FINTI has elapsed. Only one FINISH card is recognized per run; if multiple FINISH cards are encountered, the last FINISH card will be accepted.

3. RELERR   XDOT=1.E-5, X=1.E-4, etc.

A relative error may be specified for each integrator output to be used by the chosen integration routine to control error. This is used only for integration routines which are allowed to vary the integration interval to satisfy error bounds. These are RKS and MILNE. If any relative errors are specified, then the last error specified is applied to all integrators that are unspecified. If none are specified, the error is set at .005.

4. ABSERR   XDOT=1.E-4, etc.

This statement specifies an absolute error control in the same manner as in RELERR. If ABSERR is not specified, the error is set at .0005. The calculation of integration step size based on these errors is described in Chapter 9.

5. CONTIN

This statement forces the continuation of the current parameter study from the point at which it is terminated in a run. TIME is not reset to zero, and output specification, integration interval, error bounds on variables, or parameter values may be changed prior to continuing the run.

6. INTEG RKS

The name of one of the DSL integration routines appears in columns 7–72. The specified integration routine will be used for all parameter studies until some other routine is requested via another INTEG statement. The available integration routines are

RKS: Fourth-order Runge-Kutta integration method with variable integration interval. The discrepancy between integrated values and a Simpson's rule check is maintained within the error bounds.

RKSFX: Fourth-order Runge-Kutta with fixed integration interval. No error test is made.

MILNE: Fifth-order predictor-corrector method by Milne with variable integration interval.

TRAPZ: Trapezoidal integration with fixed integration interval.

ADAMS: Adams-Moulton integration method with fixed integration interval.

If the integration method is not specified by the user, the MILNE routine will be used. Details on the integration methods and the algorithms used for integration will be found in Chapter 9.

7. RESET   PRINT, GRAPH FINISH, etc.

This statement eliminates specified output features in columns 7–72 as though unspecified. When used with PRINT, SCOPE, GRAPH, RANGE, or FINISH the corresponding list index is set equal to zero. (Note that each appearance of a parameter in RESET causes a reset of the list index so that only one may appear per run.) Two other data labels may be used with RESET to set all corresponding table values to zero. These are RELERR and ABSERR. Leaving columns 7–72 on a RESET card blank is the same as specifying PRINT, SCOPE, GRAPH, and RANGE. A way of resetting structure and parameter variables is to respecify their current value rather than to use the RESET card.

4.3.6 Data and Result Output

Statements in this group are used to specify the form and timing of data output from the simulation. Several of these statements specify a data output interval, or communication interval, at which problem data are written to some output device. As the simulation proceeds, each time the independent variable (usually TIME) reaches a value of communication interval, the simulator either performs the required output function or passes control to the proper subroutine. For best results, the data output intervals should be a multiple of the specified integration size, especially for the fixed step integration routines. In the following description, the problem output control statements are listed by example, followed by a brief description of the format and function of each statement.

1. PRINT   .01, X, Y, NUMB, YDOT

The first field appearing in columns 7–72 is numeric and specifies the interval at which problem data are to appear in the program listing. All other fields contain names of program variables whose values are to be listed at each print interval. Up to 25 variables may appear in a PRINT statement. If a PRINT statement is used, the value of the independent variable (TIME) will be automatically printed at each interval. The named variables may include variables declared as integer variables via the INTGER command.

2. TITLE   MASS-SPRING-DAMPER PROBLEM

The text which appears in columns 7–72 of the TITLE statement or statements will be printed at the head of each page of the output listing, one line per TITLE statement. There may be three TITLE statements in effect at any time. The TITLE statement may not be continued over more than one line.

3. TYPE   X, XDOT, VAR1, etc.

Current variables listed in columns 7–72 will be printed on the system typewriter (1053 in the case of the 1800 and the console typewriter on the 1130). This statement should be used following an interrupt of a run or at the end of a run. If the typewriter is used during the run for printing intermediate values of the variables, excessive run time may be required for problem solution.

4. GRAPH   .05, 8, 11, TIME, XDOT
    SCOPE   .05, 8, 11, TIME, XDOT

This statement creates a plot either on the 1627 plotter or a storage scope connected to the 1800 for two variables. The first numeric field in columns 7–72 contains the plot interval of the independent variable. The next two numeric fields contain the lengths (in inches) of the horizontal and vertical axes, respectively. If the lengths are not specified, the axes will default to 12-inch horizontal and 8-inch vertical lengths. The first variable listed after the numeric field will be used as the variable for the horizontal axis. The next variable will be plotted on the vertical axis. Only one GRAPH or SCOPE card (or statement) is allowed per run. Also, in a single GRAPH statement only one independent variable is allowed. When problem solution is restarted with a new parameter value and the independent variable TIME is reset to zero the pen returns to the 0, 0 print position on the plotter. A new graph may be started by the use of data switch 5, as discussed in a later chapter.

LABEL   GRAPH OF VELOCITY VERSUS TIME

Columns 7-46 in this statement will be used to write a label below the horizontal axis. When LABEL is used it overlays and replaces the third TITLE card. If no LABEL is used, the third TITLE statement is used to provide the LABEL. The names of the variables in the GRAPH statement are printed adjacent to the horizontal and vertical axis.

SCALE   2., .06

The two numeric values in columns 7–72 specify the multiplying factors for the independent and dependent variables, respectively, appearing on the GRAPH statement. Thus, if the maximum value of TIME (FINTI) is 100 and the horizontal axis length is chosen as 10 inches, the scale factor will be .1. In the case of the dependent variable, if the maximum value is not known, a guess must be made in order to predict the scale factor, to keep the plot within the specified lengths of the axes.

RANGE   X, XDOT, Y, etc.

The minimum and maximum values reached during the simulation run for the specified variables in columns 7-72 are printed at the end of the run. These values are collected at integration step interval and not at the print interval.

In problem data output in DSL only one set of output cards PRINT, GRAPH (or SCOPE), SCALE, LABEL, and RANGE is recognized per problem. If more than one card is specified for, say, PRINT, the last in sequence will be used to print the output. TITLE cards overlay those cards of the previous data set.

4.4 Translator Commands or Pseudooperations

Translator commands are used to request special operations by the translator as it converts the DSL structure statements into the FORTRAN subroutine names UPDAT.

1. Columns 1-6 contain a 1-6 character command beginning in column 1. There are a few exceptions.

2. Columns 7-72 are used for additional detail required for certain commands.

RENAME   TIME = DISPL, DELT = DELTX, etc.

With this card names reserved by the system can be altered. The substitute names appear on all output and should be used in the user's structure statements. The following six names reserved by the system may be changed; TIME, DELT, DELMI, FINTI, DELTP, and DELTC.

INTGER   K, GO, COUNT, NO

The listed variables are declared as integers within the UPDAT subroutine and can then (and only then) be used as integers in DSL structure and/or FORTRAN statements.

MEMORY   INT(4), DELAY(2) etc.

Columns 7-72 contain function block name followed by integers enclosed in parentheses. The value of the integer indicates the number of words to be allocated to the named function as memory. The space thus allocated may be used for history. A pointer to the reserved storage will be inserted into the argument list of the function block reference. The user must initialize the values in the MEMORY blocks by inserting appropriate input values via PARAM statements. For standard DSL blocks, memory is assigned by the system and need not be the concern of the user. (Further details on MEMORY blocks will be described in DSL system features, Chapter 8.)

STORAG   IC(6), PARAM(8), etc.

Variable names appearing on this card are considered to be subscripted. The number within parentheses must be the maximum storage locations necessary to contain data for the corresponding variable. Storage variables must be implicitly REAL. Data for the subscripted variables are not allowed by PARAM, CONST, or INCON commands but only by the use of the TABLE data card.

INTGRL   CMPXP(2), MYBLOK(4), Y, YDOT

Columns 7-72 specify the names of user-supplied functions (called I blocks) which contain integrators. The number of integrators contained in each block is indicated, following the block name, within parentheses. Commas are used to separate block names. As a result of this command, for each reference to one of these blocks, the translator will reserve the required amount of storage in the integrator areas and insert a pointer into the argument list of the function block reference. The pointer may be used to find the integrator output area which was reserved. The storage and pointer are unique for each reference of the block name within the DSL problem program. For example, Y in the function LEDLG, listed in the table of DSL functions, and Y and $\dot{Y}$ in CMPXP could be specified as INTGRL if these were not standard DSL blocks.

SORT

Specifies that all DSL structure statements following this statement are to be sequenced such that inputs to the statement are either previously computed in the current iteration cycle or available as parameter data input. Variables appearing in one SORT section of the program but being computed in a later section are identified as an error. (If no SORT or NOSORT card appears, the SORT option is exercised.)

NOSORT

Specifies that all DSL structure statements following this statement are not to be sequenced (i.e., they remain procedural by using the input sequence for computation). Note that the SORT-NOSORT pair gives the user the facility to cause sections of the model to be sorted or not, depending on the function to be performed (see Chapters 5 and 6 for examples on the use of this statement).

PROCED
$$X = FCN(A, B, PAR5, IC3)$$

ENDPRO

Columns 7-72 are blanks. This statement itself is not executed but serves to delimit a section of procedural code and define its output and input names. The statements between a PROCED command and an ENDPRO command will not be sequenced, but the entire set of statements will be sorted as a unit relative to other connection statements according to the output and input names (see Chapters 5 and 6 for an example of the use of these statements).

END

Columns 7-72 are blank. Specifies the end of data input for a particular parameter study and the beginning of the actual simulation.

STOP

Columns 7–72 are blank. This card is intended to end the run immediately. It should follow the last END card. It is also used by the DSL translator to signify the end of all DSL input.

DUMP

Columns 7–72 are blank. The DUMP pseudo-op will cause a dump of all tables related to DSL language translator and sort routines. This card is intended to aid the user in debugging the program.

For the 1800,

*LOCAL (UPDAT, INTGR, AFGEN, IMPL)

For the 1130,

*LOCALSIMUL, PRINT, XYPLT, TYPIN, AFGEN, IMPL

These cards are used when the user wishes to conserve storage and allow larger space for expanding simulation model. One *LOCAL is permitted in each system, and the format must be exactly as shown above. For details on the general use of *LOCAL cards, see Reference (13).

*PROCESS OBJECT (R, LIB(R1))

This card is used in System/3, where the subroutines are to be stored on the disk pack containing the DSL programs. It specifies that the object program obtained by the FORTRAN compiler is to be stored on removable disk 1 (R1) and that the present subroutine is to replace the previous subroutine with the source name on the disk, if it is previously stored on that disk.

5 APPLICATION OF SIMULATION

5.0 Examples are Most Important

While Chapters 3 and 4 may be somewhat unappealing to the average reader who is interested in quickly solving his engineering problem, they are necessary to specify all the rules required of any computer language. The best tool of learning a computer language for a student is via examples that he finds in his own field. The purpose of Chapters 5 and 6 then is to provide as many examples as possible in the various fields of engineering and at the same time to avoid excessive redundancy in the methodology of the solution. These chapters are divided, therefore, into sections separated to some extent by engineering discipline. In each section, the simpler examples are given first, and more complex problems are illustrated in the latter part of each section.

5.1 General Engineering Problems

5.1.1 Mass-Spring Damper System

Let us revert to the example shown in Chapter 3 and Fig. 1.1 to start with the simple mass-spring damper system and follow it up with more complexity by changing the boundary conditions.

If the mass is displaced from its rest position and then released, it oscillates until the energy is dissipated by the damper.

An equation representing the system is

$$MX'' + BX' + KX = 0 \tag{5.1}$$

with the boundary conditions

$$X(0) = 0 \quad \text{and} \quad X(0) = A \tag{5.2}$$

To solve this problem the DSL statements required are shown in Table 3.9. Now suppose that instead of a single spring with a spring constant K, the system has dead spaces as shown in Fig. 5.1 and that the spring forces are functions of the displacement and the dead spaces. The equation of motion then becomes

$$MX'' + BX' + KX = F(s) \tag{5.3}$$

where $F(s)$ represents the different spring forces that act each half-cycle. When

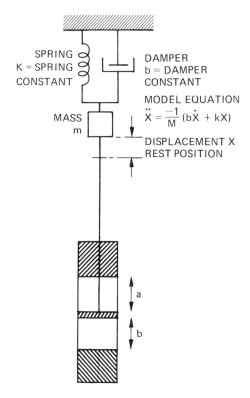

SPRING
K = SPRING
CONSTANT

DAMPER
b = DAMPER
CONSTANT

MODEL EQUATION

$$\ddot{X} = \frac{-1}{M}(b\dot{X} + kX)$$

MASS
m

DISPLACEMENT X
REST POSITION

a

b

Figure 5.1.   Mass spring damper with dead spaces.

the displacement of the mass is positive, the spring is represented by

$$F(s) = 0 \qquad \text{if } X \leqslant b$$
$$F(s) = K_2 \ (X\text{-}b) \qquad \text{if } X > b \qquad \qquad (5.4)$$

and when the displacement is negative the equation is

$$F(s) = 0 \qquad \text{if } X \geqslant a$$
$$F(s) = K_1 \ (\mid X \mid - a) \qquad \text{if } \mid X \mid < a \qquad \qquad (5.5)$$

The formulation of this problem in DSL thus involves dynamically changing a function during the problem solution. Notice that the spring force function is discontinuous. A careful evaluation of the available functions in the DSL library shown in Table 4.2 indicates that the functions FCNSW and INSW will provide the necessary discontinuous functions to complete the problem solution. The DSL statements for the solution of this problem are shown in Table 5.1, where the use of FCNSW and INSW is illustrated. Tables 5.2 and 5.3

TABLE 5.1
DSL Input for the Mass-Spring Problem with Dead
Spaces-Translator Input

```
                    *** DSL/1800 TRANSLATOR INPUT ***
TITLE MASS SPRING DAMPER WITH DEAD SPACES
      XDOT=INTGR(0.,(-B*XDOT+FOFS-K*X)/M)
      X=INTGR(XO,XDOT)
      S1=ABS(X)-A
      S2=X-B
      Y1=K1*S1
      Y2=K2*S2
      FOFS=FCNSW(X,F1,0,F2)
      F1=INSW(S1,0.,Y1)
      F2=INSW(S2,0.,Y2)
INTEG MILNE
PARAM M=9.,B=1.5,A=1.2,K=2.,K1=3.8,K2=3.6
INCON XO=10.
CONTRL DELT=0.05,FINTI=10.
PRINT .2,X,XDOT
END
STOP

OUTPUT VARIABLE SEQUENCE
   4   6   9   3   5   8   7   1   2

STORAGE USED/MAXIMUM
INTGR   2/50, IN VARS  23/300, OUT VARS   9/100, PARAMS 13/60,
 SYMBOLS 39/200, FORT WDS   99/5000, DATA WDS   87/1000
```

TABLE 5.2

Translator Output for the Mass-Spring Problem

```
SUBROUTINE UPDAT
REAL   INTGR,K      ,M       ,K1     ,K2      ,INSW
COMMON NALRM,IZZZ(833) ,TIME ,DELT ,DELMI,FINTI,DELTP,DELTC,XDOT ,
1X     ,S1     ,S2     ,Y1     ,Y2     ,FOFS ,F1     ,F2     ,M     ,B     ,A     ,
1K     ,K1     ,K2     ,X0
S2=X-B
Y2=K2*S2
F2=INSW(S2,0.,Y2)
S1=ABS(X)-A
Y1=K1*S1
F1=INSW(S1,0.,Y1)
FOFS=FCNSW(X,F1,0,F2)
XDOT=INTGR(0.,(-B*XDOT+FOFS-K*X)/M)
X=INTGR(X0,XDOT)
RETURN
END

                   *** DSL/1800 SIMULATION DATA ***
TITLE MASS SPRING DAMPER WITH DEAD SPACES
INTEG MILNE
PARAM M=9.,B=1.5,A=1.2,K=2.,K1=3.8,K2=3.6
INCON X0=10.
CONTRL DELT=0.05,FINTI=10.
PRINT .2,X,XDOT
END
```

show the computer run, which includes the UPDAT subroutine produced by the DSL translation and the results of simulation for two sets of parameters. Note that the function FCNSW must be used in conjunction with INSW to meet the boundary conditions of Eqs. (5.4) and (5.5). Note also the use of the FORTRAN function ABS to extract the absolute value of S. The results from the first set of K values show an unstable system; the second set yields a stable system, plots of which are shown in Fig. 5.2.

5.1.2 A Simple Control System

The following example is typical of problems encountered by control engineers interested in simulating complex control systems to observe their responses to known input or other disturbances. By variation of the parameters such as time constants, gain constants, and reset constants, the control engineer will be able to arrive at a satisfactory stable control system design after multiple DSL runs on the computer. The plotting features of the DSL system become extremely useful for the control engineer. His familiarity with time response plots and Nyquist plots assists him in quickly determining a satisfactory control system design.

TABLE 5.3

Simulation Output for the Mass-Spring Problem

MASS SPRING DAMPER WITH DEAD SPACES

TIME	X	XDOT
0.000E 00	0.1000E 02	0.0000E 00
0.200E 00	0.1002E 02	0.2319E 00
0.400E 00	0.1009E 02	0.4579E 00
0.600E 00	0.1020E 02	0.6796E 00
0.800E 00	0.1036E 02	0.8989E 00
0.100E 01	0.1056E 02	0.1117E 01
0.120E 01	0.1081E 02	0.1336E 01
0.140E 01	0.1110E 02	0.1557E 01
0.160E 01	0.1143E 02	0.1782E 01
0.179E 01	0.1181E 02	0.2012E 01
0.200E 01	**0.1223E 02**	0.2248E 01

TIME	X	XDOT
0.100E 02	**0.1297E 03**	0.4378E 02
0.100E 02	**0.1297E 03**	0.4378E 02

```
                          *** DSL/1800 SIMULATION DATA ***
PARAM K1=-2.3,K2=2.,K=4.
SCALE 1.,1.
LABEL X VERSUS TIME FOR MS/SP WITH DEAD SPACES
GRAPH .1,8,10,TIME,X
END
```

MASS SPRING DAMPER WITH DEAD SPACES

TIME	X	XDOT
0.000E 00	0.1000E 02	0.0000E 00
0.200E 00	0.9949E 01	-0.5019E 00
0.400E 00	0.9800E 01	-0.9830E 00
0.600E 00	0.9557E 01	-0.1439E 01
0.800E 00	0.9226E 01	-0.1868E 01
0.100E 01	0.8812E 01	-0.2267E 01
0.120E 01	0.8321E 01	-0.2633E 01
0.140E 01	0.7761E 01	-0.2964E 01
0.160E 01	0.7138E 01	-0.3258E 01
0.179E 01	**0.6460E 01**	**-0.3514E 01**
0.200E 01	**0.5734E 01**	**-0.3731E 01**

MASS **SPRING DAMPER WITH DEAD SPACES**

TIME	X	XDOT
0.100E 02	0.6198E 00	0.2380E 01
0.100E 02	0.6198E 00	0.2380E 01

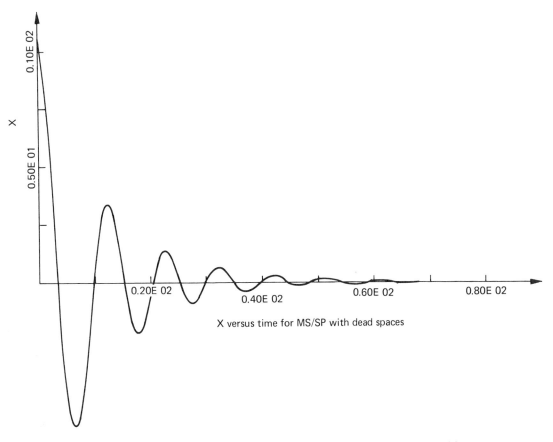

X versus time for MS/SP with dead spaces

Figure 5.2. Simulation results for a stable mass spring system with dead spaces.

Figure 5.3 shows a control system with several components containing first-order transfer functions. The system consists of a compensator, an amplifier, an exciter, a generator, and a load. The feedback is with a variable FGAIN. It is desired to observe the system response with variation in T and FGAIN.

The compensator block can be simulated using the LEDLG subroutine provided by the program as shown in Table 4.2. The exciter, the generator, and the load blocks can be simulated using the functions REALP in the same table. Alternatively, the blocks exciter and generator or the blocks load and generator can be combined to yield the Laplace transform in the form

$$\frac{1}{S^2 + 2P_1 P_2 S + P_2^2} \tag{5.6}$$

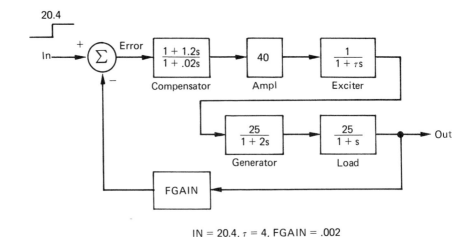

IN = 20.4, τ = 4, FGAIN = .002

Figure 5.3.   Control system (first-order transfer functions).

so that the subroutine CMPXP may be used. Another method is to use the DSL function TRNFR if we were to combine all the blocks as shown in Fig. 5.4.

Table 5.4 shows the DSL statements for simulation of the control system wherein LEDLG and REALP are used, whereas Table 5.5 shows the DSL statements where TRNFR is used. In addition, the tables show the UPDAT routines generated for the two cases. Figure 5.5 shows the response of the system output as a result of a step input, whereas Table 5.6 shows the simulation results.

In the case where the DSL function TRNFR is used, we also note that the parameters A and B are multidimensioned arrays. That is, A and B are subscripted. Table 5.5 thus also illustrates the use of TABLE and STORAG language features of DSL to handle subscripted variables. More extensive use of this facility is further described with an illustration in Chapter 6.

Note also the fact that the translator output will show, as a part of data read by the SIMX program of DSL, parameters A and B with the values of 16 and 18. These numbers are actually the indices of the subscripted variables A and B described by the STORAG statements. During execution of the simulation

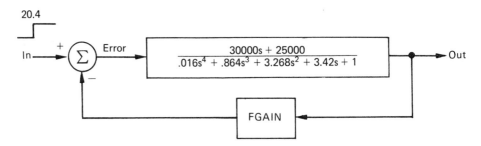

Figure 5.4.   Control system (high-order transfer function).

TABLE 5.4
DSL Input for the Control System and its
UPDAT Subroutine

```
                      *** DSL/1800 TRANSLATOR INPUT ***
     TITLE  CONTROL SYSTEM WITH 1ST ORDER TRANSFER FCNS
     PARAM INPUT=20.4,FGAIN=.002,TAU=.4
            ERROR=INPUT-FGAIN*LOAD
            COMP,Y=LEDLG(0.,1.2,.02,ERROR)
            AMPLF=40.*COMP
            EXCIT=REALP(0.,TAU,AMPLF)
            GEN=REALP(0.,2.,25.*EXCIT)
            LOAD=REALP(0.,1.,25.*GEN)
     PRINT .08,ERROR,GEN,LOAD
     CONTRL FINTI=4.,DELT=0.02
     GRAPH 0.02,8,8,TIME,LOAD
     LABEL SIMPLE CONTROL SYSTEM
     SCALE 2.,.0004
     END
     STOP


     SUBROUTINE UPDAT
     REAL   INPUT,LOAD ,LEDLG
     COMMON NALRM,IZZZ(833) ,TIME ,DELT ,DELMI,FINTI,DELTP,DELTC,Y
    1EXCIT,GEN  ,LOAD ,ERROR,COMP ,AMPLF,INPUT,FGAIN,TAU
     ERROR=INPUT-FGAIN*LOAD
     CALL LEDLG(0.,1.2,.02,ERROR,COMP,Y)
     AMPLF=40.*COMP
     EXCIT=REALP(0.,TAU,AMPLF)
     GEN=REALP(0.,2.,25.*EXCIT)
     LOAD=REALP(0.,1.,25.*GEN)
     RETURN
     END
VARIABLE ALLOCATIONS
 NALRM(IC)=7FFF       IZZZ(IC)=7FFE-7CBE   TIME(RC)=7CBC     DELT(RC)=7CBA
 DELTP(RC)=7CB4       DELTC(RC)=7CB2          Y(RC)=7CB0     EXCIT(RC)=7CAE.
 ERROR(RC)=7CA8       COMP(RC)=7CA6       AMPLF(RC)=7CA4     INPUT(RC)=7CA2

FEATURES SUPPORTED
 ONE WORD INTEGERS

CALLED SUBPROGRAMS
 LEDLG   REALP   FMPY    FLD     FSTO     FSBR

REAL CONSTANTS
 .000000E 00=0002    .120000E 01=0004    .200000E-01=0006    .400000E 02=0008
 .100000E 01=000E

CORE REQUIREMENTS FOR UPDAT
 COMMON   866  VARIABLES    2   PROGRAM    72

RELATIVE ENTRY POINT ADDRESS IS 0010 (HEX)

END OF COMPILATION
```

phase, these numbers will be used to locate their relative position in COMMON storage.

The function TRNFR must have the proper coefficients, as shown by the formulations in Table 5.7.

TABLE 5.5
DSL Input for the Control System Using the DSL
Transfer Function

```
                        *** DSL/1800 TRANSLATOR INPUT ***
TITLE CONTROL SYSTEM USING TRANSFER FUNCTION
        LOAD=TRNFR(1,4,A,B,ERROR)
        ERROR=INPUT-FGAIN*LOAD
STORAG A(2),B(5)
PARAM INPUT=20.4,FGAIN=.002
TABLE A(1-2)=25000.,30000.,B(1-5)=1.,3.42,3.268,.864,.016
CONTRL FINTI=4.,DELT=.04
PRINT .08,ERROR,LOAD
END
STOP

SUBROUTINE UPDAT
REAL LOAD ,INPUT
COMMON NALRM,IZZZ(833) ,TIME ,DELT ,DELMI,FINTI,DELTP,DELTC,LOAD ,
1ZZ001,ZZ002,ZZ003,ERROR,INPUT,FGAIN
1 ,ZZ990( 2) ,A    ( 2) ,B    ( 5)
ERROR=INPUT-FGAIN*LOAD
LOAD=TRNFR(1,4,A,B,ERROR)
RETURN
END
                        *** DSL/1800 SIMULATION DATA ***
PARAM  A    = 16 ,B    = 18
TITLE CONTROL SYSTEM USING TRANSFER FUNCTION
PARAM INPUT=20.4,FGAIN=.002
TABLE A(1-2)=25000.,30000.,B(1-5)=1.,3.42,3.268,.864,.016
CONTRL FINTI=4.,DELT=.04
PRINT .08,ERROR,LOAD
END
```

While the problem shown here considers a step input, the input can be made either sinusoidal using the standard SIN(X) function or any other function from the library (any one of the functions provided in the DSL library shown in Table 4.2), a function from the users library, or a generated function based on linear or quadratic interpolation from values supplied by the user. The last option can be implemented easily using the AFGEN and NLFGN facilities of the DSL program. These facilities will be illustrated with another example later in this chapter.

This illustration shows the ease with which DSL can be used for examining the characteristics of a proposed control system in order to design a stable control system. A large number of built-in functions shown in Table 4.2 are in fact provided for the control engineer to facilitate his use of DSL. Historically, both analog and digital computer simulation were the product of control system simulation requirements, and therefore it is not surprising that the various functions required in this discipline are provided by simulation languages. As we shall see in the examples that follow, many of these built-in functions can also

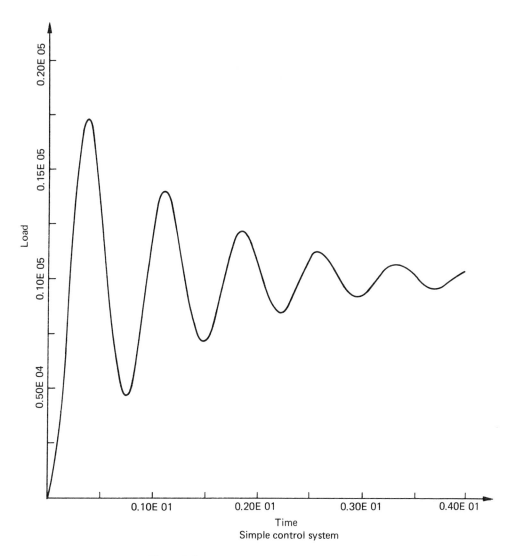

Figure 5.5. Control system response.

be used in other disciplines.

The use of PLOT and SCOPE (in the 1800) facilities is, to a control engineer, very valuable in that they provide him quickly with the expected behavior of the system under simulated disturbances, which can be used as input with the various DSL functions. In this case, the SCOPE function is considerably faster than the 1130/1800 hard copy plotter. Unfortunately, the System/3 users do not have this capability, and this author can only suggest enterprising users to find a programmer who can write an interface routine to perform a PRINT-PLOT

TABLE 5.6

Simulation Results for the Control System

```
                            *** DSL/1800 SIMULATION DATA ***
TITLE   CONTROL SYSTEM WITH 1ST ORDER TRANSFER FCNS
PARAM   INPUT=20.4,FGAIN=.002,TAU=.4
PRINT   .08,ERROR,GEN,LOAD
CONTRL  FINTI=4.,DELT=0.02
GRAPH   0.02,8,8,TIME,LOAD
LABEL   SIMPLE CONTROL SYSTEM
SCALE   2.,,.0004
END
```

CONTROL SYSTEM WITH 1ST ORDER TRANSFER

TIME	ERROR	GEN	LOAD
0.000E 00	0.2040E 02	0.0000E 00	0.0000E 00
0.799E-01	0.1759E 02	0.1674E 04	0.1400E 04
0.159E 00	0.8270E 01	0.3078E 04	0.6064E 04
0.239E 00	-0.3247E 01	0.3168E 04	0.1182E 05
0.319E 00	-0.1175E 02	0.2061E 04	0.1607E 05
0.399E 00	-0.1399E 02	0.3806E 03	0.1719E 05
0.479E 00	-0.9774E 01	-0.1082E 04	0.1508E 05
0.559E 00	-0.1715E 02	-0.1728E 04	0.1105E 05
0.639E 00	0.6253E 01	-0.1382E 04	0.7073E 04
0.719E 00	0.1073E 02	-0.3161E 03	0.4834E 04
0.799E 00	0.1023E 02	0.9241E 03	0.5083E 04
0.879E 00	**0.5591E 01**	**0.1785E 04**	**0.7404E 04**
0.959E 00	**-0.6881E 00**	**0.1946E 04**	**0.1054E 05**

CONTROL SYSTEM WITH 1ST ORDER TRANSFER

TIME	ERROR	GEN	LOAD
0.399E 01	-0.2383E 00	0.4600E 03	0.1031E 05
0.399E 01	-0.2383E 00	0.4600E 03	0.1031E 05

TABLE 5.7

DSL Standard TRNFR Block Form

$$Y = \text{TRNFR}(N, M, A, B, X)$$

$A = $ Numer. coeff. vector

$B = $ Denom. coeff. vector

Mth Order Transfer Function

$$\frac{Y}{X} = \frac{\displaystyle\sum_{i=0}^{N} A_{(N-i+1)} S^{N-i}}{\displaystyle\sum_{i=0}^{M} B_{(M-i+1)} S^{M-i}} \quad (N < M)$$

function on a System/3 printer device. For System/3 Model 6, the visual display tube (2265) may be used for this purpose. A discussion of program conversion for System/3 in Chapter 11 will provide further details.

5.1.3 Heat Exchanger Control

Figure 5.6 shows a proposed scheme to develop a control system for a heat exchanger. The heat exchanger fluid temperature is measured by a thermocouple and is controlled by adjustment of a valve which regulates the flow of steam. Steam is used as the heating medium. It is proposed that a three-mode controller be designed for regulation of the heat exchanger. T_m is the temperature measured by the thermocouple for the fluid, and T_s is the set point value. The thermocouple measuring system itself acts as a first-order system, acting on T_0, the temperature of the fluid at any given instant. The relation between the temperature of the heating medium T_h and T_0, the heated fluid in the heat exchanger, is also assumed to be first-order. The control valve acts on a voltage V_c, with a second-order relationship to the temperature T_h.

Thus, changes in T_0, reflected in T_m, the measured temperature, show up as a deviation from the set point value T_s. The error is to be used by the controller to produce a new value of V_c to return the fluid temperature to the desired

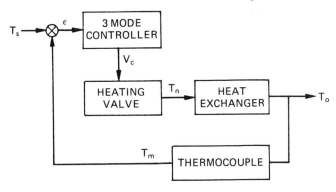

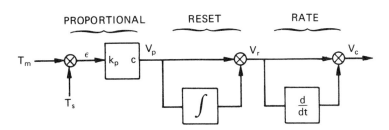

Figure 5.6.   Temperature controller for heat exchanger.

value. The three-mode controller is to be designed to perform close control of T_0, both during transient and steady-state conditions.

The proportional section of the controller will provide gain of the error signal; reset will provide control voltage V_C with an integral of the error voltage to reduce the steady-state error signal to zero. The rate section will provide the control voltage with a derivative of the output of the reset section of the control in order to anticipate temperature changes and smooth the transient response of the temperature control system.

The system equations both in differential form and in transfer function form are as follows:

Differential form   *Transfer function*

1. Heat exchanger

$$\dot{T}_0 + \frac{1}{\tau_e} T_0 = \frac{1}{\tau_e} T_h \qquad\qquad \frac{T_0}{T_h}(S) = \frac{1}{\tau_e S + 1} \qquad (5.7)$$

2. Heating valve

$$\ddot{T}_h + W_n 2\xi \dot{T}_h + W_n^2 T_h = \frac{W_n^2 V_c}{C} \qquad \frac{T_h}{V_c}(S) = \frac{1/C}{(S^2/W_n^2) + (2\xi/W_n)S + 1} \qquad (5.8)$$

3. Thermocouple

$$\dot{T}_m + \frac{1}{\tau_t} T_m = \frac{1}{\tau_t} T_0 \qquad\qquad \frac{T_m}{T_0}(S) = \frac{1}{\tau_t S + 1} \qquad (5.9)$$

4. Controller
   a. Proportional

$$K_P(T_s - T_m)(C) = V_P \qquad\qquad \frac{V_P}{T_s - T_m} = K_P \cdot C \qquad (5.10)$$

   b. Reset

$$V_P + \frac{1}{K_r} \int_0^t V_P \, dt = V_r \qquad\qquad \frac{V_r}{V_P}(S) = \frac{K_r S + 1}{K_r S} \qquad (5.11)$$

   c. Rate

$$\frac{1}{\alpha} V_r + \frac{1}{\alpha K_d} \int_0^t (V_r - V_c) \, dt = V_c \qquad \frac{V_c}{V_r}(S) = \frac{K_d S + 1}{\alpha K_d S + 1} \qquad (5.12)$$

Range of parameters and variables:

---

1. **Controller**
    K_P = proportional gain $0 < K_P \leqslant 10$
    K_r = reset time $\infty \geqslant K_r \geqslant 1$ second
    K_d = rate time $\infty \geqslant K_d \geqslant .1$ second
    α = compensation ratio $1/10$
    T_s = set point temperature $0 \leqslant T_s \leqslant 100°$
    V_P = proportional voltage $0 \leqslant V_P \leqslant 10$ volts
    V_r = reset voltage $0 \leqslant V_r \leqslant 10$ volts

2. **Exchanger**
    T_0 = output temperature   0 to 100°F
    T_h = input temperature   0 to 100°F
    τ_e = time constant   2 seconds

3. **Thermocouple**
    T_m = thermocouple output   0 to 100°F
    τ_t = time constant   1 second

4. **Valve**
    W_n = natural frequency $\sqrt{18}$ radians per second
    V_c = valve input voltage   0 to 10 volts
    ξ = damping factor   .7
    C = control constant   1 volt/10°F

With the ranges of parameters and variables defined above, the problem is to develop a DSL program to investigate the operation of a control system with various values of proportional gain, reset, and rate constants in regulating the temperature T_0. Initial values for the constants are

$$K_p = 5, \; K_r = 10, \text{ and } K_d = 2. \tag{5.13}$$

The set point temperature T_s is set at 20°F. Figure 5.7 shows an analog block diagram for the system for those users interested in analog simulation. The DSL structure and data statements are shown in Table 5.8. The UPDAT subroutine is shown as compiled by FORTRAN in Table 5.9.

The reader should especially note the use of NOSORT-SORT section where FORTRAN statements are inserted. The purpose of this portion is to provide a one-time calculation of the parameters required in the solution of the heating valve equation. A switch key SWT is used for this purpose, initially set by DSL to the value by CONTRL to 1 and then reset during first pass execution of the user's program to 2. Subsequent passes through the calculations will bypass the initial calculations. Note that in DSL, SWT has to

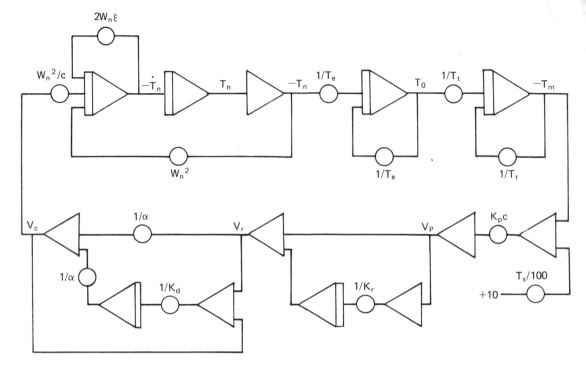

Figure 5.7. Analog block diagram for the heat exchanger problem.

be declared as an integer, while in normal FORTRAN one could simply use ISWT to make this variable an integer. However, in DSL if ISWT were used, it would also require declaration of ISWT as an INTGER, since DSL assumes all variables to be REAL, i.e., with a decimal point and not whole integers. Table 5.10 shows the results of the simulation for this problem.

An interesting variation of this problem for the reader to try is to have a more complex relationship between T_0 and T_h, which may exist in many heat exchangers, and attempt to find out if a three-mode controller will be able to regulate the heat exchanger. This exercise will prove useful for many industrial heat exchanger designs which involve multipass heat exchange.

5.1.4 Study of Transient Response in a Nonlinear System

Figure 5.8 shows a gate and driver system, used in the stations of the San Francisco Bay Area Rapid Transit (BART) System. The problem, as taken from Reference (6), consists of computing the angular motion of a gate alternately due to action of a motor and a brake, through a gear train with a

TABLE 5.8
DSL Translator Input for the Heat Exchanger Problem

```
            *** DSL/1800 TRANSLATOR INPUT ***
TITLE   THREE MODE TEMPERATURE CONTROL
INTGER SWT
PARAM DAMP=.7,TAUE=2.,TAUT=1.,C=.1,ALPHA=.1,KR=10.,KD=2.,KP=5.
NOSORT
        GO TO (1,2),SWT
      1 WN=SQRT(18.)
        WNSQ=18.
        D2=(DAMP+DAMP)*WN
        WN2=WNSQ/C
        KP2=KP*C
        SWT=2
      2 CONTINUE
SORT
INCON THDTO=0.,THO=0.,TOO=0.,TMO=0.,TS=20.
*       EXCHANGER CALCULATIONS
        TO=INTGR(TOO,(TH-TO)/TAUE )
*       VALVE CALCULATIONS
        THDOT=INTGR(THDTO,WN2*VC-D2*THDOT-WNSQ*TH)
        TH=INTGR(THO,THDOT)
*       THERMOCOUPLE CALCULATIONS
        TM=INTGR(TMO,(TO-TM)/TAUT )
*       THREE MODE CONTROLLER EVALUATION
*       PROPORTIONAL SECTION
        VP=KP2*(TS-TM)
*       RESET SECTION
        VR=VP-INTGR(O.,VP)/KR
*       RATE SECTION
        VC=(VR-INTGR(O.,VR+VC)/KD)/ALPHA
CONTRL DELT=.001,FINTI=1.,SWT=1
INTEG RKS
RELERR TH=.001
ABSERR TH=.01
PRINT .1,TO,THDOT,TH,TM,VP,VR,VC
END
STOP
```

transmission ratio of 30:1. The load consists of the system inertia J and a gravity force W, acting on the center of gyration point CG, which is a distance R away from the pivot point.

The motor and brake torques are given by the manufacturer in the form of curves of torque versus speed, as shown in Fig. 5.9.

The gate, initially closed ($\theta_0 = -15$), is to be opened by energizing the motor (until an angle is reached), followed by applying the brake. If impact at the final position ($\theta_f = +15$) is to be minimized, what must the brake angle be? What is the total transit time and the best brake angle to use? Compute drive shaft speed and load shaft angle versus time as well as phase trajectories.

TABLE 5.9

Heat Exchanger Problem UPDAT Subroutine from DSL

```
      SUBROUTINE UPDAT
      REAL    KR    ,KD   ,KP   ,KP2  ,INTGR
      INTEGER      SWT
      COMMON NALRM,IZZZ(833)  ,TIME ,DELT ,DELMI,FINTI,DELTP,DELTC,TO    ,
     1ZZ001,ZZ002,THDOT,TH    ,TM   ,WN    ,WNSQ ,D2    ,WN2  ,KP2   ,VP   ,
     1VR    ,VC    ,DAMP ,TAUE ,TAUT ,C     ,ALPHA,KR    ,KD   ,KP    ,THDTO,
     1THO   ,TOO   ,TMO  ,TS   ,INZZO,SWT
      GO TO (1,2),SWT
    1 WN=SQRT(18.)
      WNSQ=18.
      D2=(DAMP+DAMP)*WN
      WN2=WNSQ/C
      KP2=KP*C
      SWT=2
    2 CONTINUE
      TO=INTGR(TOO,(TH-TO)/TAUE )
      VP=KP2*(TS-TM)
      VR=VP-INTGR(0.,VP)/KR
      VC=(VR-INTGR(0.,VR+VC)/KD)/ALPHA
      THDOT=INTGR(THDTO,WN2*VC-D2*THDOT-WNSQ*TH)
      TH=INTGR(THO,THDOT)
      TM=INTGR(TMO,(TO-TM)/TAUT )
      RETURN
      END
VARIABLE ALLOCATIONS
 NALRM(IC)=7FFF        IZZZ(IC)=7FFE-7CBE   TIME(RC)=7CBC      DELT(RC)=7CBA
 DELTP(RC)=7CB4        DELTC(RC)=7CB2         TO(RC)=7CB0    ZZ001(RC)=7CAE
   TH(RC)=7CA8           TM(RC)=7CA6          WN(RC)=7CA4     WNSQ(RC)=7CA2
  KP2(RC)=7C9C           VP(RC)=7C9A          VR(RC)=7C98       VC(RC)=7C96
 TAUT(RC)=7C90            C(RC)=7C8E       ALPHA(RC)=7C8C       KR(RC)=7C8A
THDTO(RC)=7C84          THO(RC)=7C82         TOO(RC)=7C80      TMO(RC)=7C7E
  SWT(IC)=7C7A

STATEMENT ALLOCATIONS
 1    =0012  2    =0033

FEATURES SUPPORTED
 ONE WORD INTEGERS

CALLED SUBPROGRAMS
 INTGR   FSQRT   FADD    FSUB    FMPY    FDIV    FLD    FSTO    FSBR

REAL CONSTANTS
 .180000E 02=0006    .000000E 00=0008

INTEGER CONSTANTS
    2=000A

CORE REQUIREMENTS FOR UPDAT
 COMMON    902  VARIABLES     6  PROGRAM    146

RELATIVE ENTRY POINT ADDRESS IS 000B (HEX)

END OF COMPILATION
```

TABLE 5.10
Simulation Output for the Heat Exchanger Problem

THREE MODE TEMPERATURE CONTROL

TIME	TO	THDOT	TH	TM	VP	VR	VC
0.000E 00	0.0000E 00	0.0000E 00	0.0000E 00	0.0000E 00	0.1000E 02	0.1000E 02	0.1000E 03
0.999E-01	0.1106E 01	0.9682E 03	0.6067E 02	0.2884E-01	0.9985E 01	0.9885E 01	0.5570F 02
0.199E 00	0.6481E 01	0.9116E 03	0.1600E 03	0.3537E 00	0.9823E 01	0.9623E 01	0.2769E 02
0.299E 00	0.1587E 02	0.4522E 03	0.2299E 03	0.1367E 01	0.9316E 01	0.9021E 01	0.8120E 01
0.399E 00	0.2698E 02	-0.8780E 02	0.2479E 03	0.3282E 01	0.8358E 01	0.7974E 01	-0.6922E 01
0.499E 00	0.3713E 02	-0.5509E 03	0.2150E 03	0.6046E 01	0.6976E 01	0.6515E 01	-0.1872E 02
0.599E 00	0.4416E 02	-0.8674E 03	0.1427E 03	0.9376E 01	0.5311E 01	0.4789E 01	-0.2726E 02
0.699E 00	0.4668E 02	-0.1013E 04	0.4722E 02	0.1284E 02	0.3576E 01	0.3009E 01	-0.3208E 02
0.799E 00	0.4419E 02	-0.9930E 03	-0.5442E 02	0.1598E 02	0.2006E 01	0.1412F 01	-0.3277E 02
0.899E 00	0.3705E 02	-0.8276E 03	-0.1465E 03	0.1835E 02	0.8202E 00	0.2118E 00	-0.2936E 02
0.100E 01	0.2627E 02	-0.5544E 03	-0.2163E 03	0.1964E 02	0.1797E 00	-0.4332E 00	-0.2259E 02
0.100E 01	0.2627E 02	-0.5544E 03	-0.2163E 03	0.1964E 02	0.1797E 00	-0.4332E 00	-0.2259E 02

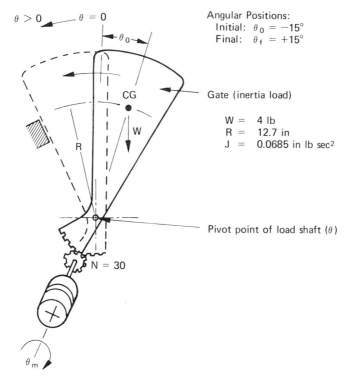

Angular Positions:
Initial: $\theta_0 = -15°$
Final: $\theta_f = +15°$

Gate (inertia load)

W = 4 lb
R = 12.7 in
J = 0.0685 in lb sec²

Pivot point of load shaft (θ)

Figure 5.8. Gate and driver system.

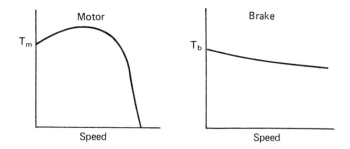

Figure 5.9.   Torque curves for gate-driver system.

The torque balance is described by

$$T_d - T_g = J\ddot{\theta}_m \qquad \text{(at the load shaft)}, \qquad (5.14)$$

where $T_g = \dfrac{WR}{N} \sin \theta$ is torque due to gravity, and

$$T_d = \begin{cases} k_m T_m(\theta) & \text{when the motor is on} \\ -k_b T_b(\theta) & \text{when the brake is on} \end{cases} \qquad (5.15)$$

Figure 5.10 shows a functional diagram of the system; somewhat different from the normal analog block diagram, it nevertheless serves the purpose of understanding the system behavior.

The AFGEN blocks provide the tool to describe the dependency of T_m and T_b on speed (W_m). To give these functions to the DSL system, several points are chosen on the X-coordinate of the curves in Fig. 5.10 and the Y-coordinates are evaluated. The X- and Y-coordinates are then provided to DSL in an AFGEN statement, as monotonically increasing values X1, Y1, X2, Y2, etc.

Table 5.11 shows the DSL statements for this problem. The AFGEN block use is illustrated. Note that the FINISH statement is used to terminate simulation, as an alternative to terminating at FINTI, the simulation finish time. The UPDAT subroutine generated by DSL is also shown in the table. The computer output on the printer is shown in Table 5.12. Notice the error generated on the printout. This error is a result of RPM going to less than zero in a large step increment. However, the results are valid up to the point where the error occurred. The plots for gate (RPM) versus time as well as phase trajectories are shown in Figs. 5.11 and 5.12. Notice that multiple runs repeatedly use the same plotted coordinates, since at the end of simulation the plotter is returned to its 0,0 position.

From inspecting the figures, a first conclusion might be that a smooth gate motion could be expected if a brake angle of $\theta = 11.5$ were used. Also, it would take approximately 340 milliseconds to complete the motion.

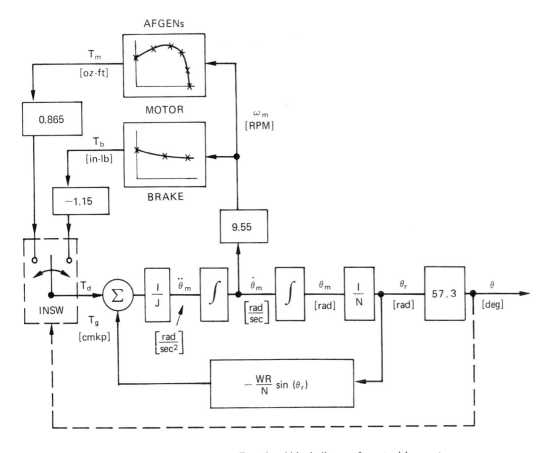

Figure 5.10. Functional block diagram for gate-driver system.

The model used for this system, as crude as it is, would be difficult to analyze in other ways and would become rapidly prohibitive if such elements as backlash in the gears, delay in the motor and clutch response, and dry and viscous damping were introduced. In digital simulation, however, complexity can be easily increased as model features change. Note that magnitude and time scaling are hardly ever needed and that computing elements do not have sign inversions and scale factors connected with them as in analog simulation.

5.1.5 A Focus Control System
for the Electron Beam

In the next example, taken from Reference (6), the design of a control system for focusing an electron beam will be illustrated.

The ability to keep an electron beam well focused on a specimen is an

TABLE 5.11
DSL Statements for the Gate and Driver System
and its UPDAT Subroutine

```
TITLE GATE CONTROL SYSTEM FOR BART
PARAM J=.00788,N=30.,W=1.81,R=32.3,THO=-7.86
AFGEN MOTOR=0.,3.05,200.,3.15,400.,3.2,500.,3.15,600.,3.05,...
      700.,2.8,800.,2.2,900.,1.2,980.,0.
AFGEN BRAKE=0.,20.,750.,15.,1500.,10.
      ACCMT=(TD-TG)/J
      SPEED=INTGR(0.,ACCMT)
      THETM=INTGR(THO,SPEED)
      THETR=THETM/N
      THETA=THETR*57.3
      RPM=9.55*SPEED
       TG=-W*R*SIN(THETR)/N
      TM=AFGEN(MOTOR,RPM)
      TB=AFGEN(BRAKE,RPM)
      TD=INSW(THETA-B,0.865*TM,-1.15*TB)
CONTRL FINTI=.5,DELMI=0.00001,DELT=0.001
RELERR THETA=0.001,RPM=0.001
FINISH THETA=16.,RPM=-0.01
PRINT .005,ACCMT,RPM,THETA,TG,TD
GRAPH .01,10,8,TIME,RPM
LABEL GATE MOTION - B=12/10/8 DEGREES
PARAM B=12.
SCALE 20.,.01
END
PARAM B=10.
END
LABEL PHASE TRAJECTORIES
PARAM B=12.
GRAPH .01,8,8,THETA,RPM
SCALE .2,.008
END
PARAM B=8.
END
STOP


SUBROUTINE UPDAT
REAL  J    ,N     ,MOTOR,INTGR,INSW
COMMON NALRM,IZZZ(833) ,TIME ,DELT  ,DELMI,FINTI,DELTP,DELTC,SPEED,
1THETM,ACCMT,THETR,THETA,RPM  ,TG    ,TM    ,TB    ,TD    ,J     ,N     ,
1W    ,R    ,THO  ,MOTOR,BRAKE,B
 THETR=THETM/N
  TG=-W*R*SIN(THETR)/N
 RPM=9.55*SPEED
 TB=AFGEN(BRAKE,RPM)
 TM=AFGEN(MOTOR,RPM)
 THETA=THETR*57.3
 TD=INSW(THETA-B,0.865*TM,-1.15*TB)
 ACCMT=(TD-TG)/J
 SPEED=INTGR(0.,ACCMT)
 THETM=INTGR(THO,SPEED)
 RETURN
 END
```

TABLE 5.12
Simulation Results for the Gate and Driver System

GATE CONTROL SYSTEM FOR BART

TIME	ACCMT	RPM	THETA	TG	TD
0.000E 00	0.2707E 03	0.0000E 00	-0.1501E 02	0.5047E 00	0.2638E 01
0.509E-02	0.2715E 03	0.1319E 02	-0.1500E 02	0.5045E 00	0.2643E 01
0.907E-02	0.2721E 03	0.2353E 02	-0.1499E 02	0.5040E 00	0.2648E 01
0.149E-01	0.2731E 03	0.3894E 02	-0.1495E 02	0.5028E 00	0.2655E 01
0.199E-01	0.2740E 03	0.5201E 02	-0.1490E 02	0.5013E 00	0.2660E 01
0.249E-01	0.2749E 03	0.6512E 02	-0.1485E 02	0.4994E 00	0.2666E 01
0.299E-01	0.2760E 03	0.7827E 02	-0.1477E 02	0.4970E 00	0.2672E 01
0.349E-01	0.2770E 03	0.9148E 02	-0.1469E 02	0.4942E 00	0.2677E 01
0.399E-01	0.2782E 03	0.1047E 03	-0.1459E 02	0.4910E 00	0.2683E 01
0.449E-01	0.2794E 03	0.1180E 03	-0.1448E 02	0.4873E 00	0.2689E 01

GATE CONTROL SYSTEM FOR BART

TIME	ACCMT	RPM	THETA	TG	TD
0.249E 00	0.3063E 03	0.7169E 03	0.2345E 01	-0.7974E-01	0.2334E 01
0.254E 00	0.2999E 03	0.7313E 03	0.3069E 01	-0.1043E 00	0.2259E 01
0.259E 00	0.2937E 03	0.7455E 03	0.3808E 01	-0.1294E 00	0.2185E 01
0.264E 00	0.2878E 03	0.7594E 03	0.4560E 01	-0.1549E 00	0.2113E 01
0.269E 00	0.2821E 03	0.7730E 03	0.5326E 01	-0.1809E 00	0.2042E 01
0.274E 00	0.2767E 03	0.7864E 03	0.6106E 01	-0.2072E 00	0.1973E 01
0.279E 00	0.2715E 03	0.7994E 03	0.6899E 01	-0.2340E 00	0.1905E 01
0.284E 00	0.2612E 03	0.8122E 03	0.7705E 01	-0.2612E 00	0.1797E 01
0.289E 00	0.2513E 03	0.8244E 03	0.8524E 01	-0.2888E 00	0.1691E 01
0.294E 00	0.2419E 03	0.8362E 03	0.9354E 01	-0.3167E 00	0.1589E 01
0.299E 00	0.2330E 03	0.8475E 03	0.1019E 02	-0.3449E 00	0.1491E 01
0.304E 00	0.2246E 03	0.8584E 03	0.1104E 02	-0.3734E 00	0.1397E 01
0.309E 00	0.2167E 03	0.8690E 03	0.1191E 02	-0.4022E 00	0.1305E 01
0.314E 00	-0.2104E 04	0.7813E 03	0.1274E 02	-0.4298E 00	-0.1700E 02
0.319E 00	-0.2200E 04	0.6786E 03	0.1347E 02	-0.4540E 00	-0.1779E 02
0.324E 00	-0.2289E 04	0.5853E 03	0.1402E 02	-0.4722E 00	-0.1851E 02
0.329E 00	-0.2410E 04	0.4585E 03	0.1461E 02	-0.4916E 00	-0.1948E 02
0.334E 00	-0.2523E 04	0.3408E 03	0.1501E 02	-0.5048E 00	-0.2038E 02
0.339E 00	-0.2641E 04	0.2175E 03	0.1529E 02	-0.5140E 00	-0.2133E 02
0.344E 00	-0.2766E 04	0.8842E 02	0.1544E 02	-0.5190E 00	-0.2232E 02

DSL ERR 33
DSL ERR 33
DSL ERR 33
DSL ERR 33
DSL ERR 33
DSL ERR 33
DSL ERR 33
DSL ERR 33
DSL ERR 33
DSL ERR 33

0.349E 00	-0.2852E 04	-0.4643E 02	0.1546E 02	-0.5197E 00	-0.2300E 02
0.349E 00	-0.2852E 04	-0.4643E 02	0.1546E 02	-0.5197E 00	-0.2300E 02

SIMULATION HALTED, RPM   = -0.4643E 02

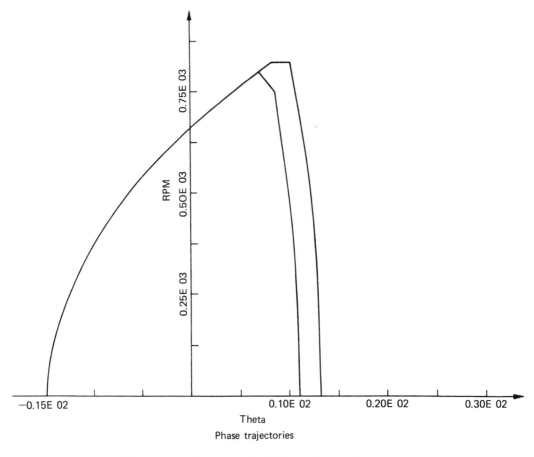

Phase trajectories

Figure 5.11.   Plot output from DSL for the gate-driver system.

important requirement when unattended, long-term operation is demanded. As the gun potential changes, due to filament supply variations, and as lens power changes, due to lens supply variations, the focal point of the beam near the target moves about, leaving the image more or less in focus at the target plane. The long-term beam diameter variations caused by insufficient supply regulation as well as thermal expansion of gun and lens assembly can be compensated for by current adjustment in the final lens of the column. Figure 5.13 illustrates the beam geometry during focusing.

5.1.5.1   Problem Formulation

Given the expected extent of defocusing in a certain period of time, the time interval during which the refocusing has to be accomplished, and the

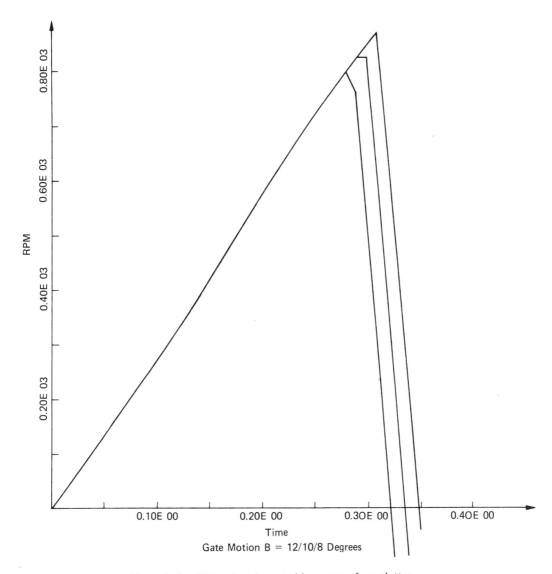

Figure 5.12 DSL output for gate-driver system from plotter.

accuracy required to approximate the best possible focus, the engineering problem is to design an automatic control system capable of achieving the focusing with a minimum of circuitry and a maximum of reliability. A small focusing lens serves as the system actuator to force the beam of electrons to converge over the desired range of working distances. A diode and a test target constitute the transducer for the beam, which is deflected at constant velocity over the target.

Figure 5.14 shows how a chopped target current signal $I_T(t)$ can be obtained by sweeping the beam across a hole pattern in the target.

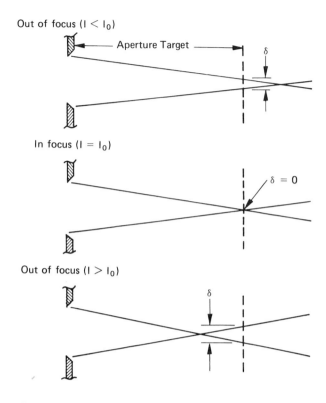

Figure 5.13   Beam geometry during focusing for an electron beam.

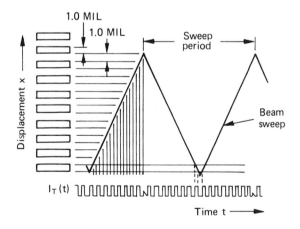

Figure 5.14. Illustration of test signal generation.

5.1.5.2 System Description

With the given actuator, it is possible to control the spot size ϕ of the beam in the target plan, as long as a signal related to ϕ can be derived from the detector output $I_T(t)$.

For a beam that originates from a perfectly stable gun and lens system, a relationship between spot size and focusing lens current (normalized by the amount of current I_0 that focuses the beam) can be derived from the basic relation as expressed in Eq. (5.16). This characteristic relationship is plotted in Fig. 5.15.

$$\phi = \sqrt{\phi_0^2 + (\phi_m^2 - \phi_0^2)\left(1 - \frac{I^2}{I_0^2}\right)^2} \tag{5.16}$$

ϕ_0 and ϕ_m here are the smallest possible spot size (at focus) and the maximum spot size, obtained for zero lens current. Note that the spot size is independent of the polarity of the current.

The basic concept underlying the feedback control system described here is the minimization of $\phi(I)$ by electronic means. One part of the circuitry generates a function H, inversely proportional to ϕ, and maximizes it. The other part either uses an error signal to adjust the lens current (during refocusing) or holds the lens current constant (during beam utilization periods).

By operating on the target current signal with time differentiation, rectification, and peak detection, the signal H is generated (see Fig. 5.16). Figure 5.17 shows how ϵ is related to spot size ϕ. From inspection it becomes obvious that the problem of focusing reduces to maximizing H (or setting its current

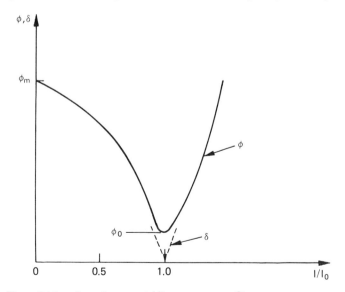

Figure 5.15. Spot size, actual (ϕ), and idealized (δ) vs. lens current.

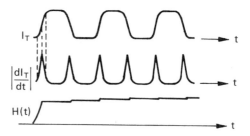

Figure 5.16. Generation of a continuous function $H(t) \sim 1/\phi(t)$ by differentiation, and peak detection of 1_T.

derivative ϵ to zero). Differentiation with respect to focus control lens current I_{FCL} is achieved by constantly varying the lens current from a value above to a value below the quiescent lens current and by generating the difference of their respective H values $\epsilon = H_{hi} - H_{lo}$. As indicated in the schematic system diagram (Fig. 5.18), this is done by superimposing on the lens current a square-wave dither current and driving a sampler ahead of two peak detectors synchronously with the dither. The difference of the two peak detector outputs (special hold circuits) then is the desired error ϵ. Approximate integration of the error signal by a filter circuit accomplishes the generation of a smooth signal proportional to the desired lens drive current I_{FCL}. Upon conclusion of each refocusing cycle, the dither is removed, and the lens current is kept constant by opening the loop inside an analog storage circuit. The control system loop is closed via the electron beam column since the lens

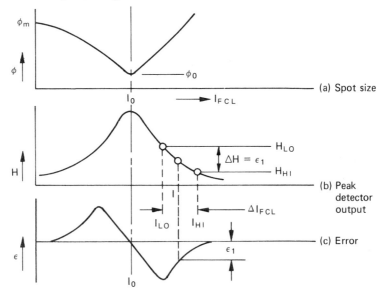

Figure 5.17. Derivation of an error function $\epsilon(1)$ from $\phi(1)$.

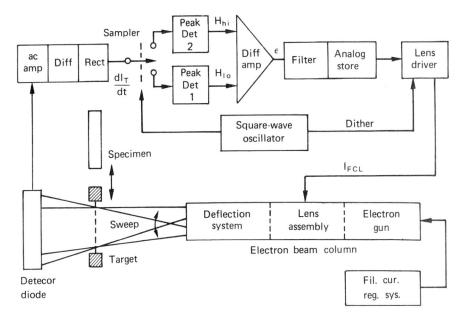

Figure 5.18. Schematic diagram of electron beam column and control systems.

current acts upon the beam and the detector diode picks up the changing beam current, containing information on the spot size in the target plane.

A mathematical representation of the system is illustrated in Figs. 5.19 and 5.20. The parabolic relationship between lens current and spot size is shown in one nonlinear block and the conversion from ϕ to H in another. The effect of dither and subsequent synchronous detection is represented as a finite difference operation Δ/Δ_L. Two predominant linear time constants due to the peak detectors and the integrating filter are shown as the linear plant characteristics. A third one, due to eddy current losses in the focus control lens, is physically present but mathematically negligible and, therefore, is omitted. A summing point, which accomplishes summation of a quiescent lens current I_Q and a servo current I_s, represents the final link of the loop.

Realistic representation of system disturbances which manifest themselves in changes of ϕ_0 and ϕ_m are indicated in conjunction with the $\phi(I_L)$ block. Note also the interdependence of I_0 and ϕ_m characterized for a particular lens used by the constant $K = .00035$ micron per square milliampere.

5.1.5.3 Steady-State Analysis

In a simplified block diagram, the two nonlinearities and the finite difference operator can be lumped together as one nonlinear function $\epsilon(I)$, termed the S-curve and shown as curve (c) in Fig. 5.17. The locus of operating points in the steady state can be seen to be the straight load line,

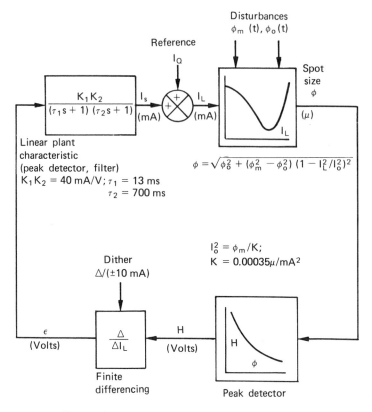

Figure 5.19. Block diagram of focus control system.

obtained (upon setting S = 0) from the simple relationship

$$I_s = K_1 K_2 \epsilon = I_L - I_Q \tag{5.16a}$$

$$\epsilon = \frac{1}{K_1 K_2}(I_L - I_Q) \tag{5.17}$$

The intersections of this line with the curves of the $\epsilon(I_L)$ family represent all the trajectory end points in the phase plane (ϵ versus I_L) shown in Fig. 5.21, which is obtained from simulation plots with DSL. The steady-state accuracy of focusing is thus seen to be very high since the load line slope is very small and the slopes of the S-curves are very large in their linear center sections.

5.1.5.4 Simulation and Dynamic Analysis

The digital simulation approach for this control system can be seen at a glance from Fig. 5.20, the simulation diagram on which is based the list of DSL input statements shown in Table 5.13. Two REALPL blocks are used to

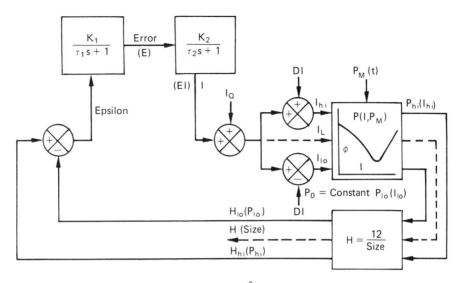

$K_1 = 1.0$ $K_2 = 40$ $I_Q = 220$ $P_0 = 1.2$ $P_M = KI_0{}^2$
$\tau_1 = 0.013$ $\tau_2 = 0.70$ DI $= 10$ K $= 0.00035$ $I_0 = I_L(t) = 210,230,210,220$ $I_{FCL} = I + 180$

Figure 5.20.   Simulation diagram of focus control system.

represent the real pole transfer function from ϵ to 1. The finite differencing of $H(I)$ is done by adding to and subtracting from the lens current I_L the dither half-amplitude D1, generating the resulting spot sizes (ϕ_{hi} and ϕ_{lo}) and their inverses H and taking the difference of the high and the low current H. The relationship between ϕ and H can be seen through a physical analysis to be hyperbolic in nature. Equation (5.18) was obtained by curve-fitting the experimental data:

$$H = \frac{12}{\phi} \text{ (volts per micron)} \tag{5.18}$$

The minimum spot size ϕ_0 was treated here as a changing constant, while the variations in ϕ_m, (and, therefore, I_0) were simulated by specifying I_0 as a series of step functions:

$$I_0 = 210*\text{STEP}(0.) + 20*\text{STEP}(0.2) - 20*\text{STEP}(.5) + 10*\text{STEP}(.8) \tag{5.19}$$

A 40-milliampere shift of the $\phi(I_L)$ characteristic into the negative current direction was experienced on the electron beam column due to interference of the focus control lens and the final main lens; it can be taken care of by specifying I_0 to be 220 milliamperes (180 milliamperes, the actual value needed in the circuit, plus 40 milliamperes due to the curve shift). The total current in the focus control lens may then be defined to be $I_{FCL} = I + 180$ for plotting purposes only.

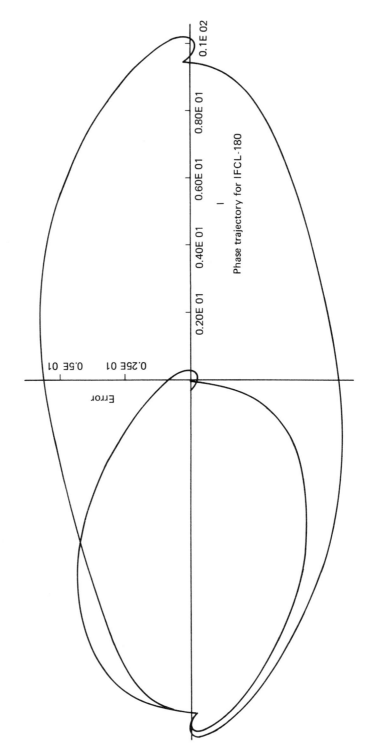

Figure 5.21. Phase plane plot for electron beam focus control system.

98

TABLE 5.13
DSL Input for the Electron Beam Focus
Control System

```
                     *** DSL/1800 TRANSLATOR INPUT ***
TITLE ELECTRON BEAM FOCUS CONTROL SYSTEM
      E=REALP(IC1,TOU1,EPSLN)
      ERROR=K1*E
      EI=REALP(IC2,TOU2,ERROR)
      I=K2*EI
      IL=I+IQ
      IHI=IL+DI
      ILO=IL-DI
      IO=210.*STEP(0.)+20.*STEP(0.2)-20.*STEP(0.5)+10.*STEP(0.8)
      IO2=IO*IO
      PM=K*IO2
      DPP=PM*PM-PO*PO
      PHI=SQRT(PO*PO+DPP*(1.-IHI*IHI/IO2)**2)
      PLO=SQRT(PO*PO+DPP*(1.-ILO*ILO/IO2)**2)
      SIZE=SQRT(PO*PO+DPP*(1.-IL*IL/IO2)**2)
      H=12./SIZE
      HHI=12./PHI
      HLO=12./PLO
      EPSLN=HHI-HLO
      IFCL=I+180.
PARAM K1=1.,K2=40.,TOU1=.013,TOU2=0.7,DI=10.,PO=1.2,K=.00035,IQ=220.
CONTRL DELT=.001,FINTI=1.,DELMI=0.00001
INCON IC1=0.,IC2=0.
PRINT .01,I,IL,SIZE,H,EPSLN,ERROR,EI,PHI,PLO,IFCL
GRAPH .005,10,8,TIME,SIZE
SCALE 10.,2.
LABEL ELECTRON BEAM FOCUS CONTROL
END
RESET GRAPH
RESET PRINT
LABEL PHASE TRAJECTORY FOR IFCL-180
GRAPH .005,10,5,I,ERROR
SCALE 1.,,.8
END
STOP

OUTPUT VARIABLE SEQUENCE
   8   9   4   5   7  10  11  13  17   6  12  16  18   1   2   3  14  15  19

STORAGE USED/MAXIMUM
INTGR   2/50,  IN VARS  39/300,  OUT VARS  19/100,  PARAMS 16/60,  SYMBOLS  51/200,
```

Table 5.14 shows the UPDAT subroutine produced by DSL as well as the data statements created. Table 5.15 shows portions of the print output from simulation. Notice that the print format is substantially different from other examples. This occurs because the number of print variables referenced are larger than a normal tabular format will permit.

Simultaneous plots of spot size and current as functions of time are obtained by the first GRAPH instruction, and phase plane trajectories for the run are obtained by the second one. See Figs. 5.21 and 5.22. These graphs are obtained after experimentation and design optimization in parallel with the simulation effort. They agreed very closely with the experimental step response

```
PAGE    2                    DSL/1800  (DIGITAL  SIMULATION  LANGUAGE)

      SUBROUTINE UPDAT
      REAL  IC1  ,K1    ,IC2  ,I    ,K2   ,IL   ,IQ   ,IHI  ,ILO  ,IO   ,
     1IO2  ,K    ,IFCL
      COMMON NALRM,IZZZ(833) ,TIME ,DELT ,DELMI,FINTI,DELTP,DELTC,E     ,
     1EI   ,ERROR,I    ,IL   ,IHI  ,ILO  ,IO   ,IO2  ,PM   ,DPP  ,PHI   ,
     1PLO  ,SIZE ,H    ,HHI  ,HLO  ,EPSLN,IFCL ,K1   ,K2   ,TOU1 ,TOU2  ,
     1DI   ,PO   ,K    ,IQ   ,IC1  ,IC2
      IO=210.*STEP(0.)+20.*STEP(0.2)-20.*STEP(0.5)+10.*STEP(0.8)
      IO2=IO*IO
      I =K2*EI
      IL=I+IQ
      ILO=IL-DI
      PM=K*IO2
      DPP=PM*PM-PO*PO
      PLO=SQRT(PO*PO+DPP*(1.-ILO*ILO/IO2).**2)
      HLO=12./PLO
      IHI=IL+DI
      PHI=SQRT(PO*PO+DPP*(1.-IHI*IHI/IO2)**2)
      HHI=12./PHI
      EPSLN=HHI-HLO
      E =REALP(IC1,TOU1,EPSLN)
      ERROR=K1*E
      EI=REALP(IC2,TOU2,ERROR)
      SIZE=SQRT(PO*PO+DPP*(1.-IL*IL/IO2)**2)
      H=12./SIZE
      IFCL=I+180.
      RETURN
      END



                    *** DSL/1800 SIMULATION DATA ***
TITLE ELECTRON BEAM FOCUS CONTROL SYSTEM
PARAM K1=1.,K2=40.,TOU1=.013,TOU2=0.7,DI=10.,PO=1.2,K=.00035,IQ=220.
CONTRL DELT=.001,FINTI=1.,DELMI=0.00001
INCON IC1=0.,IC2=0.
PRINT .01,I,IL,SIZE,H,EPSLN,ERROR,EI,PHI,PLO,IFCL
GRAPH .005,10,8,TIME,SIZE
SCALE 10.,2.
LABEL ELECTRON BEAM FOCUS CONTROL
END
```

curves obtained from the electron beam recorder. Note that plots of spot size versus time were generated for which no transducer exists in the machine and which are, therefore, not directly obtainable.

The manner in which the phase plane plots (Fig. 5.22) were obtained on the IBM 1130 will be of special interest to the reader. Notice that the 0, 0 point for X, Y is in the center of the plot. As opposed to plotting facilities in DSL for IBM 7090 or Systems 360/370, the 1130/1800 system prepares the

TABLE 5.15
Simulation Results for the Electron Beam Focus Control System

ELECTRON BEAM FOCUS CONTROL SYSTEM

TIME	=	0.0000E 00	I	=	0.0000E 00	IL	=	0.2200E 03	SIZE	=	0.1921E 01	H	=	0.6245E 01

TIME = 0.0000E 00   I = 0.0000E 00   IL = 0.2200E 03   SIZE = 0.1921E 01   H = 0.6245E 01
                     EPSLN = -0.6360E 01   ERROR = 0.0000E 00   EI = 0.0000E 00   PHI = 0.3296E 01
                     PLO = 0.1199E 01   IFCL = 0.1800E 03

TIME = 0.9077E-02   I = -0.9153E 00   IL = 0.2190E 03   SIZE = 0.1813E 01   H = 0.6615E 01
                     EPSLN = -0.6141E 01   ERROR = -0.3155E 01   EI = -0.2288E-01   PHI = 0.3160E 01
                     PLO. = 0.1207E 01   IFCL = 0.1790E 03

TIME = 0.1999E-01   I = -0.3390E 01   IL = 0.2166E 03   SIZE = 0.1551E 01   H = 0.7732E 01
                     EPSLN = -0.4965E 01   ERROR = -0.4524E 01   EI = -0.8476E-01   PHI = 0.2800E 01
                     PLO = 0.1297E 01   IFCL = 0.1766E 03

TIME = 0.2999E-01   I = -0.5884E 01   IL = 0.2141E 03   SIZE = 0.1345E 01   H = 0.8917E 01
                     EPSLN = -0.3265E 01   ERROR = -0.4253E 01   EI = -0.1471E 00   PHI = 0.2451E 01
                     PLO = 0.1470E 01   IFCL = 0.1741E 03

TIME = 0.3999E-01   I = -0.7945E 01   IL = 0.2120E 03   SIZE = 0.1237E 01   H = 0.9696E 01
                     EPSLN = -0.1732E 01   ERROR = -0.3241E 01   EI = -0.1986E 00   PHI = 0.2177E 01
                     PLO = 0.1656E 01   IFCL = 0.1720E 03

TIME = 0.4999E-01   I = -0.9340E 01   IL = 0.2106E 03   SIZE = 0.1203E 01   H = 0.9967E 01
                     EPSLN = -0.6798E 00   ERROR = -0.2090E 01   EI = -0.2335E 00   PHI = 0.2001E 01
                     PLO = 0.1797E 01   IFCL = 0.1706E 03

TIME = 0.5999E-01   I = -0.1011E 02   IL = 0.2098E 03   SIZE = 0.1200E 01   H = 0.9999E 01
                     EPSLN = -0.9709E-01   ERROR = -0.1138E 01   EI = -0.2527E 00   PHI = 0.1908E 01
                     PLO = 0.1879E 01   IFCL = 0.1698E 03

TIME = 0.6999E-01   I = '-0.1041E 02   IL = 0.2095E 03   SIZE = 0.1201E 01   H = 0.9987E 01
                     EPSLN = 0.1345E 00   ERROR = -0.4974E 00   EI = -0.2603E 00   PHI = 0.1871E 01
                     PLO = 0.1912E 01   IFCL = 0.1695E 03

ELECTRON BEAM FOCUS CONTROL SYSTEM

TIME = 0.9599E 00   I = -0.2461E 00   IL = 0.2197E 03   SIZE = 0.1200E 01   H = 0.9995E 01
                     EPSLN = 0.1434E-01   ERROR = 0.9864E-02   EI = -0.6153E-02   PHI = 0.1945E 01
                     PLO = 0.1950E 01   IFCL = 0.1797E 03

TIME = 0.9699E 00   I = -0.2367E 00   IL = 0.2197E 03   SIZE = 0.1200E 01   H = 0.9995E 01
                     EPSLN = 0.7137E-02   ERROR = 0.1008E-01   EI = -0.5917E-02   PHI = 0.1946E 01
                     PLO = 0.1949E 01   IFCL = 0.1797E 03

TIME = 0.9799E 00   I = -0.2285E 00   IL = 0.2197E 03   SIZE = 0.1200E 01   H = 0.9995E 01
                     EPSLN = 0.8831E-03   ERROR = 0.6543E-02   EI = -0.5713E-02   PHI = 0.1947E 01
                     PLO = 0.1948E 01   IFCL = 0.1797E 03

TIME = 0.9899E 00   I = -0.2228E 00   IL = 0.2197E 03   SIZE = 0.1200E 01   H = 0.9995E 01
                     EPSLN = -0.3456E-02   ERROR = 0.2108E-02   EI = -0.5571E-02   PHI = 0.1948E 01
                     PLO = 0.1947E 01   IFCL = 0.1797E 03

TIME = 0.9999E 00   I = -0.2196E 00   IL = 0.2197E 03   SIZE = 0.1200E 01   H = 0.9996E 01
                     EPSLN = -0.5928E-02   ERROR = -0.1699E-02   EI = -0.5490E-02   PHI = 0.1949E 01
                     PLO = 0.1947E 01   IFCL = 0.1797E 03

TIME = 0.9999E 00   I = -0.2196E 00   IL = 0.2197E 03   SIZE = 0.1200E 01   H = 0.9996E 01
                     EPSLN = -0.5928E-02   ERROR = -0.1699E-02   EI = -0.5490E-02   PHI = 0.1949E 01
                     PLO = 0.1947E 01   IFCL = 0.1797E 03

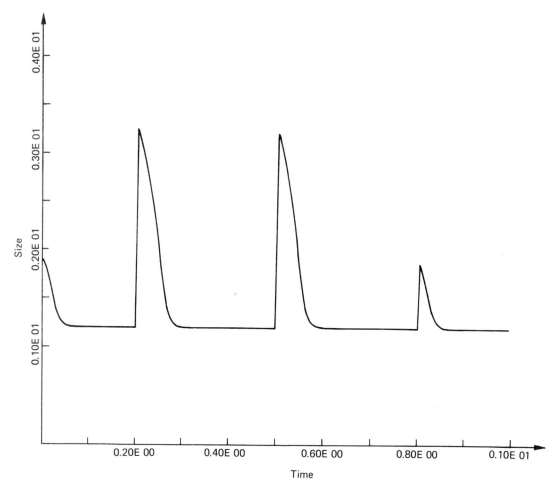

Figure 5.22.   Step response for electron beam focus control system.

ordinates prior to simulation start and therefore will not be able to handle negative values of ordinates. If the user is aware of his ranges of the variables, then he must accordingly scale the variables in plotting as well as, in this case, center the plotter pen to be able to plot negative values of the X- and Y-coordinates. If this is not done, at execution time the user will find the plotter pen hitting the bottom for negative values of Y-coordinate and overwriting some previous plots for the negative X-coordinate values.

A comparison of the phase plane plots with the 1130 DSL simulation and the results in the referenced article, where simulation was performed on an IBM 7090 (plotting completed after completion of simulation), will show the reader the difference.

5.1.6 Aircraft Arresting Gear Problem

This example is taken from the *MIDAS Programming Guide,* published by the Analog Computation Division of Wright-Patterson Air Force Base, Ohio; it also appears in the DSL user's guide (17).

Equations:		Constants:
$m_3 \ddot{y}_3 = f_{k_2} - f_D$ $y_3(0) = 0, \dot{y}_3(0) = 0$		$m_1 = 1400$ slugs
$m_2 \ddot{y}_2 = 2f_{k_1} - f_{k_2}$ $y_2(0) = 0, \dot{y}_2(0) = 0$		$m_2 = 45.28$ slugs
$m_1 \ddot{x} = -2f_{k_1} \sin\theta$ $x(0) = 0, \dot{x}(0) = 290$ ft/sec		$m_3 = 20$ slugs
$f_{k_2} = k_2(y_2 - y_3) = 0$ For $y_2 \geqslant y_3$		$k_1 = 4550$ lb/ft
$f_{k_1} = k_1(y_1 - 2y_2) = 0$ $y_2 < y_3$		$k_2 = 25{,}300$ lb/ft
$f_D = f(y_3) \cdot (\dot{y}_3)^2$ For $y_1 \geqslant 2y_2$		$h = 125$ ft
$y_1 = \sqrt{x^2 + h^2} - h$ $y_1 < 2y_2$		
$\sin\theta = \dfrac{x}{h + y_1} = \dfrac{x}{\sqrt{x^2 + h^2}}$ $f(y_3) =$ arbitrary linear function of y_3 representing damping coefficient		

The system shown above with its set of differential equations and their boundary condition, is illustrated in Fig. 5.23. Three second-order differential equations have to be solved simultaneously to determine the characteristics of the system. The corresponding analog block diagram used for analog simulation is also reproduced here in Fig. 5.24. The input relays are shown by crossed lines and are used to account for the physical fact that the cables can only transmit tensile forces. Thus, when $y_2 \geqslant y_3$, there is a force in tension equal to $k_2(y_2 - y_3)$ between

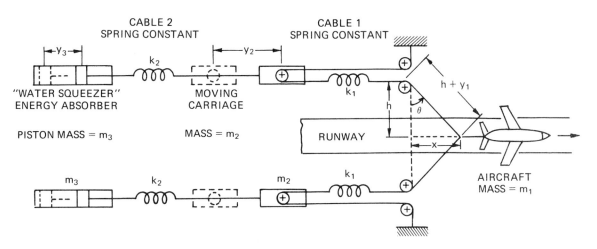

Figure 5.23.   Aircraft arresting gear system.

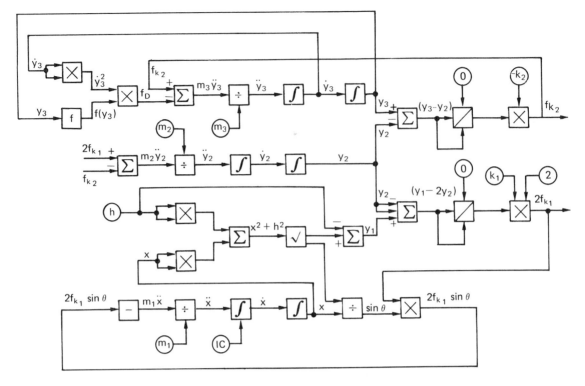

Figure 5.24.   Block diagram for the aircraft arresting gear system.

the moving carriage and the piston of the water squeezer. When $y_2 < y_3$, no compressive force exists. The damping coefficient, $f(y_3)$, is approximated by a straight line.

The DSL program shown in Table 5.16 demonstrates the use of procedural block PROCED, a linear function generator AFGEN, an implicit function IMPL, and the addition of a multiple output block at execution time RESOL. The blocks DELAY, DERIV, and DEBUG are shown here to demonstrate their use. Results of the simulation are shown in Fig. 5.25. Note that in order to solve the problem, prior to execution of the translated simulation program, the subroutine RESOL is compiled as shown in Table 5.17 and stored in the user subroutine library user area (UA).

The illustration also shows how the problem can be terminated by satisfying either the condition of FINTI (final value of time) or a variable reaching a certain value in the FINISH statement.

Note that the second run restarts at time = 0, changes over the initial condition, suppresses RANGE and GRAPH, and integrates with the use of the fourth-order Runge-Kutta variable step integration method. The run is set to terminate either at finish time or at ANGLE reaching 1.427.

TABLE 5.16
DSL Translator Input for the Aircraft Arresting Gear Problem

```
                    *** DSL/1800 TRANSLATOR INPUT ***
TITLE AIRCRAFT ARRESTING GEAR PROBLEM
      SQR1=SQRT(X**2+H**2)
      Y12=SQR1-H-2.*Y2
      FK1=INSW(Y12,0.,K1*Y12)
      XDOT=INTGR(XDOTO,-2.*FK1*(X/SQR1)/M1)
      X=INTGR(0.,XDOT)
PROCED FD=FCNY3(Y3,Y3DOT)
      IF(Y3-324.)4,4,5
    4 FD=AFGEN(FOFY3,Y3)*Y3DOT**2
      GO TO 6
    5 FD=90.*Y3DOT**2
    6 CONTINUE
ENDPRO
      Y23=Y2-Y3
      FK2=INSW(Y23,0.,K2*Y23)
      Y3DOT=INTGR(0. ,(FK2-FD)/M3)
      Y3=INTGR(0.,Y3DOT)
      Y2=INTGR(0.,INTGR(0.,(2.0*FK1-FK2)/M2) )
      ANGLE=ATAN(X/H)
      THETA=IMPL(THETO,ERROR,FOFTH)
      SINTH,COSTH=RESOL(THETA)
      R=H*COSTH+X*SINTH
      FOFTH=(X*COSTH-SINTH+R*THETA)/R
      Y3DEL=DELAY(50,2.,Y3DOT)
      Y3DER=DERIV(0.,Y3)
      WW=DEBUG(3,1.)
PARAM H=125.,THETO=0.,ERROR=1.E-3
CONST K1=4550.,K2=25300.,M1=1400.,M2=45.28,M3=20.
INCON XDOTO=290.
AFGEN FOFY3= 0.,8.33,30.,4.,60.,1.6,120.,5.2,150.,5.20,180.,6.6,210.,...
      8.3,240.,10.7,270.,16.,282.,21.,294.,28.,306.,41.,312.,50.,...
      324.,90.,999.,90.
CONTRL DELT=.002,DELMI=1.E-10 ,FINTI=8.
INTEG MILNE
RELERR Y3DOT=1.E-4
PRINT 0.1,Y3DOT,Y3,XDOT,X,DELT
GRAPH .05,8,6,TIME,Y3DOT
LABEL Y3DOT VS. TIME (ARRESTING GEAR)
SCALE 1.,.05
RANGE Y3DOT,Y3,XDOT,X,DELT
END
TITLE ARRESTING GEAR PROBLEM RUNGA-KUTTA INTEGRATION
INCON XDOTO=200.
CONTRL FINTI=6.,DELT=.002
PRINT .5,Y3DOT,Y3,Y2,THETA,XDOT,X,Y3DEL,Y3DER,DELT,ANGLE
FINISH ANGLE=1.427
INTEG RKS
RESET RELERR,ABSERR,GRAPH,RANGE
RELERR XDOT=1.E-3
ABSERR XDOT=1.E-2
END
STOP
```

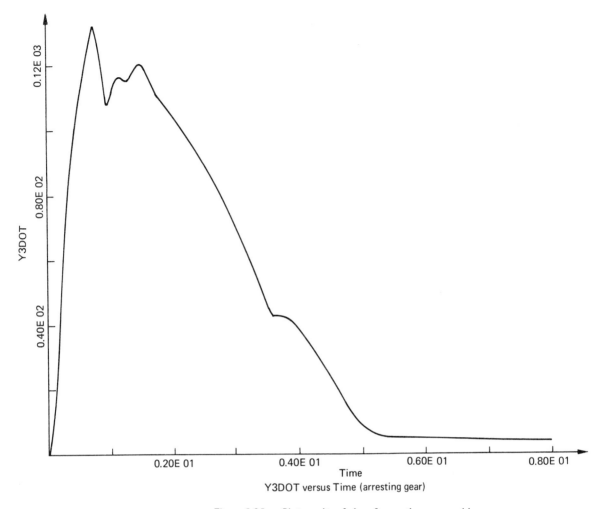

Y3DOT versus Time (arresting gear)

Figure 5.25.   Plot results of aircraft arresting gear problem.

The simulation results shown in Tables 5.18, 5.19, and 5.20 illustrate the output when the DSL function DEBUG is used. Note that all the variables for these integration steps starting at a specified value of time (or the independent variable) are printed out by the program. In addition, the RANGE statement of DSL used here prints the minimum and maximum value of the variables specified.

5.1.7   Pilot Ejection Study, an Example in Iterative Solution Using Console Switches

The purpose of this investigation is to determine the trajectory of a pilot ejected from a moving fighter aircraft in the air and to ascertain whether he will strike the vertical stabilizer of the aircraft as the aircraft continues to move under

TABLE 5.17
UPDAT Subroutine for the Aircraft Arresting Gear
Problem and Compilation of RESOL Subroutine
Prior to Execution of Simulation

```
      SUBROUTINE UPDAT
      REAL  INSW ,K1     ,INTGR,M1    ,K2    ,M3    ,M2    ,IMPL
      COMMON NALRM,IZZZ(833) ,TIME ,DELT ,DELMI,FINTI,DELTP,DELTC,XDOT ,
     1X     ,Y3DOT,Y3    ,ZZOO1,Y2    ,SQR1 ,Y12   ,FK1    ,FD    ,Y23   ,FK2   ,
     1ANGLE,SINTH,COSTH,R     ,THETA,Y3DEL,Y3DER,WW     ,H     ,THETO,ERROR,
     1K1    ,K2    ,M1    ,M2    ,M3    ,XDOTO,FOFY3
      SQR1=SQRT(X**2+H**2)
      Y12=SQR1-H-2.*Y2
      FK1=INSW(Y12,0.,K1*Y12)
      XDOT=INTGR(XDOTO,-2.*FK1*(X/SQR1)/M1)
      X=INTGR(0.,XDOT)
      IF(Y3-324.)4,4,5
    4 FD=AFGEN(FOFY3,Y3)*Y3DOT**2
      GO TO 6
    5 FD=90.*Y3DOT**2
    6 CONTINUE
      Y23=Y2-Y3
      FK2=INSW(Y23,0.,K2*Y23)
      Y3DOT=INTGR(0. ,(FK2-FD)/M3)
      Y3=INTGR(0.,Y3DOT)
      Y2=INTGR(0.,INTGR(0.,(2.0*FK1-FK2)/M2) )
      ANGLE=ATAN(X/H)
9100  THETA=IMPL(899,THETO,ERROR,FOFTH)
      IF(NALRM) 9200, 9200, 9300
9300  CONTINUE
      CALLRESOL(THETA,SINTH,COSTH )
      R=H*COSTH+X*SINTH
      FOFTH=(X*COSTH-SINTH+R*THETA)/R
      GO TO 9100
9200  CONTINUE
      Y3DEL=DELAY(794,50,2.,Y3DOT)
      Y3DER=DERIV(788,0.,Y3)
      WW=DEBUG(787,3,1.)
      RETURN
      END

   // FOR
   *ONE WORD INTEGERS
   *LIST ALL
         SUBROUTINE RESOL(THETA,SINTH,COSTH)
         SINTH=SIN(THETA)
         COSTH=COS(THETA)
         RETURN
         END
   FEATURES SUPPORTED
    ONE WORD INTEGERS

   CALLED SUBPROGRAMS
    FSIN    FCOS    FSTO    SUBIN

   CORE REQUIREMENTS FOR RESOL
    COMMON      0  VARIABLES      0  PROGRAM     22

   RELATIVE ENTRY POINT ADDRESS IS 0000 (HEX)

   END OF COMPILATION

   // DUP

   *DELETE            RESOL
   CART ID 1021   DB ADDR  2DA0    DB CNT   0003

    *STORE     WS  UA  RESOL
    CART ID 1021   DB ADDR  3000    DB CNT   0003
```

TABLE 5.18
Simulation Results for the Aircraft Arresting
Gear Problem

```
                    *** DSL/1800 SIMULATION DATA ***
TITLE AIRCRAFT ARRESTING GEAR PROBLEM
PARAM H=125.,THETO=0.,ERROR=1.E-3
CONST K1=4550.,K2=25300.,M1=1400.,M2=45.28,M3=20.
INCON XDOT0=290.
AFGEN FOFY3= 0.,8.33,30.,4.,60.,1.6,120.,5.2,150.,5.20,180.,6.6,210.,,...
        8.3,240.,10.7,270.,16.,282.,21.,294.,28.,306.,41.,312.,50.,,...
        324.,90.,999.,90.
CONTRL DELT=.002,DELMI=1.E-10 ,FINTI=8.
INTEG MILNE
RELERR Y3DOT=1.E-4
PRINT 0.1,Y3DOT,Y3,XDOT,X,DELT
GRAPH .05,8,6,TIME,Y3DOT
LABEL Y3DOT VS. TIME (ARRESTING GEAR)
SCALE 1.,.05
RANGE Y3DOT,Y3,XDOT,X,DELT
END
```

```
AIRCRAFT ARRESTING GEAR PROBLEM


       TIME         Y3DOT         Y3          XDOT          X           DELT
    0.000E 00    0.0000E 00    0.0000E 00   0.2900E 03   0.0000E 00   0.1999E-02
    0.999E-01    0.8147E 01    0.1580E 00   0.2898E 03   0.2899E 02   0.3999E-02
    0.199E 00    0.4859E 02    0.3001E 01   0.2890E 03   0.5795E 02   0.7999E-02
    0.299E 00    0.7483E 02    0.9364E 01   0.2874E 03   0.8678E 02   0.7999E-02
    0.399E 00    0.8855E 02    0.1749E 02   0.2851E 03   0.1154E 03   0.7999E-02
    0.499E 00    0.1075E 03    0.2730E 02   0.2820E 03   0.1437E 03   0.1599E-01
    0.599E 00    0.1208E 03    0.3879E 02   0.2788E 03   0.1718E 03   0.7999E-02
    0.699E 00    0.1293E 03    0.5136E 02   0.2760E 03   0.1995E 03   0.1599E-01
    0.799E 00    0.1296E 03    0.6454E 02   0.2738E 03   0.2270E 03   0.7999E-02
    0.899E 00    0.1145E 03    0.7666E 02   0.2724E 03   0.2543E 03   0.7999E-02
```

```
TIME    0.1006E 01   DELT   0.7999E-02   DELMI   0.1000E-09   FINTI   0.8000E 01   DELTP   0.1000E 00
DELTC   0.0000E 00   XDOT   0.2700E 03   X       0.2833E 03   Y3DOT   0.1088E 03   Y3      0.8828E 02
ZZ001   0.1158E 03   Y2     0.8991E 02   SQR1    0.3097E 03   Y12     0.4903E 01   FK1     0.2231E 03
FD      0.3903E 05   Y23    0.1631E 01   FK2     0.4126E 05   ANGLE   0.1155E 01   SINTH   0.9999E 00
COSTH   0.3528E-02   R      0.2838E 03   THETA   0.1567E 01   Y3DEL   0.0000E 00   Y3DER   0.1086E 03
WW      0.1000E 01   H      0.1250E 03   THETO   0.0000E 00   ERROR   0.1000E-02   K1      0.4550E 04
K2      0.2530E 05   M1     0.1400E 04   M2      0.4528E 02   M3      0.2000E 02   XDOT0   0.2900E 03
FOFY3   0.3946E-41


TIME    0.1006E 01   DELT   0.7999E-02   DELMI   0.1000E-09   FINTI   0.8000E 01   DELTP   0.1000E 00
DELTC   0.0000E 00   XDOT   0.2700E 03   X       0.2833E 03   Y3DOT   0.1088E 03   Y3      0.8828E 02
ZZ001   0.1158E 03   Y2     0.8991E 02   SQR1    0.3097E 03   Y12     0.4902E 01   FK1     0.2230E 05
FD      0.3906E 05   Y23    0.1632E 01   FK2     0.4131E 05   ANGLE   0.1155E 01   SINTH   0.9999E 00
COSTH   0.3528E-02   R      0.2838E 03   THETA   0.1567E 01   Y3DEL   0.0000E 00   Y3DER   0.1087E 03
WW      0.1000E 01   H      0.1250E 03   THETO   0.0000E 00   ERROR   0.1000E-02   K1      0.4550E 04
K2      0.2530E 05   M1     0.1400E 04   M2      0.4528E 02   M3      0.2000E 02   XDOT0   0.2900E 03
FOFY3   0.3946E-41


TIME    0.1006E 01   DELT   0.7999E-02   DELMI   0.1000E-09   FINTI   0.8000E 01   DELTP   0.1000E 00
DELTC   0.0000E 00   XDOT   0.2700E 03   X       0.2833E 03   Y3DOT   0.1088E 03   Y3      0.8828E 02
ZZ001   0.1158E 03   Y2     0.8991E 02   SQR1    0.3097E 03   Y12     0.4902E 01   FK1     0.2230E 05
FD      0.3906E 05   Y23    0.1632E 01   FK2     0.4131E 05   ANGLE   0.1155E 01   SINTH   0.9999E 00
COSTH   0.3528E-02   R      0.2838E 03   THETA   0.1567E 01   Y3DEL   0.0000E 00   Y3DER   0.1088E 03
WW      0.1000E 01   H      0.1250E 03   THETO   0.0000E 00   ERROR   0.1000E-02   K1      0.4550E 04
K2      0.2530E 05   M1     0.1400E 04   M2      0.4528E 02   M3      0.2000E 02   XDOT0   0.2900E 03
FOFY3   0.3946E-41
    0.999E 00    0.1080E 03    0.8752E 02   0.2702E 03   0.2815E 03   0.7999E-02
    0.109E 01    0.1157E 03    0.9881E 02   0.2669E 03   0.3083E 03   0.7999E-02
    0.119E 01    0.1159E 03    0.1104E 03   0.2631E 03   0.3348E 03   0.1599E-01
    0.129E 01    0.1150E 03    0.1219E 03   0.2587E 03   0.3609E 03   0.3999E-02
    0.139E 01    0.1184E 03    0.1336E 03   0.2538E 03   0.3866E 03   0.1599E-01
    0.149E 01    0.1203E 03    0.1456E 03   0.2487E 03   0.4117E 03   0.1599E-01
    0.159E 01    0.1166E 03    0.1574E 03   0.2437E 03   0.4363E 03   0.7999E-02
    0.169E 01    0.1124E 03    0.1689E 03   0.2386E 03   0.4604E 03   0.1599E-01
    0.179E 01    0.1097E 03    0.1800E 03   0.2333E 03   0.4840E 03   0.7999E-02
    0.189E 01    0.1074E 03    0.1909E 03   0.2278E 03   0.5071E 03   0.1599E-01
    0.199E 01    0.1050E 03    0.2015E 03   0.2220E 03   0.5296E 03   0.1599E-01
    0.209E 01    0.1029E 03    0.2119E 03   0.2160E 03   0.5515E 03   0.3999E-02
```

TABLE 5.19
Simulation Results for the Aircraft Arresting Gear
Problem with Terminal Output from DEBUG

AIRCRAFT ARRESTING GEAR PROBLEM

TIME		Y3DOT		Y3		XDOT		X		DELT
0.499E	01	0.7760E	01	0.3635E	03	0.1126E	02	0.8437E	03	0.1999E-02
0.509E	01	0.6270E	01	0.3642E	03	0.1100E	02	0.8448E	03	0.3999E-02
0.519E	01	0.5594E	01	0.3647E	03	0.1080E	02	0.8458E	03	0.1999E-02
0.529E	01	0.5334E	01	0.3653E	03	0.1062E	02	0.8469E	03	0.1999E-02
0.539E	01	0.5233E	01	0.3658E	03	0.1045E	02	0.8480E	03	0.3999E-02
0.549E	01	0.5161E	01	0.3663E	03	0.1028E	02	0.8490E	03	0.3999E-02
0.559E	01	0.5078E	01	0.3668E	03	0.1012E	02	0.8500E	03	0.1999E-02
0.569E	01	0.4999E	01	0.3673E	03	0.9963E	01	0.8510E	03	0.1999E-02
0.579E	01	0.4918E	01	0.3678E	03	0.9810E	01	0.8520E	03	0.3999E-02
0.589E	01	0.4842E	01	0.3683E	03	0.9662E	01	0.8530E	03	0.1999E-02
0.599E	01	0.4762E	01	0.3688E	03	0.9519E	01	0.8539E	03	0.1999E-02
0.609E	01	0.4694E	01	0.3693E	03	0.9380E	01	0.8548E	03	0.1999E-02
0.619E	01	0.4626E	01	0.3697E	03	0.9244E	01	0.8558E	03	0.1999E-02
0.629E	01	0.4566E	01	0.3702E	03	0.9113E	01	0.8567E	03	0.1999E-02
0.639E	01	0.4487E	01	0.3706E	03	0.8986E	01	0.8576E	03	0.1999E-02
0.649E	01	0.4436E	01	0.3711E	03	0.8862E	01	0.8585E	03	0.1999E-02
0.659E	01	0.4369E	01	0.3715E	03	0.8741E	01	0.8593E	03	0.3999E-02
0.669E	01	0.4306E	01	0.3719E	03	0.8624E	01	0.8602E	03	0.1999E-02
0.679E	01	0.4261E	01	0.3724E	03	0.8510E	01	0.8611E	03	0.1999E-02
0.689E	01	0.4190E	01	0.3728E	03	0.8399E	01	0.8619E	03	0.9999E-03
0.699E	01	0.4148E	01	0.3732E	03	0.8291E	01	0.8627E	03	0.9999E-03
0.709E	01	0.4079E	01	0.3736E	03	0.8185E	01	0.8635E	03	0.1999E-02
0.719E	01	0.4036E	01	0.3740E	03	0.8083E	01	0.8644E	03	0.1999E-02
0.729E	01	0.3987E	01	0.3744E	03	0.7982E	01	0.8652E	03	0.3999E-02
0.739E	01	0.3932E	01	0.3748E	03	0.7885E	01	0.8659E	03	0.3999E-02
0.749E	01	0.3892E	01	0.3752E	03	0.7790E	01	0.8667E	03	0.1999E-02
0.759E	01	0.3840E	01	0.3756E	03	0.7697E	01	0.8675E	03	0.9999E-03
0.769E	01	0.3792E	01	0.3760E	03	0.7606E	01	0.8682E	03	0.9999E-03
0.779E	01	0.3743E	01	0.3763E	03	0.7518E	01	0.8690E	03	0.9999E-03
0.789E	01	0.3706E	01	0.3767E	03	0.7431E	01	0.8697E	03	0.3999E-02
0.799E	01	0.3665E	01	0.3771E	03	0.7347E	01	0.8705E	03	0.9999E-03
0.799E	01	0.3665E	01	0.3771E	03	0.7347E	01	0.8705E	03	0.9999E-03

VARIABLE	MINIMUM	MAXIMUM
Y3DOT	0.0000E 00	0.1324E 03
Y3	0.0000E 00	0.3771E 03
XDOT	0.7347E 01	0.2900E 03
X	0.0000E 00	0.8705E 03
DELT	0.9999E-03	0.1599E-01

his seat. Several combinations of speed and altitude are required to be investigated since the drag on the pilot, which causes his relative horizontal motion with respect to the aircraft, is a function of both air density and velocity. The ejection system is so devised that it causes the pilot and his seat to travel along rails at a specified exit velocity V_E, at an angle several combinations backward from vertical. The seat becomes disengaged from the rails at $Y = Y_1$.

TABLE 5.20
Simulation Results for the Aircraft Arresting Gear
Problem with FINISH Facility of DSL and More
Extensive Variable Printout

```
                    *** DSL/1800 SIMULATION DATA ***
TITLE ARRESTING GEAR PROBLEM RUNGA-KUTTA INTEGRATION
INCON XDOT0=200.
CONTRL FINTI=6.,DELT=.002
PRINT .5,Y3DOT,Y3,Y2,THETA,XDOT,X,Y3DEL,Y3DER,DELT,ANGLE
FINISH ANGLE=1.427
INTEG RKS
RESET RELERR,ABSERR,GRAPH,RANGE
RELERR XDOT=1.E-3
ABSERR XDOT=1.E-2
END
```

ARRESTING GEAR PROBLEM RUNGA-KUTTA INTEG


```
TIME  =  0.0000E 00   Y3DOT =  0.0000E 00   Y3    =  0.0000E 00   Y2    =  0.0000E 00   THETA =  0.0000E 00
                      XDOT  =  0.2000E 03   X     =  0.0000E 00   Y3DEL =  0.0000E 00   Y3DER =  0.0000E 00
                      DELT  =  0.1999E-02   ANGLE =  0.0000E 00

TIME  =  0.5000E 00   Y3DOT =  0.5781E 02   Y3    =  0.1506E 02   Y2    =  0.1595E 02   THETA =  0.1560E 01
                      XDOT  =  0.1974E 03   X     =  0.9960E 02   Y3DEL =  0.0000E 00   Y3DER =  0.5775E 02
                      DELT  =  0.3954E-01   ANGLE =  0.6728E 00

TIME  =  0.9999E 00   Y3DOT =  0.8396E 02   Y3    =  0.5241E 02   Y2    =  0.5305E 02   THETA =  0.1565E 01
                      XDOT  =  0.1909E 03   X     =  0.1966E 03   Y3DEL =  0.0000E 00   Y3DER =  0.8384E 02
                      DELT  =  0.4481E-01   ANGLE =  0.1004E 01
```

```
TIME   0.1017E 01   DELT   0.7103E-01   DELMI  0.1000E-09   FINTI  0.6000E 01   DELTP  0.5000E 00
DELTC  0.0000E 00   XDOT   0.1907E 03   X      0.2000E 03   Y3DOT  0.8454E 02   Y3     0.5390E 02
ZZ001  0.8295E 02   Y2     0.5452E 02   SQR1   0.2358E 03   Y12    0.1842E 01   FK1    0.8381E 04
FD     0.1492E 05   Y23    0.6148E 00   FK2    0.1555E 05   ANGLE  0.1012E 01   SINTH  0.9999E 00
COSTH  0.4999E-02   R      0.2006E 03   THETA  0.1565E 01   Y3DEL  0.0000E 00   Y3DER  0.8428E 02
WW     0.1000E 01   H      0.1250E 03   THETO  0.0000E 00   ERROR  0.1000E-02   K1     0.4550E 04
K2     0.2530E 05   M1     0.1400E 04   M2     0.4528E 02   M3     0.2000E 02   XDOT0  0.2000E 03
F0FY3  0.3946E-41
```

```
TIME   0.1017E 01   DELT   0.7103E-01   DELMI  0.1000E-09   FINTI  0.6000E 01   DELTP  0.5000E 00
DELTC  0.0000E 00   XDOT   0.1907E 03   X      0.2000E 03   Y3DOT  0.8452E 02   Y3     0.5391E 02
ZZ001  0.8298E 02   Y2     0.5452E 02   SQR1   0.2358E 03   Y12    0.1823E 01   FK1    0.8297E 04
FD     0.1491E 05   Y23    0.6124E 00   FK2    0.1549E 05   ANGLE  0.1012E 01   SINTH  0.9999E 00
COSTH  0.4999E-02   R      0.2006E 03   THETA  0.1565E 01   Y3DEL  0.0000E 00   Y3DER  0.8452E 02
WW     0.1000E 01   H      0.1250E 03   THETO  0.0000E 00   ERROR  0.1000E-02   K1     0.4550E 04
K2     0.2530E 05   M1     0.1400E 04   M2     0.4528E 02   M3     0.2000E 02   XDOT0  0.2000E 03
F0FY3  0.3946E-41
```

```
TIME   0.1035E 01   DELT   0.7103E-01   DELMI  0.1000E-09   FINTI  0.6000E 01   DELTP  0.5000E 00
DELTC  0.0000E 00   XDOT   0.1905E 03   X      0.2034E 03   Y3DOT  0.8500E 02   Y3     0.5541E 02
ZZ001  0.8337E 02   Y2     0.5600E 02   SQR1   0.2387E 03   Y12    0.1754E 01   FK1    0.7983E 04
FD     0.1421E 05   Y23    0.5858E 00   FK2    0.1482E 05   ANGLE  0.1019E 01   SINTH  0.9999E 00
COSTH  0.4916E-02   R      0.2040E 03   THETA  0.1565E 01   Y3DEL  0.0000E 00   Y3DER  0.8495E 02
WW     0.1000E 01   H      0.1250E 03   THETO  0.0000E 00   ERROR  0.1000E-02   K1     0.4550E 04
K2     0.2530E 05   M1     0.1400E 04   M2     0.4528E 02   M3     0.2000E 02   XDOT0  0.2000E 03
F0FY3  0.3946E-41
```

```
TIME  =  0.1499E 01   Y3DOT =  0.8333E 02   Y3    =  0.9350E 02   Y2    =  0.9449E 02   THETA =  0.1567E 01
                      XDOT  =  0.1855E 03   X     =  0.2909E 03   Y3DEL =  0.0000E 00   Y3DER =  0.8365E 02
                      DELT  =  0.5108E-01   ANGLE =  0.1165E 01

TIME  =  0.1999E 01   Y3DOT =  0.8367E 02   Y3    =  0.1346E 03   Y2    =  0.1361E 03   THETA =  0.1568E 01
                      XDOT  =  0.1749E 03   X     =  0.3812E 03   Y3DEL =  0.0000E 00   Y3DER =  0.8385E 02
                      DELT  =  0.9517E-01   ANGLE =  0.1254E 01

TIME  =  0.2500E 01   Y3DOT =  0.7741E 02   Y3    =  0.1750E 03   Y2    =  0.1765E 03   THETA =  0.1568E 01
                      XDOT  =  0.1627E 03   X     =  0.4657E 03   Y3DEL =  0.5781E 02   Y3DER =  0.7739E 02
                      DELT  =  0.5371E-01   ANGLE =  0.1308E 01
```


ARRESTING GEAR PROBLEM RUNGA-KUTTA INTEG


```
TIME  =  0.5999E 01   Y3DOT =  0.1829E 02   Y3    =  0.3463E 03   Y2    =  0.3474E 03   THETA =  0.1569E 01
                      XDOT  =  0.2539E 02   X     =  0.8136E 03   Y3DEL =  0.5481E 02   Y3DER =  0.1829E 02
                      DELT  =  0.2974E-01   ANGLE =  0.1418E 01

TIME  =  0.5999E 01   Y3DOT =  0.1829E 02   Y3    =  0.3463E 03   Y2    =  0.3474E 03   THETA =  0.1569E 01
                      XDOT  =  0.2539E 02   X     =  0.8136E 03   Y3DEL =  0.5481E 02   Y3DER =  0.1829E 02
                      DELT  =  0.2974E-01   ANGLE =  0.1418E 01
```

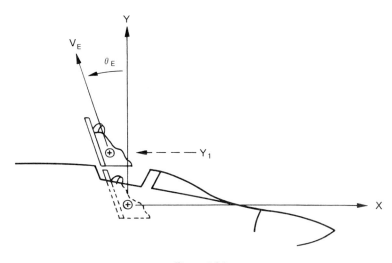

Figure 5.26.

Figure 5.26 illustrates the geometry of the ejection and the aircraft. The ejection trajectory is illustrated in Fig. 5.27.

Once the pilot and seat combination leaves the rails, it follows a ballistic trajectory which can be calculated; however, since it is the relative motion of the pilot with respect to the aircraft (which is assumed to fly level at constant speed) that is important, we can formulate the equations so as to obtain this motion directly. This phase of the ejection is shown in Fig. 5.27.

The governing equations of motion along the various directions are given as follows:

$$\dot{X} = V\cos\theta - V_A \qquad \theta = 0 \text{ for } 0 \leqslant Y \leqslant Y_1$$

$$\dot{Y} = V\sin\theta \qquad\qquad = \frac{-(g\cos\theta)}{V} \text{ for } Y > Y_1$$

$$\dot{V} = 0 \qquad\qquad 0 \leqslant Y \leqslant Y_1 \qquad\qquad (5.20)$$

$$\qquad = -\frac{D}{M} - g\sin\theta \qquad Y > Y_1$$

$$D = \tfrac{1}{2}\rho C_D S V^2$$

The values of constants (for all cases) and parameters required for the solution are

$$m = 7 \text{ slugs} \qquad\qquad Y_1 = 40 \text{ feet}$$

$$g = 32.2 \text{ feet per square second} \qquad V_E = 40 \text{ feet per second} \qquad (5.21)$$

$$C_D = 1 \qquad\qquad \theta_E = 15°$$

$$S = 10 \text{ square feet}$$

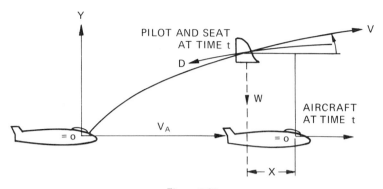

Figure 5.27.

The initial values of V and θ (the pilot's initial velocity vector at the moment of leaving the cockpit rails) are given by

$$V(0) = \left[(V_A - V_E \sin\theta_E)^2 + (V_E \cos\theta_E)^2 \right]^{1/2}$$

$$\theta(0) = \tan^{-1} \frac{V_E \cos\theta_E}{V_A - V_E \sin\theta_E}$$

(5.22)

and further

$$X(0) = Y(0) = 0$$

A run is to be terminated when any one of these conditions occurs: $X < -60$ feet (pilot beyond vertical stabilizer) or $Y > 30$ feet (pilot above tail)

The following quantities are to be printed every .02 second of the independent variable t:

$$t, \dot{V}, V, \theta, X, Y$$

The following four run cases are required to be investigated:

Case	Velocity, V_A (ft/sec)	Density, ρ (slugs/ft^3)	Condition
1	900	2.3769 × 10^{-3}	Sea level
2	900	.2238 × 10^{-3}	60,000 ft
3	500	.2238 × 10^{-3}	60,000 ft
4	500	2.3769 × 10^{-3}	Sea level

The DSL structure statements for the problem as formulated above are shown in Table 5.21, whereas the corresponding UPDAT subroutine generated is shown in Table 5.22. Note again the use of a procedural block to define a function using FORTRAN statements intermixed with DSL structure statements. The use of the FINISH statement to terminate the problem solution for conditions other than the time reaching FINTI value is also given in Table 5.21.

Table 5.22 shows how, prior to the execution of the translated program, the function ROTAT is defined as a subroutine and stored in the user area, as was the case in the previous example (Section 5.1.6).

Note that the INSW block from the DSL program library could have been used to produce the same results as the statements in PROCED block by defining

$$YGEY1 = INSW(Y-Y1, 0., 1.)$$

Furthermore, a PROCED block could have replaced the calling of ROTAT by putting the FORTRAN statements in the PROCED block.

TABLE 5.21

Input for Pilot Ejection Study-Implicit Structure

```
                    *** DSL/1800 TRANSLATOR INPUT ***
TITLE DSL 1130 PILOT EJECTION STUDY. IMPLICIT STRUCTURE
CONST M=7.,CD=1.,S=10.,Y1=4.,THETD=15.,VA=900.,RHO=2.3769E-3,VE=40.
*
      XDOT=V*COS(THETA)-VA
      X=INTGR(0.,XDOT)
      YDOT=V*SIN(THETA)
      Y=INTGR(0.,YDOT)
PROCED YGEY1=PRO( Y,Y1)
      YGEY1=1.
      IF(Y-Y1) 1,2,2
    1 YGEY1=0.
    2 CONTINUE
ENDPRO
      VDOT=YGEY1*(-D/M-G*SIN(THETA))
      V=INTGR(VZERO,VDOT)
CONST G=32.2
      THDOT=YGEY1*(-G*COS(THETA)/V )
      THETA=INTGR(THETO,THDOT)
      D=0.5*RHO*CD*S*V*V
      VX1,VY=ROTAT(VE,THETE)
      VX=VA-VX1
      THETE=THETD/57.3
      VZERO=SQRT(VX*VX+VY*VY)
      THETO= ATAN(VY/VX)
FINISH X=-60.,Y=30.
CONTRL FINTI=4.,DELT=.01
*    CALCULATE VALUES OF THETA IN DEGREES
      THTDG=THETA*57.3
PRINT .02,VDOT,V,THTDG,X,Y
END
PARAM RHO=.2238E-3
END
PARAM VA=500.
END
PARAM RHO=2.3769E-3
END
STOP
```

TABLE 5.22
UPDAT Subroutine for Pilot Ejection Study-
Implicit Structure

```
     SUBROUTINE UPDAT
     REAL  M      ,INTGR
     COMMON NALRM,IZZZ(833) ,TIME ,DELT ,DELMI,FINTI,DELTP,DELTC,X
    1Y     ,V      ,THETA,XDOT ,YDOT ,YGEY1,VDOT ,THDOT,D     ,VX1  ,VY
    1VX    ,THETE,VZERO,THETO,THTDG,M     ,CD   ,S     ,Y1    ,THETD,VA
    1RHO   ,VE     ,G
     XDOT=V*COS(THETA)-VA
     X=INTGR(0.,XDOT)
     YDOT=V*SIN(THETA)
     Y=INTGR(0.,YDOT)
     D=0.5*RHO*CD*S*V*V
     YGEY1=1.
     IF(Y-Y1) 1,2,2
   1 YGEY1=0.
   2 CONTINUE
     VDOT=YGEY1*(-D/M-G*SIN(THETA))
     THETE=THETD/57.3
     CALL ROTAT(VE,THETE,VX1,VY)
     VX=VA-VX1
     VZERO=SQRT(VX*VX+VY*VY)
     V=INTGR(VZERO,VDOT)
     THDOT=YGEY1*(-G*COS(THETA)/V )
     THETO= ATAN(VY/VX)
     THETA=INTGR(THETO,THDOT)
     THTDG=THETA*57.3
     RETURN
     END

     // FOR ROTAT
     *LIST ALL
     *ONE WORD INTEGERS
          SUBROUTINE ROTAT( VE,THETE,VX1,VY)
          VX1=VE*SIN(THETE)
          VY= VE*COS(THETE)
          RETURN
          END

     // DUP

     *STORE     WS  UA  ROTAT
```

The FINISH statement terminates the run whenever the value of X passes through −60 or whenever the value of Y passes through 30.

The four END cards indicate that four runs are to be made. Note that the data for run 2 are identical to those for run 1 with the exception of RHO, which has a new value of .0002238. Only one parameter card reflecting this change is required.

The simulation results for this case are shown in Table 5.23, where the

TABLE 5.23
Simulation Results for the Implicit Structure Pilot Ejection Study

```
                    *** DSL/1800 SIMULATION DATA ***
TITLE DSL 1130 PILOT EJECTION STUDY. IMPLICIT STRUCTURE
CONST M=7.,CD=1.,S=10.,Y1=4.,THETD=15.,VA=900.,RHO=2.3769E-3,VE=40.
CONST G=32.2
FINISH X=-60.,Y=30.
CONTRL FINTI=4.,DELT=.01
PRINT .02,VDOT,V,THTDG,X,Y
END
```

DSL 1130 PILOT EJECTION STUDY. IMPLICIT

TIME	VDOT	V	THTDG	X	Y
0.000E 00	0.0000E 00	0.8904E 03	0.2486E 01	0.0000E 00	0.0000E 00
0.199E-01	0.0000E 00	0.8904E 03	0.2486E 01	-0.2070E 00	0.7727E 00
0.358E-01	0.0000E 00	0.8904E 03	0.2486E 01	-0.3711E 00	0.1385E 01
0.558E-01	0.0000E 00	0.8904E 03	0.2486E 01	-0.5781E 00	0.2157E 01
0.799E-01	0.0000E 00	0.8904E 03	0.2486E 01	-0.8281E 00	0.3090E 01
0.999E-01	0.0000E 00	0.8904E 03	0.2486E 01	-0.1035E 01	0.3863E 01
0.119E 00	-0.1275E 04	0.8664E 03	0.2448E 01	-0.1462E 01	0.4621E 01
0.139E 00	-0.1204E 04	0.8416E 03	0.2405E 01	-0.2399E 01	0.5344E 01
0.159E 00	-0.1138E 04	0.8182E 03	0.2360E 01	-0.3816E 01	0.6034E 01
0.179E 00	-0.1077E 04	0.7960E 03	0.2315E 01	-0.5689E 01	0.6693E 01
0.199E 00	-0.1021E 04	0.7751E 03	0.2268E 01	-0.7991E 01	0.7321E 01
0.219E 00	-0.9695E 03	0.7552E 03	0.2220E 01	-0.1070E 02	0.7920E 01
0.239E 00	-0.9216E 03	0.7363E 03	0.2170E 01	-0.1379E 02	0.8492E 01
0.259E 00	-0.8772E 03	0.7183E 03	0.2119E 01	-0.1726E 02	0.9036E 01
0.279E 00	-0.8359E 03	0.7011E 03	0.2067E 01	-0.2107E 02	0.9555E 01
0.299E 00	-0.7974E 03	0.6848E 03	0.2014E 01	-0.2522E 02	0.1004E 02
0.319E 00	-0.7616E 03	0.6692E 03	0.1960E 01	-0.2969E 02	0.1051E 02
0.339E 00	-0.7280E 03	0.6543E 03	0.1904E 01	-0.3446E 02	0.1096E 02
0.359E 00	-0.6967E 03	0.6401E 03	0.1847E 01	-0.3953E 02	0.1138E 02
0.379E 00	-0.6673E 03	0.6265E 03	0.1789E 01	-0.4487E 02	0.1179E 02
0.399E 00	-0.6398E 03	0.6134E 03	0.1729E 01	-0.5048E 02	0.1217E 02
0.419E 00	-0.6139E 03	0.6008E 03	0.1669E 01	-0.5634E 02	0.1253E 02
0.439E 00	-0.5896E 03	0.5888E 03	0.1606E 01	-0.6245E 02	0.1287E 02
0.439E 00	-0.5896E 03	0.5888E 03	0.1606E 01	-0.6245E 02	0.1287E 02

SIMULATION HALTED, X   = -0.6245E 02
```

problem termination as X passes the value of –60 is illustrated. Only one of the four parameter runs is illustrated for brevity.

### 5.1.7.1 Determining Safe Ejection Altitude Using Man-Machine Interaction

Now let us pose the problem in a somewhat different manner. The time history of each run is not necessary if one is interested only in determining the envelope of safe ejection as a function of aircraft altitude and velocity. Only the
```

TABLE 5.24

Variation of Air Density with Altitude

h (K ft)	ρ (×10³)	h (K ft)	ρ (×10³)
0.	2.377	15.	1.497
1.	2.308	20.	1.267
2.	2.241	30.	.891
4.	2.117	40.	.587
6.	1.987	50.	.364
10.	1.755	60.	.2238

TABLE 5.25

*Translator Input for the Pilot Ejection Study—
Iterative Solution*

```
              *** DSL/1800 TRANSLATOR INPUT ***
TITLE PILOT EJECTION STUDY. ITERATED SOLUTION
INTGER SW
PARAM H=0.,VA=300.
CONTRL FINTI=2.,DELT=.01
CONST M=7.,G=32.2,CD=1.,S=10.,Y1=4.,THETD=15.,VE=40.
PARAM SW=1
AFGEN DNSTY=0.,2.377E-3,1000.,2.308E-3,2000.,2.241E-3,4000.,2.117E-3,...
      6000.,1.987E-3,10000.,1.755E-3,15000.,1.497E-3,20000.,1.267E-3,...
      30000.,.891E-3,40000.,.587E-3,50000.,.364E-3,60000.,.2238E-3
INTEG MILNE
NOSORT
      GO TO(1,2),SW
    1 SW=2
      THETE=.01745*THETD
      VX=VA-VE*SIN(THETE)
      VY=VE*COS(THETE)
      VO=SQRT(VX*VX+VY*VY)
      THETO=ATAN( VY/VX)
      RHOP=AFGEN(DNSTY,H)*CD*S
    2 CONTINUE
SORT
FINISH X=-30.
      X=INTGR(0.,V*COS(THETA)-VA)
      Y=INTGR(0.,V*SIN(THETA))
      YGEY1=INSW(Y-Y1,0.,1.)
      VDOT=YGEY1*(-D/M-G*SIN(THETA))
      V=INTGR(VO,VDOT)
      THDOT=YGEY1*(-G*COS(THETA)/V)
      THETA=INTGR(THETO,THDOT)
      D=0.5*RHOP*V**2
PRINT 0.05,X,Y
END
PRINT 0.01,X,Y
END
STOP
```

value of altitude and velocity V_A that result in "near" misses are of interest. The air density, which is a function of altitude, is shown in Table 5.24.

The problem then may be run as follows: Choose a large value of print interval and print only the values of Y. Use data switch 2 to perform the next input from the keyboard entry (see Table 7.4, Data Switch Control). Obtain the value of Y at the end of the first run with a first guess of a startup value of H. If the value of Y is less than 20, add 100 to the value of H, type in the new value of H in a PARAM statement, and repeat the run. Continue this procedure until $Y \geqslant 20$, at which time either enter commands for complete printouts and plots via the keyboard or switch back to the card reader (by turning off data switch 2) and rerun the problem with the last value of H to obtain the desired output.

This procedure is shown in Tables 5.25 and 5.26, where the DSL structure statements as well as the input to the console typewriter are shown for the problem solution with the operator interaction.

Note the use of AFGEN to describe density as a function of height H—both its definition and the data specification. Note in addition that the problem statements are restructured—with initial calculations performed only at simulation start. The FORTRAN logic in this case is now inserted in a NOSORT block since no blocks are to be defined. Note, however, that at each new value of H in the iterative solution the parameter SW must be reset to 1. With each new value of velocity V_A, the iterative solution is repeated to obtain that value of H which results in the Y value exceeding 20. The user can then manually plot the safe altitude of ejection

TABLE 5.26
Showing the Input from the 1130 Console
Keyboard for the Iterated Solution of Pilot
Ejection Study

```
PARAM H=100.,SW=1
END
PARAM H=500.,SW=1
END
PARAM H=2000.,SW=1
END
PARAM H=5000.,SW=1
END
PARAM H=10000.,VA=500.,SW=1
END
PARAM H=30000.,SW=1
END
PARAM H=32000.,SW=1
END
PARAM VA=700.,H=40000.,SW=1
END
PARAM H=45000.,SW=1
END
PARAM VA=900.,H=50000.,SW=1
END
PARAM H=52000.,SW=1
END
```

H versus velocity. An interesting exercise for the user will be to use the plot facilities of DSL and the NLFGN (nonlinear function generator) function with data obtained from the results to prepare the plot. The NLFGN function will provide some smoothing of the curve between the points.

The results indicate that as the velocity increases the safe height also increases rapidly, so that the pilot must either slow down the aircraft, change his plane of travel direction, or be at a sufficient altitude before he can safely eject from the aircraft.

5.1.8 Cooling Fin Design

The next problem is encountered in designing a heat exchanger which is carrying a hot fluid cooled in deep outer space. The heat exchanger has cooling fins attached to the tube to increase the heat exchange area. The heat is conducted along the fins and then transmitted to the outer space by means of radiation. It is desired to find a suitable thickness of cooling fin for a given length and width of fin to dissipate a specified amount of heat. Figure 5.28 shows the geometry of the cooling fin, with heat transfer occurring by conduction along (direction x) the length of the fin which is in equilibrium with the heat dissipated by radiation from

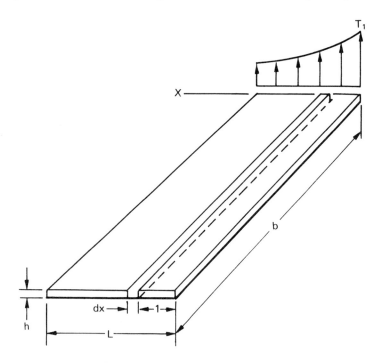

Figure 5.28. Geometry of the radiating fin.

the surfaces of the fin to the outer space. If one neglects the heat radiated from the tip of the fin, the equation describing this heat equilibrium is

$$kbh \frac{d^2 T_1}{dx^2} = 2 \, w\epsilon b \left(T_1^4 - T_2^4 \right) \tag{5.23}$$

where

T_1	= absolute Fahrenheit temperature at any position along the fin, $^\circ R$	
T_2	= absolute Fahrenheit temperature of surrounding space $0^\circ R$	
w	= Stefan-Boltzmann constant, $.173(10)^{-3}$ Btu/(ft)^2 (hr)($^\circ R$)^4	
ϵ	= emissivity of fin, .8	
b	= width of fin in z-direction, .5 ft	(5.24)
h	= thickness of fin, $.005 \leqslant h \leqslant .01$ ft	
k	= thermal conductivity of fin, 25 Btu/(hr) (ft) ($^\circ R$)	
$(T_1)_0$	= constant temperature of fin at root end ($x = 0$), $2000^\circ R$	
L	= length of fin, .25 ft	

Neglecting any radiation from the very small perimetrical surfaces of the fin and the curvature of the root area of the fin, we may write

$$\left[-kbh \left(\frac{dT_1}{dx} \right)_x \right] - \left[-kbh \left(\frac{dT_1}{dx} \right)_{x+dx} \right] = \sigma\epsilon_1 (2b \, dx)\left(T_1^4 - T_2^4 \right) \tag{5.25}$$

from which

$$kbh \frac{(dT_1/dx)_{x+dx} - (dT_1/dx)_x}{dx} = 2\sigma\epsilon_1 b \left(T_1^4 - T_2^4 \right) \tag{5.26}$$

or

$$\frac{d^2 T_1}{dx^2} = \frac{2\sigma\epsilon_1}{kh} \left(T_1^4 - T_2^4 \right) \tag{5.27}$$

Equation (5.27) is the differential equation defining the thermal equilibrium state of the fin, the solution of which will yield the temperature distribution along the fin. With the given parameter values inserted, Eq. (5.27) becomes

$$\frac{d^2 T_1}{dx^2} = .011 \times 10^{-8} \times \frac{T_1^4}{h} \tag{5.28}$$

where h is left a variable for programming purposes.

The boundary values for solution of the equation are

$$\begin{array}{lll} \text{at } x = 0, & T_1 = 2000^\circ R & \\ \text{at } x = .25, & \dfrac{dT_1}{dx} = 0 & \end{array} \tag{5.29}$$

TABLE 5.27

DSL Input for Cooling Fin Design

```
                              *** DSL/1800 TRANSLATOR INPUT ***
      TITLE COOLING FIN DESIGN FOR SPACE VEHICLE
            TERM=2.*SIGMA*E*(TEMP**4)
            TDT=INTGR(TDTO,TDT2)
            TDT2=TERM/(K*H)
            T=INTGR(TO,TDT)
            TDTO=-QO/(K*H*WIDTH)
            Q=-TDT*K*H*WIDTH
            TEMP=LIMIT(0.,TO,T)
      PRINT .05,Q,TDT,T
      CONST SIGMA=.173E-8,E=.8,K=25.
      PARAM WIDTH=.5,H=0.02
      INCON QO=1000.,TO=2000.
      CONTRL DELT=.0001,FINTI=.25
      INTEG RKS
      RENAME TIME=X
      END
      LABEL TEMPERATURE GRADIENT ON FIN
      SCALE 40.,,.004
      GRAPH .001,10,8,X,T
      PRINT .01,TDT,T,Q
      END
      STOP
```

The heat flux $Q = -kbh(dT/dx)$ is specified as 10,000 BTUs per hour at $x = 0$, and it is desired to find the value of h which satisfies the boundary condition $x = .25$.

As in the case of the pilot ejection study problem, a trial and error solution is called for. The two equations to be solved, coded in DSL, are shown in Table 5.27. The procedure for solving this problem is to turn on data switch 2 after the program execution begins (i.e., the solution of the problem with the first value of h begins), examine the value of dT/dx, make a correction to a second estimate of h, type through the keyboard a new value of h, type END, and repeat the problem solution. When a sufficiently low value of dT/dx is found, simply repeat the solution after inserting a more extensive PRINT statement, and if desired use GRAPH facility to observe, for example, the temperature gradient across the fin width.

Figure 5.29 and Table 5.28 show the output from the program with the above procedure.

5.2 Two-Point Boundary-Value Problems in Fluid Mechanics

In Chapter 3, we discussed the problem of a special case of a non-Newtonian fluid flow past a wedge described by

$$nF'''(F'')^{n-1} + \frac{1-M+2Mn}{n+1} FF'' = M(F'^2 - 1) \qquad (5.30)$$

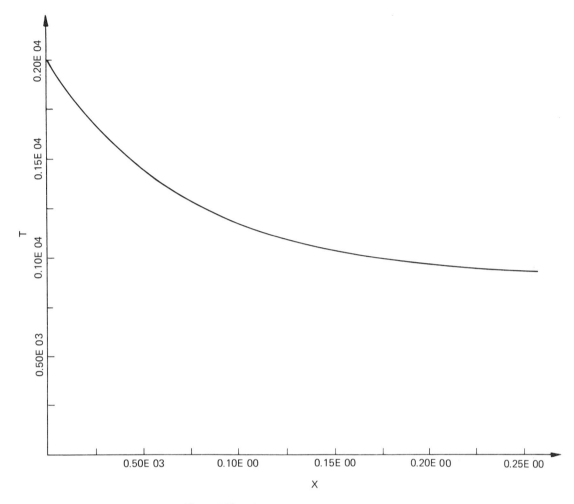

Figure 5.29.   Temperature gradient on fin.

wherein the boundary conditions were given as follows: at $\eta = 0$, $F = F' = 0$, and $F'' = X$ for $m = 1$, with n set to 1 for Newtonian flow.

However, in actual practice, the initial boundary conditions for F'' are not given; instead it is specified that $F' = 1$ when $\eta = \infty$. Furthermore, for non-Newtonian fluids, n is a fraction with a value between 0 and 1.

Having illustrated the two-point boundary condition problem solution in other disciplines, the reader should have no difficulty solving this problem using DSL. Table 5.29 shows the DSL code for the solution. Note that the FINISH facility of DSL is used to terminate the problem. Thus, when F' reaches a value of 1, F''

TABLE 5.28
Simulation Results for the Cooling Fin Design with
Initial and Final Values of Thickness h

```
                        *** DSL/1800 SIMULATION DATA ***
TITLE COOLING FIN DESIGN FOR SPACE VEHICLE
PRINT .05,Q,TDT,T
CONST SIGMA=.173E-8,E=.8,K=25.
PARAM WIDTH=.5,H=0.02
INCON Q0=1000.,T0=2000.
CONTRL DELT=.0001,FINTI=.25
INTEG RKS
END


COOLING FIN DESIGN FOR SPACE VEHICLE


     X              Q             TDT              T
 0.000E 00   0.1000E 04   -0.4000E 04   0.2000E 04
 0.427E-01   0.1609E 03   -0.6439E 03   0.1903E 04
 0.999E-01  -0.9265E 03    0.3706E 04   0.1987E 04
 0.149E 00  -0.2031E 04    0.8125E 04   0.2282E 04
 0.199E 00  -0.3138E 04    0.1255E 05   0.2799E 04
 0.249E 00  -0.4245E 04    0.1698E 05   0.3538E 04
 0.249E 00  -0.4245E 04    0.1698E 05   0.3538E 04

                        *** DSL/1800 SIMULATION DATA ***
PARAM H=0.0046177
END


COOLING FIN DESIGN FOR SPACE VEHICLE


     X             TDT             T              Q
 0.000E 00   -0.1732E 05   0.2000E 04   0.1000E 04
 0.999E-02   -0.1406E 05   0.1843E 04   0.8115E 03
 0.199E-01   -0.1165E 05   0.1715E 04   0.6729E 03
 0.299E-01   -0.9829E 04   0.1608E 04   0.5673E 03
 0.399E-01   -0.8399E 04   0.1518E 04   0.4848E 03
 0.139E 00   -0.2372E 04   0.1052E 04   0.1369E 03
 0.149E 00   -0.2090E 04   0.1030E 04   0.1206E 03
 0.159E 00   -0.1830E 04   0.1010E 04   0.1056E 03
 0.169E 00   -0.1588E 04   0.9935E 03   0.9170E 02
 0.179E 00   -0.1362E 04   0.9788E 03   0.7862E 02
 0.189E 00   -0.1147E 04   0.9662E 03   0.6624E 02
 0.199E 00   -0.9433E 03   0.9558E 03   0.5444E 02
 0.209E 00   -0.7468E 03   0.9473E 03   0.4310E 02
 0.219E 00   -0.5564E 03   0.9408E 03   0.3211E 02
 0.229E 00   -0.3704E 03   0.9362E 03   0.2138E 02
 0.239E 00   -0.1874E 03   0.9334E 03   0.1082E 02
 0.249E 00   -0.5910E 01   0.9324E 03   0.3411E 00
 0.249E 00   -0.5910E 01   0.9324E 03   0.3411E 00
```

TABLE 5.29
Translator Input and Simulation Output for
Solution of Non-Newtonian Flow Past a Wedge

```
                    *** DSL/1800 TRANSLATOR INPUT ***
TITLE NONNEWTONIAN FLOW PAST A WEDGE
INTEG MILNE
      F=INTGR(0.,FDT)
      FDT=INTGR(0.,F2DT)
      F2DT=INTGR(X,F3DT)
      F3DT=((FDT*FDT-1.)*M-(1.-M+2.*M*N)/(N+1.)*F*F2DT)/(N*F2DT**(N-1.))
RELERR FDT=.0001
PARAM M=1.,N=.2,X=3.2628
CONTRL DELT=0.002,DELMI=0.00001,FINTI=100.
FINISH F2DT=0.0001
RENAME TIME=ETA
PRINT .5,F,FDT,F2DT
LABEL PLOT OF FDT VERSUS ETA IN WEDGE NON-NEWTONIAN FLOW
GRAPH .1,10,8,ETA,FDT
SCALE .5,6.
END
STOP
```

NONNEWTONIAN FLOW PAST A WEDGE

ETA	F	FDT	F2DT
0.000E 00	0.0000E 00	0.0000E 00	0.3262E 01
0.500E 00	0.2332E 00	0.7216E 00	0.5360E 00
0.100E 01	0.6389E 00	0.8712E 00	0.1601E 00
0.150E 01	0.1089E 01	0.9240E 00	0.6850E-01
0.200E 01	0.1558E 01	0.9489E 00	0.3586E-01
0.250E 01	0.2036E 01	0.9628E 00	0.2132E-01
0.300E 01	0.2520E 01	0.9713E 00	0.1381E-01
0.165E 02	0.1591E 02	0.9971E 00	0.1432E-03
0.170E 02	0.1641E 02	0.9971E 00	0.1554E-03
0.175E 02	0.1690E 02	0.9972E 00	0.1147E-03
0.180E 02	0.1741E 02	0.9973E 00	0.1462E-03
0.181E 02	0.1750E 02	0.9973E 00	0.9533E-04

SIMULATION HALTED, F2DT = 0.9533E-04

reaches close to zero and this is the variable used for termination of the run. Again, with the use of data switch 2, the user can iterate the problem solution for different values of F'' in the PARAM statement. The case illustrated in Fig. 5.30 is for $M = 1$, which represents stagnation flow. It is left as an exercise for the reader to arrive at the boundary conditions for $F''(0)$ for several different values of n and M.

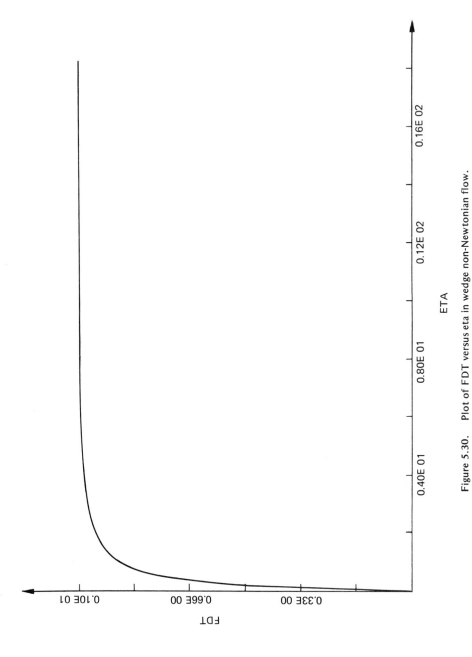

Figure 5.30.   Plot of FDT versus eta in wedge non-Newtonian flow.

6 PROBLEMS IN THE FIELD OF CHEMISTRY AND CHEMICAL ENGINEERING

6.0   Introduction

We shall begin this chapter with a very common problem in chemical engineering, which is illustrated in the study of dynamics of chemical reactors and chemical systems in almost all the papers and books published to date on that subject. The problem is the examination of chemical reaction in a stirred tank reactor. This system—although, truly speaking, semicontinuous—will be treated as a continuous reactor system here.

6.1   Reaction Kinetics in a Stirred Tank

Figure 6.1 shows a typical stirred tank reactor, when the liquid level is maintained at a certain level. The mixing is considered ideal: i.e., the time for the portion of the arriving fluid to thoroughly disperse in the tank is much smaller than the time it takes for completion of reaction.

The chemical reaction occuring in the tank is either endothermic or exothermic, requiring coils in the reactor to transfer heat to or from another outside fluid. Control of the reactor is generally accomplished by regulating the flow of the ingredients added to the reactor and controlling the temperature of the reactor by means of the flow of a cooling or heating fluid in the coils.

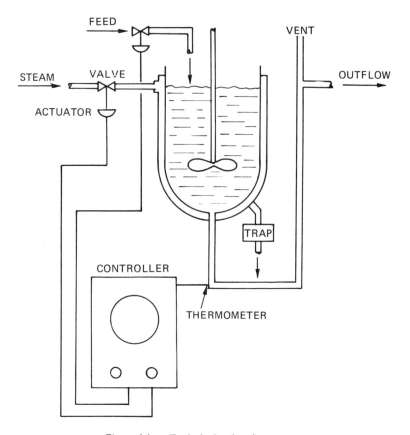

Figure 6.1. Typical stirred tank reactor.

The chemist may be interested in finding out the rate of reaction in the tank and correlating it to the concentration of the various reaction raw materials and products to determine the reaction rate expression. He may run the stirred tank reactor in a continuous mode as described above, or he may just use the reactor in a batch mode, wherein all the reactants are transferred to the reactor at once and the concentration gradient of a product is observed with respect to time, with isothermal conditions maintained in the reactor. The chemical engineer, on the other hand, may be interested in finding out what the total surface area of cooling coils in the reactor should be or what maximum rate of conversion he can obtain without seriously affecting other properties of the reaction mixture such as the density and the viscosity of the product.

To start with the simplest example, let us consider the reaction

$$X \rightarrow Y \tag{6.1}$$

The differential equations describing the material and energy balance for this system are

$$\frac{dc}{dt} = F(c_0 - c) + A \exp\left(\frac{-E}{RT}\right)c \qquad (6.2)$$

where c is the concentration of reactant X in the reactor; c_0 the concentration of X in the incoming stream; F a function of the flow rate and reactor volume; $A, E,$ and R constants in the reaction rate expression; and T the temperature of the reaction mixture. Note that the reverse reaction is neglected and that a simple first-order reaction rate dependency is assumed. The energy balance is given by

$$\frac{dT}{dt} = G(T_0 - T) + HA \exp\left(\frac{-E}{RT}\right)c + U(T - T_c) \qquad (6.3)$$

where T_0 is the temperature of the incoming stream; T_C the temperature of fluid in the cooling coils, U a function of the heat transfer coefficient and the heat transfer area for the coils, H a function of the heat of reaction, and G a function of the flow rate and heat capacity of the incoming stream.

For isothermal reactions, or reactions where H is very small, Eq. (6.2) is the only one to be solved and has an analytical solution. While this is a trivial case, the solution of Eq. (6.2) becomes complex when multiple reactions are involved or if the reaction rate R is a function of concentration of more than one component and reversible reactions are involved. Analytical integration is not possible, especially when the reaction rate is expressed as fractional or whole powers of the concentrations of reactor components. For example, if we were to examine the reaction

$$I + 2J = L + \tfrac{1}{2}M + N \qquad (6.4)$$

and the rate expression R is given by

$$r = k_1 c(I)c(J)^2 - k_2 c(L)c(M)^{1/2}c(N) \qquad (6.5)$$

numerical integration becomes necessary to obtain the expression for concentration as a function of time.

The use of DSL will be very helpful in this illustration. One very direct application is in determining the reaction kinetics with various possible combinations for the reaction rate expression. The chemist may be able to examine the various reaction hypotheses using DSL and compare the simulation results with his laboratory data. The graph features of DSL will immediately give him, to his scale, plots of concentrations of components versus time, for comparison of his laboratory measurements of concentration/time data with the simulation results, and will determine rapidly which of the hypotheses is most sound.

To illustrate the usefulness of DSL here, let us take an example from the book *Chemical Engineering Kinetics* by J.M. Smith (McGraw-Hill, New York, 1958) p. 118. The isothermal reactions between acetic acid and ethyl alcohol proceed as follows:

$$CH_3COOH + C_2H_5OH \underset{k_1}{\overset{k}{\rightleftharpoons}} CH_3COOC_2H_5 + H_2O \qquad (6.6)$$

with the rate expressed as

$$r = kC_H C_{OH} - k' C_E C_W \qquad (6.7)$$

The conversion of acetic acid, X, which is of interest is then expressed, based on the initial given concentrations, in terms of

$$r = k \times 4 \times (1-X)(10.8-4X) - k' \times 4X(18+4X) \qquad (6.8)$$

The problem may be to determine the conversion when the rate r becomes zero, or it may be to find the best values of k and k' to match a particular set of laboratory data, with the reaction rate expression assumed as in Eq. (6.7).

The DSL coding for this example and simulation results for the initial conditions of the example are shown in Table 6.1, and Fig. 6.2, with one set of values for k and k'. Although in this example the rate expression has been in the form of conversion X, one could have used the concentrations of the other components, with the relation of their concentrations to C specified algebraically. This would make it easier to express the rate expression if it were of the form of Eq. (6.5).

In this example, the chemist can try various values of k and k', or he can express the reaction rate as

$$r = k\left(C_H C_{OH} - \frac{1}{K} C_E C_W \right) \qquad (6.9)$$

and since K, the equilibrium constant, is known from thermodynamic principles, he has only to vary k to obtain a fit to his laboratory measurements. The reader will note that for the batch reactor case here $F = 0$, since there is no flow of reactants once the tank has been charged with the reaction mixture.

6.2 The Continuous Stirred Tank Reactor Problem

In the example above, we considered only a special case of Eqs. (6.2) and (6.3) by assuming that there is no continuous flow in the reactor ($F = 0$).

TABLE 6.1
Simulation Input and Output for Reaction Kinetics

```
                        *** DSL/1800 TRANSLATOR INPUT ***
            TITLE STIRRED TANK REACTOR KINETICS
            *       EXPRESS RATE    OF REACTANT DEPLETION
                    CDOT=F*(CO-C) - RATE
                    C=INTGR(CX,CDOT)
                    X=1.-CX/4.
                    RATE=K*4.*(1.-X)*(10.8-4.*X)-KP*4.*X*(18.+4.*X)
            PARAM F=0.,CX=4.0,K=.000476,KP=.000163,CO=4.0
            CONTRL DELT=.01,FINTI=120.
            RELERR C=.001
            INTEG MILNE
            PRINT 1.,C
            GRAPH 0.5,12,8,TIME,C
            LABEL  ACETIC ACID REACTION RATE
            SCALE .1,2.
            END
            STOP
```

STIRRED TANK REACTOR KINETICS   STIRRED TANK REACTOR KINETICS

TIME	C		TIME	C
0.000E 00	0.4000E 01		0.100E 03	0.1943E 01
0.100E 01	0.3979E 01		0.101E 03	0.1923E 01
0.200E 01	0.3958E 01		0.102E 03	0.1902E 01
0.300E 01	0.3938E 01		0.103E 03	0.1881E 01
0.400E 01	0.3917E 01		0.104E 03	0.1861E 01
0.500E 01	0.3897E 01		0.105E 03	0.1840E 01
0.600E 01	0.3876E 01		0.106E 03	0.1820E 01
0.700E 01	0.3856E 01		0.107E 03	0.1799E 01
0.800E 01	0.3835E 01		0.108E 03	0.1779E 01
0.900E 01	0.3814E 01		0.109E 03	0.1758E 01
0.100E 02	0.3794E 01		0.110E 03	0.1737E 01
0.110E 02	0.3773E 01		0.111E 03	0.1717E 01
0.120E 02	0.3753E 01		0.112E 03	0.1696E 01
0.130E 02	0.3732E 01		0.113E 03	0.1676E 01
0.140E 02	0.3712E 01		0.114E 03	0.1655E 01
0.150E 02	0.3691E 01		0.115E 03	0.1635E 01
0.160E 02	0.3670E 01		0.116E 03	0.1614E 01
0.170E 02	0.3650E 01		0.117E 03	0.1594E 01
0.180E 02	0.3629E 01		0.118E 03	0.1573E 01
0.190E 02	0.3609E 01		0.119E 03	0.1552E 01
0.200E 02	0.3588E 01		0.120E 03	0.1532E 01
0.210E 02	0.3568E 01		0.120E 03	0.1532E 01
0.220E 02	0.3547E 01			
0.230E 02	0.3527E 01			
0.240E 02	0.3506E 01			
0.250E 02	0.3485E 01			

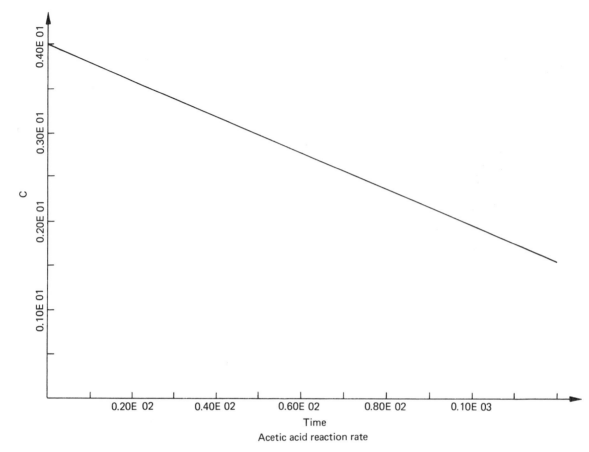

Figure 6.2.   Plot output for reaction kinetics problem.

In industrial practice, the reactor is continuously fed with reactants, the products are removed, the reactor temperature is controlled by cooling or heating coils, and in a large number of cases, such as in the polymer and synthetic rubber industry, a series of these reactors is used in sequence. The product from one reactor goes to the next reactor as a feed.

For a single reactor then, Eqs. (6.2) and (6.3) need to be solved. If the reaction rate expression is of a more complex form, as in Eq. (6.9), than the simple form in Eq. (6.2), the DSL coding is as shown in Table 6.2. Note that the rate is for conversion of reactant A, which is also directly proportional to conversion of B, and produces C and D according to the chemical reaction

$$A + B \rightleftharpoons C + D \tag{6.10}$$

The values in the example of Table 6.2 are given for a fictitious chemical reaction. As an exercise, the reader may use his own kinetic data to run this problem.

TABLE 6.2
DSL Input for Continuous Stirred Tank Reactor
Dynamic Simulation

```
            *** DSL/1800 TRANSLATOR INPUT ***
TITLE CSTR DYNAMIC STUDY
      RATE=K*(CA*CB-CC*CD/KEQ)
      CADOT=F*(CAO-CA)- RATE
      CBDOT= F*(CBO-CB)-RATE
      CCDOT= F*(CCO -CC)+RATE
      CDDOT=F*(CDO-CD)+RATE
      KEQ=ALPHA*EXP(-EEQ/T)
      K=ARHEN*EXP(-ACTEN/R/T)
*     PERFORM ENERGY BALANCE
      TDOT=(G*(TO-T)+H*RATE+U*(TCOOL-T))/HTCAP
      T=INTGR(TO,TDOT)
      CA=INTGR(CAST, CADOT)
      CB=INTGR(CBST,CBDOT)
      CC=INTGR(CCST,CCDOT)
      CD=INTGR(CDST,CDDOT)
ABSERR CA=.01,T=.01
PARAM CAST=.5,CBST=.5,CAO=.5,CBO=.5,ARHEN=4.4E3, ACTEN=10000.,...
      ALPHA=1.7E6 ,EEQ=5000.,TO=373.,TCOOL=298.,H=1000.,R=1.987,...
      G=100.,U=100.,F=110.,HTCAP=150.
INTEG MILNE
CONTRL DELT=.1,FINTI=100.
PRINT 2.,CA,CB,CC,CD,T
END
STOP
PARAM NOT INPUT, SET=0  CCO
PARAM NOT INPUT, SET=0  CDO
PARAM NOT INPUT, SET=0  CCST
PARAM NOT INPUT, SET=0  CDST

OUTPUT VARIABLE SEQUENCE
   6   7   1   8   9   2  10   3  11   4  12   5  13

STORAGE USED/MAXIMUM
INTGR   5/50,  IN VARS  47/300,  OUT VARS  13/100,  PARAMS 26/60, SYMBOLS  55
```

Note that the initial values of those concentrations which are zero need not be specified, since the DSL program sets them to zero together with the warning message as seen in Table 6.2.

Notice how easy it is to use any reaction rate expression, with whole or fractional powers on the concentration which may be included in the coding. In addition, for example, an engineer interested in designing the cooling coils may make DSL runs, with selected values of the variables U, TCOOL, etc., so as to find the right values to sustain the temperature of the reaction.

An extension of the illustration above is the case of multiple continuous stirred tank reactors in series where the output of one reactor becomes an input to the next reactor.

In effect this problem would involve solution of several simultaneous differential equations, which are treated in other examples later.

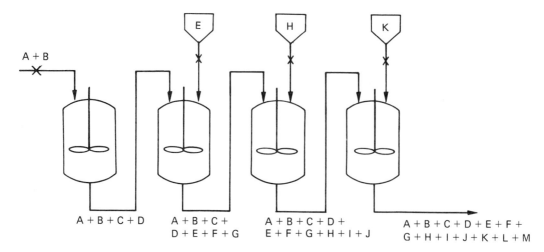

Figure 6.3. A series of continuous stirred tank reactors for polymer production.

As an exercise the reader may solve a two-reactor problem as illustrated in Fig. 6.3. Reactor 1 has pure components A and B entering and products C and D leaving together with unconverted A and B. In reactor 2, products from reactor 1 are fed together with a new reactant E, which forms products F and G based on the following reactions:

$$C + D + E \rightleftharpoons F + G \qquad (6.11)$$

Assume that no further conversion of A and B takes place. Note that the concentrations of reactants C and D entering the second reactor must be computed after addition of the reactant E. Also, the initial conditions for the second reactor may assume no time delay in transport from one reactor to another. This assumption can be removed by providing a known transport lag in a second part of the reader exercise. The solution of this problem is left to the reader.

6.3 An Exothermic Continuous Stirred Tank Reactor

A slight variation of the problem illustrated in Table 6.2 can be found in Reference (22), where transients for a first-order exothermic reaction taking place in a continuous stirred tank reactor are described.

The mass balance for this illustration is given by

TABLE 6.3
Parameter Values for the Exothermic CSTR Problem

V	=	reactor volume = 2000 cm^3
F	=	feed flow rate = 10 cm^3/sec
k	=	reaction constant = 7.86 × 10^{12}
E	=	energy of activation = 22,500
ρ	=	density of solution = 1 g/cm^3
c_p	=	heat capacity of solution = 1 cal/(g) (°C)
ΔH	=	heat of reaction = −10,000 cal/mole
U	=	heat transfer coefficient for cooling coil = 1.356 cal/(°C) (sec)
T'	=	average cooling water temperature = 350°K
T_0	=	T_∞ = 300°K
c_0	=	c_∞ = 5.0 × 10^{-3} mole/liter

$$V\frac{dc}{dt} = Fc_0 - Fc - Vk \exp\left(\frac{-E}{RT}\right)c \qquad (6.12)$$

and the corresponding heat balance is given by

$$V\rho c_p \frac{dT}{dt} = F\rho c_p(T_0 - T) - \Delta H V k \exp\left(\frac{-E}{RT}\right)c - U(T - T') \qquad (6.13)$$

The steady-state values of the temperature and the concentration are given or may be easily calculated from Eqs. (6.11) and (6.12) when the derivatives are zero. The value of the various parameters are shown in Table 6.3. Using both Runge-Kutta and Milne integration methods, let us evaluate the transient characteristics of this reactor and compare the results with the numerical tables in Reference (22).

Table 6.4 shows the DSL code for this example as well as the results of simulation in the form of tables of concentration and temperature as a function of time. Note that these results are in agreement with the results in the reference.

As an exercise, the reader may change the value of the heat transfer coefficient and/or the reaction rate dependency to other than the first-order dependency shown in the illustration. It may be worthwhile to try various powers of c in the rate expression and observe their effect on transient behavior of the reactor.

6.4 Tubular Reactors

Next, let us investigate the simulation of continuous tubular reactors, which may have chemical reaction occurring on the walls of the reactor or on the surfaces of the catalyst with which the reactor tube may be packed.

TABLE 6.4
DSL Input and Simulation Output for Exothermic
CSTR Problem

*** DSL/1800 TRANSLATOR INPUT ***

```
TITLE EXHOTHERMIC REACTION IN CSTR
INTEG MILNE
      C=INTGR(CO,CDT)
      CDT=(F*CO-F*C-V*K*EXP(-E/R/T)*C)/V
      T=INTGR(TO,TDT)
      TDT=(F*RHO*CP*(TO-T)-DELH*V*K*EXP(-E/R/T)*C-U*(T/TPR))/(V*RHO*CP)
PARAM V=2000.,F=10.,K=7.86E12,E=22500.,RHO=1.,CP=1.,DELH=-10000.,...
      U=1.356,TPR=350.,TO=300.,CO=.005,R=1.986
CONTRL DELT=1.,FINTI=2000.
ABSERR T=0.1,C=0.00000001
RELERR C=0.001
PRINT 100.,C,T
END
INTEG RKS
END
STOP
```

EXHOTHERMIC REACTION IN CSTR

TIME	C	T
0.000E 00	0.4999E-02	0.3000E 03
0.100E 03	0.4867E-02	0.3012E 03
0.200E 03	0.4771E-02	0.3022E 03
0.300E 03	0.4700E-02	0.3029E 03
0.400E 03	0.4647E-02	0.3034E 03
0.500E 03	0.4607E-02	0.3038E 03
0.600E 03	0.4576E-02	0.3041E 03
0.700E 03	0.4552E-02	0.3043E 03
0.800E 03	0.4534E-02	0.3045E 03
0.900E 03	0.4520E-02	0.3046E 03
0.100E 04	0.4509E-02	0.3047E 03
0.110E 04	0.4501E-02	0.3048E 03
0.120E 04	0.4494E-02	0.3049E 03
0.130E 04	0.4489E-02	0.3049E 03
0.140E 04	0.4485E-02	0.3050E 03
0.150E 04	0.4481E-02	0.3050E 03
0.160E 04	0.4479E-02	0.3050E 03
0.170E 04	0.4477E-02	0.3051E 03
0.180E 04	0.4475E-02	0.3051E 03
0.190E 04	0.4474E-02	0.3051E 03
0.200E 04	0.4473E-02	0.3051E 03
0.200E 04	0.4473E-02	0.3051E 03

Consider the case of an isothermal homogeneous tubular reactor wherein a chemical reactant flows. For the steady-state case with no back mixing the material balance equation is

$$\frac{1}{Pe}\frac{d^2f}{dz^2} - \frac{df}{dz} - Rf^n = 0 \qquad (6.14)$$

where f = fraction of reactant remaining = $f(z)$
   z = dimensionless axial distance parameter obtained by dividing axial distance by reactor length L
   L = reactor length
   Pe = Peclet number = uL/D
   u = linear velocity of fluid
   D = Effective axial diffusion coefficient
   R = constant involving reaction constant
   n = integer depending on chemical reaction

The effect of backmixing is included in the first term of the equation. For Pe = ∞ (or D = 0) the fluid would be in plug flow, and for Pe = 0 (D = ∞) the fluid would be completely mixed. The boundary conditions associated with Eq. (6.14) are given by

$$\frac{1}{Pe}\frac{df}{dz} = 1, \quad z = 0, \text{ inlet to reactor}$$
$$\frac{df}{dz} = 0, \quad z = 1, \text{ outlet of reactor} \qquad (6.15)$$

Note that the case n = 1.0 specifies a first-order chemical reaction; the system of equations is then linear, and an analytical solution can be obtained with little difficulty. A choice of

$$Pe = 1.0$$
$$R = 2.0$$

is made here for convenience, and the problem under examination is to calculate f as a function of z.

Equation (6.15) indicates that the boundary conditions are specified at two ends of the reactor tube. The solution of Eq. (6.14) thus will require a trial and error method wherein an initial guess of f at z = 0 is made, after which df/dz at z = 0 is evaluated from Eqs. (6.14) and (6.15) integrated to obtain a value of df/dz at z = 1. A correction on $f(0)$ is made based on the deviation of f' at z =1 from zero. Multiple integration runs can be made with various guesses of $f(0)$ in a single step using DSL as shown in Table 6.5, where the reaction is assumed to be first-order; that is, n = 1.

As an exercise, the reader may solve the problem for a second-order chemical reaction with n = 2, and for the more enthusiastic reader, R may be

TABLE 6.5
DSL Input and Simulation Output for the Tubular Reactor

```
                          *** DSL/1800 TRANSLATOR INPUT ***
TITLE ISOTHERMAL TUBULAR REACTOR SIMULATION
INTEG RKS
INTGER N
      F=INTGR(FO,FDT)
      FDT=INTGR(FDTO,F2DT)
      F2DT=PE*(FDT+R*(F**N) )
      FDTO=PE*(F-1.)
FINISH F2DT=0.
PARAM N=1,PE=1.,R=2.,FO=0.522
CONTRL DELT=0.001,FINTI=1.
RELERR F=0.0001
PRINT 0.01,F,FDT,F2DT
END
STOP
```

ISOTHERMAL TUBULAR REACTOR SIMULATION

TIME	F	FDT	F2DT
0.000E 00	0.5219E 00	-0.4780E 00	0.5659E 00
0.907E-02	0.5176E 00	-0.4728E 00	0.5624E 00
0.199E-01	0.5125E 00	-0.4667E 00	0.5583E 00
0.273E-01	0.5091E 00	-0.4626E 00	0.5556E 00

TIME	F	FDT	F2DT
0.999E 00	0.2950E 00	0.3011E-01	0.6201E 00
0.999E 00	0.2950E 00	0.3011E-01	0.6201E 00

```
                       *** DSL/1800 SIMULATION DATA ***
PARAM FO=.518987
END
```

ISOTHERMAL TUBULAR REACTOR SIMULATION

TIME	F	FDT	F2DT
0.000E 00	0.5189E 00	-0.4810E 00	0.5569E 00
0.499E-01	0.4956E 00	-0.4536E 00	0.5375E 00
0.999E-01	0.4736E 00	-0.4272E 00	0.5199E 00
0.149E 00	0.4528E 00	-0.4016E 00	0.5041E 00
0.749E 00	0.2958E 00	-0.1257E 00	0.4660E 00
0.799E 00	0.2902E 00	-0.1021E 00	0.4782E 00
0.849E 00	0.2856E 00	-0.7783E-01	0.4935E 00
0.899E 00	0.2824E 00	-0.5270E-01	0.5121E 00
0.949E 00	0.2804E 00	-0.2656E-01	0.5343E 00
0.999E 00	0.2797E 00	0.7895E-03	0.5603E 00

expressed as

$$R = A \exp\left(\frac{-E}{R_1 T} \right) \tag{6.15a}$$

and T may itself be expressed as a linear, quadratic or cubic function of z, thus solving the case of a nonisothermal reactor. A second variation may be to have T obtained from a separate energy balance equation which must be solved simultaneously with the mass balance, as in the case of the continuous stirred tank reactor solution discussed earlier. These variations are left as exercises for the reader, since in subsequent examples we shall illustrate these types of solutions by solving a more complex problem of ammonia reactor simulation, where both mass and energy balance are to be solved for a packed bed reactor with the chemical reaction considerably more involved than that in Eq. (6.14).

6.5 Critical Bed Diameter of an Adiabatic Tubular Reactor

The following illustration is from Reference (22). For an exothermic catalytic reaction, it is important to have the design of the bed diameter such that the maximum temperature reached in the reactor is below the temperature at which the catalyst is destroyed. To determine this diameter, the following heat balance equation is to be solved:

$$\Delta H + k \left(\frac{\partial^2 T}{\partial r^2} + \frac{1}{r} \frac{\partial T}{\partial r} \right) - G c_p \frac{\partial T}{\partial x} = 0 \tag{6.16}$$

where T = temperature
   ΔH = volumetric rate of heat release
   k = thermal conductivity of fluid phase
   r = radial distance from center of bed
   x = axial distance from bed entrance
   G = superficial mass velocity
   c_p = specific heat of fluid stream

Now consider the temperature profile along the axis of the reactor bed. As x increases, the temperature rises to a maximum and then declines to that at the end of the bed, because the reaction progresses faster at higher temperatures, which are attained by the adsorption of the heat of the reaction by the reactants. The temperature drops as the reactants are consumed along the length of the reactor. At the maximum point

$$\frac{\partial T}{\partial x} = 0 \qquad (6.17)$$

Further, there will be a maximum temperature at all other radial positions. However, the maximum with the largest magnitude is near the center line and is called the peak temperature, as indicated by

$$\left(\frac{\partial T}{\partial x}\right)_{x=x_p} = 0 \qquad (6.17a)$$

If the assumption is made that the maxima for all radial profiles occur at the same peak point, the flow term in Eq. (6.15) can be dropped. Thus,

$$\left[\Delta H + k\left(\frac{\partial^2 T}{\partial r^2} + \frac{1}{r}\frac{\partial T}{\partial r}\right)\right]_p = 0$$

$$\left[\Delta H + k\left(\frac{d^2 T}{dr^2} + \frac{1}{r}\frac{dT}{dr}\right)\right]_p = 0 \qquad (6.18)$$

Equation (6.16), a second-order ordinary differential equation, is explicitly defined. For ease in computation, ΔH can be assumed linear in T, that is,

$$\Delta H = a + bT \qquad (6.18a)$$

so that Eq. (6.16) becomes

$$\frac{d^2 T}{dr^2} + \frac{1}{r}\frac{dT}{dr} + \frac{a+bT}{k} = 0 \qquad (6.19)$$

The boundary conditions which fit the physical situation are

$$\frac{dT}{dr} = 0 \qquad \text{at } r = 0, \text{ bed center line}$$

$$T = T_w \qquad \text{at } r = R, \text{ wall temperature} \qquad (6.19a)$$

A convenient set of numbers for the various parameters is

$$
\begin{array}{lll}
D & = & 2 \text{ feet} \\
T_w & = & 200°\text{F} \\
K_p & = & 200 \text{ Btu/(hour)(foot)(}°\text{F)} \\
a & = & 400 \text{ Btu/(hour)(foot)} \\
b & = & 600 \text{ Btu/(hour)(foot)(}°\text{F)}
\end{array} \qquad (6.20)
$$

By using Eqs. (6.18) and (6.19) the radial temperature distribution at the axial position of the peak temperature can be calculated. The equations under consideration form a boundary-value problem involving a linear second-order ordinary differential equation.

Equation (6.17) has an analytical solution; however, if the heat of reaction term were truly represented by $A \exp(-B/T)$, a numerical solution would be required.

Table 6.6 shows the DSL code for the solution of this problem for linear variation of ΔH.

Table 6.7 shows the temperature gradient along the radius of the reactor bed with maximum temperature of $528°F$, for the linear case. One other case is also shown where a different boundary value for T_0 is assumed. Note that the example here is really a two-point boundary-value problem since the second boundary condition is at the tube wall.

As an exercise, the reader should use a temperature dependency for the reaction constant and a nonlinear dependency for the heat of reaction to solve this problem. Note that the results presented here are in agreement with those in Reference (22).

TABLE 6.6
DSL Input for the Adiabatic Reactor Problem

```
                          *** DSL/1800 TRANSLATOR INPUT ***
TITLE  CRITICAL DIAMETER OF AN ADIABATIC REACTOR
INTEG MILNE
       TDT=INTGR(TDT0,T2DT)
       T=INTGR(T0,TDT)
       T2DT=-DELHR/KG-TDT/TIME
       DELHR=A+B*T
PARAM KG=200.,A=400.,B=600.,FINTI=1.,TDT0=0.,T0=528.
CONTRL DELT=0.001,FINTI=1.
PRINT 0.05,T,TDT
END
STOP
```

6.6 A Two-Point Boundary-Value Problem in Transient Heat Conduction

This example is from Reference (22). A one-dimensional transient heat conduction in a metal bar is of interest, where the heat conduction equation is given by

$$\rho c_p \frac{\partial T}{\partial t} = \frac{\partial}{\partial x}\left(k \frac{\partial T}{\partial x}\right) \tag{6.21}$$

ρ and c_p are the density and heat capacity of the solid, which are assumed constant, whereas the thermal conductivity k is a linear function of temperature.

TABLE 6.7
Adiabatic Reactor Simulation Results

CRITICAL DIAMETER OF AN ADIABATIC REACT

TIME	T	TDT
0.000E 00	0.5280E 03	0.0000E 00
0.499E-01	0.5270E 03	-0.3961E 02
0.999E-01	0.5240E 03	-0.7900E 02
0.749E 00	0.3274E 03	-0.4778E 03
0.799E 00	0.3031E 03	-0.4938E 03
0.849E 00	0.2780E 03	-0.5071E 03
0.899E 00	0.2524E 03	-0.5177E 03
0.949E 00	0.2263E 03	-0.5255E 03
0.999E 00	0.1999E 03	-0.5305E 03
0.999E 00	0.1999E 03	-0.5305E 03

*** DSL/1800 SIMULATION DATA ***
PARAM TO=550.
END
CRITICAL DIAMETER OF AN ADIABATIC REACT

TIME	T	TDT
0.000E 00	0.5500E 03	0.0000E 00
0.499E-01	0.5489E 03	-0.4126E 02
0.999E-01	0.5458E 03	-0.8228E 02
0.749E 00	0.3410E 03	-0.4976E 03
0.799E 00	0.3157E 03	-0.5143E 03
0.849E 00	0.2896E 03	-0.5282E 03
0.899E 00	0.2629E 03	-0.5393E 03
0.949E 00	0.2357E 03	-0.5474E 03
0.999E 00	0.2082E 03	-0.5526E 03
0.999E 00	0.2082E 03	-0.5526E 03

The partial differential equation (6.21) can be reduced to an ordinary differential equation by a transformation of the independent variable as follows:

$$s = \frac{x}{\alpha\sqrt{2t}} \qquad \text{where } \alpha^2 = \frac{k_0}{\rho c_p} \qquad (6.21a)$$

so that Eq. (6.21) becomes

$$-k_0 s \frac{dT}{ds} = \frac{d}{ds}\left(k\frac{dT}{ds}\right) \qquad (6.22)$$

Now if the thermal conductivity k is expressed as a linear function of

TABLE 6.8
Transient Heat Conduction Results

```
                    *** DSL/1800 TRANSLATOR INPUT ***
TITLE TRANSIENT HEAT CONDUCTION- 2 PT BDRY VALUE PROBLEM
INTEG RKS
      T=INTGR(TO,TDT)
      TDT=INTGR(TDTO,T2DT)
      T2DT=(-B*TDT*TDT-TDT*TIME)/(1.+B*T)
PARAM TO=0.,TDTO=342.,B=0.001
FINISH TDT=0.0001
PRINT 0.1,T,TDT
RELERR T=0.0001
CONTRL DELT=0.001,FINTI=20.
END
STOP
```

```
           TRANSIENT HEAT CONDUCTION- 2 PT BDRY VAL


              TIME         T           TDT
            0.000E 00  0.0000E 00  0.3420E 03
            0.100E 00  0.3358E 02  0.3292E 03
            0.199E 00  0.6579E 02  0.3147E 03
            0.273E 00  0.8865E 02  0.3032E 03
            0.589E 01  0.4006E 03  0.8544E-03
            0.600E 01  0.4006E 03  0.5587E-03
            0.609E 01  0.4006E 03  0.3627E-03
            0.619E 01  0.4006E 03  0.2338E-03
            0.629E 01  0.4006E 03  0.1496E-03
            0.639E 01  0.4006E 03  0.9511E-04
            0.639E 01  0.4006E 03  0.9511E-04

         SIMULATION HALTED, TDT    =  0.9511E-04
```

```
                        *** DSL/1800 SIMULATION DATA ***
INTEG MILNE
END
```

```
           TRANSIENT HEAT CONDUCTION- 2 PT BDRY VAL


              TIME         T           TDT
            0.000E 00  0.0000E 00  0.3420E 03
            0.100E 00  0.3358E 02  0.3292E 03
            0.200E 00  0.6579E 02  0.3147E 03
            0.300E 00  0.9649E 02  0.2989E 03
            0.400E 00  0.1255E 03  0.2822E 03
            0.500E 00  0.1529E 03  0.2648E 03
            0.600E 01  0.4006E 03  0.5434E-03
            0.610E 01  0.4006E 03  0.3456E-03
            0.619E 01  0.4006E 03  0.2223E-03
            0.630E 01  0.4006E 03  0.1379E-03
            0.640E 01  0.4006E 03  0.8700E-04
            0.640E 01  0.4006E 03  0.8700E-04

         SIMULATION HALTED, TDT    =  0.8700E-04

PARAM TDTO=300.
END
```

```
           TRANSIENT HEAT CONDUCTION- 2 PT BDRY VAL


              TIME         T           TDT
            0.000E 00  0.0000E 00  0.3000E 03
            0.100E 00  0.2951E 02  0.2899E 03
            0.200E 00  0.5793E 02  0.2781E 03
            0.300E 00  0.8510E 02  0.2649E 03
            0.400E 00  0.1108E 03  0.2506E 03
            0.540E 01  0.3540E 03  0.4084E-02
            0.550E 01  0.3540E 03  0.2718E-02
            0.560E 01  0.3540E 03  0.1798E-02
            0.569E 01  0.3540E 03  0.1173E-02
            0.580E 01  0.3540E 03  0.7640E-03
            0.590E 01  0.3540E 03  0.4870E-03
            0.600E 01  0.3540E 03  0.3117E-03
            0.610E 01  0.3540E 03  0.1940E-03
            0.619E 01  0.3540E 03  0.1215E-03
            0.630E 01  0.3540E 03  0.7410E-04
            0.630E 01  0.3540E 03  0.7410E-04

         SIMULATION HALTED, TDT    =  0.7410E-04
```

temperature,

$$k = k_0 (1 + bt) \tag{6.22a}$$

Eq. (6.21) reduces to

$$(1+bT)\frac{d^2T}{ds^2} + b\left(\frac{dT}{ds}\right)^2 + s\frac{dT}{ds} = 0 \tag{6.23}$$

with the boundary conditions

$$\begin{aligned} T &= T_0 \text{ as } s \longrightarrow \infty \\ T &= 0, \text{ at } s = 0 \end{aligned} \tag{6.23a}$$

Clearly, this is a two-point boundary condition problem with added difficulty, since the second boundary condition is at an infinite value of s. From a practical standpoint, it is best to stop the solution when T/T_0 reaches a constant value.

Table 6.8 shows the DSL code for solution of this problem. Notice the use of FINISH instead of the use of FINTI. The value of s at which the solution should be stopped is indeterminate so that FINTI cannot be specified accurately. Again, multiple guesses of dT/ds may be used in one run to get an interpolation scheme to correctly guess the first derivative value for $s = 0$.

The reader may be interested in using different methods of integration such as Milne, fourth-order Runge-Kutta, and Adams-Moulton, which are offered in DSL, not only to compare the results of the four methods but also to compare these results with the tabular results for this problem in Reference (22).

6.7 Simulation of an Ammonia Synthesis Reactor

In the previous sections we have discussed both design and operational simulation of simple tubular and stirred tank reactors with first- and second-order reactions. However, the chemical industry in practice employs a large number of catalytic reactors which may have multiple beds, may have built-in heat exchangers, and may also require steam recycling. The reactions taking place on the catalyst surface may also have a much more complex reaction formulation than the formulations discussed in the previous examples. The production of ammonia employs a typical reactor of this nature, in which the synthesis reaction,

$$N_2 + 3H_2 = 2NH_3$$

takes place at pressures of 2000-3000 pounds per square inch and temperatures of 650-900°F. A typical reactor employed in modern ammonia plants is shown

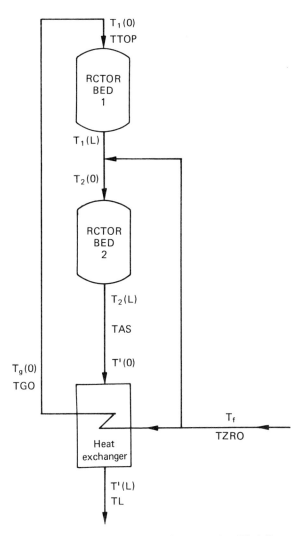

Figure 6.4. A typical ammonia reactor-simplified diagram.

in Fig. 6.4. In the reactor, multiple catalyst beds are encased, with an internal heat exchanger which heats the incoming reactor gases by using the sensible heat of the reactor products. Furthermore, the gases leaving each catalyst bed are mixed with some cooler gases prior to entering subsequent beds.

This arrangement of the reactor is useful in optimizing exothermic reaction of hydrogen conversion to produce ammonia. In this example we shall illustrate how the simulation of such a reactor can be facilitated with the use of DSL.

The simulation equations are comprised of material, energy, and pressure balance equations for each bed. Their derivation is discussed in Reference (30),

TABLE 6.9
Various Relationships Required for Ammonia
Reactor Simulation

$$K = \exp\left(\frac{-\Delta F^0}{RT}\right) = \exp\left\{\frac{1}{R} \cdot \left[\frac{9184.0}{T} - \right.\right.$$

$$\left.\left. 7.2944 \ln T + 0.34996 \times 10^2 T + 0.016781 \times 10^{-5} T^2 - 0.03875 \times 10^{-9} T^3 + 23.050\right]\right\}$$

$$K_V(T,p) = (1.7343 - 8.143 \times 10^{-4} p)$$

$$+ (5.714 \times 10^{-7} p - 2.6714 \times 10^{-3}) T$$

$$+ 2.0 \times 10^{-6} T^2$$

Heat Relationships for Heat Capacities of Various Gases:[a]

Gas	Heat Capacity Equations
Hydrogen	$(C_p)_1 = 6.952 - 0.04576 \times 10^{-2} T + 0.09563 \times 10^{-5} T^2 - 0.2079 \times 10^{-9} T^3$
Nitrogen	$(C_p)_2 = 6.903 - 0.03753 \times 10^{-2} T + 0.1930 \times 10^{-5} T^2 - 0.6861 \times 10^{-9} T^3$
Ammonia	$(C_p)_3 = 6.5846 - 0.61251 \times 10^{-2} T + 0.23663 \times 10^{-5} T^2 - 1.5981 \times 10^{-9} T^3 + [96.1678 - 0.067571 p + (-0.2225 + 1.6847 \times 10^{-4} p) T + (1.289 \times 10^{-4} - 1.0095 \times 10^{-7} p) T^2]$
Methane	$(C_p)_4 = 4.750 + 1.200 \times 10^{-2} T + 0.3030 \times 10^{-5} T^2 - 2.630 \times 10^{-9} T^3$
Argon	$(C_p)_5 = 4.9675$

[a] Temperature range: $500° - 900°K$
Pressure range: 200–1000 atm.

$$\Delta H = -9184 - 7.2949 T + 0.34996 \times 10^{-2} T^2$$
$$+ 0.03356 \times 10^{-5} T^3 - 0.11625 \times 10^{-9} T^4 - (6329.3 - 3.1619 p) +$$
$$(14.3595 + 4.4552 \times 10^{-3} p) \times T - T^2 (8.3395 \times 10^{-3} + 1.928 \times 10^{-6} p) -$$
$$51.21 + 0.14215 p$$

and they will be simply reproduced here. The material balance for hydrogen is given by

$$-\frac{dn_1}{dz} = 3f \times 9.8 \times 10^{-6} \left(\frac{300}{P}\right)^{0.63} \exp\left[\frac{-24092.2}{T} + 33.56\right]$$

$$\cdot \left[\left(\frac{K}{K_v}\right)^2 P^{1.5} n_2 n_1^{1.5} / (n_3 n_t^{1.5}) - n_3 n_t^{0.5} / (n_1^{1.5} P^{0.5})\right] A_i \quad (6.24)$$

where the subscripts on n refer, respectively, to moles of hydrogen, nitrogen, ammonia, and total moles, K and K_v are reaction equilibrium parameters which are functions of temperature T and pressure P. The various relationships for K and K_v will be found in Table 6.9; A_i is the area of the ith catalyst bed.

The heat balance is given by

$$\frac{dT}{dz} = \frac{2}{3} \frac{dn_1}{dz} \frac{\Delta H}{\sum n_i c_{p_i}} \quad (6.25)$$

where C_{p_i} is the heat capacity of the ith component in the reaction mixture and ΔH is the heat of reaction. Both c_p and ΔH are functions of temperature and pressure, and their relationships may be found in Table 6.9. The heat exchanger equations which yield the temperature of the gases entering the first catalyst bed are given by

$$T'(L) = T'(0) + \frac{\beta}{\alpha}[T_f - T_g(0)] \quad (6.26)$$

and

$$T_g(0) = \frac{T_f[1 - (\beta/\alpha)] + T'(0)\{1 - \exp[1 - (\beta/\alpha)]U'a'L'\}}{(\beta/\alpha) - \exp[1 - (\beta/\alpha)]U'a'L'} \quad (6.27)$$

where T_f is the converter feed temperature and $T'(0)$ is the feed temperature after heat exchange.

$$\alpha = \sum_{i=1}^{5} n_i(L_2)C_{p_i}(T')_{av} \quad \text{and} \quad \beta = \delta \sum_{i=1}^{5} F_{d_i} C_{p_i}(T'_g)_{av} \quad (6.28)$$

U' is the heat transfer coefficient, $a'L'$ the total area of heat transfer, and F_{d_i} the feed rate of component i to the synthesis converter.

Stoichiometry yields the material balance for components other than hydrogen:

$$n_2 = n_2(0) - \tfrac{1}{3}\left[\, n_1(0) - n_1 \,\right] \qquad \text{for nitrogen}$$

$$n_5 = n_5(0) + \tfrac{2}{3}\left[\, n_1(0) - n_1 \,\right] \qquad \text{for ammonia}$$

$$n_4 = n_4(0) \qquad\qquad\qquad\quad \text{for methane} \tag{6.29}$$

$$n_3 = n_3(0) \qquad\qquad\qquad\quad \text{for argon}$$

where $n_i(0)$ represents moles of component i at $z = 0$. Methane and argon are part of the reactor feed as inerts.

In addition reaction products leaving each bed are mixed with part of the reactor main feed to cool the mixture before it enters a subsequent bed.

Two things must be noted at the outset. One which is apparent from Fig. 6.4 is that a part of the reactor feed described as quench is diverted to mix with the reactants moving from one bed to the next.

Second, the determination of $T_g(0)$ in Eq. (6.27) involves the use of α and β, which are functions of temperature, since c_p is a function of temperature. An average temperature may be used for determining c_p with little error; however, the use of the average temperature requires knowledge of $T_g(0)$ and $T(L)$ in Eq. (6.28) prior to its determination. The use of the implicit function IMPL provided by DSL will be helpful in resolving that difficulty.

The most difficult task in the simulation of the ammonia reactor is that to determine the temperature $T_1(0)$, the temperature of the reaction product at the exit of the last reactor bed, $T_2(L)$, must be specified. However, the temperature $T_2(L)$ depends on the reactions in the various beds, which require specification of temperature $T_1(L)$. The ammonia synthesis reactor simulation thus is a two-point boundary-value problem which involves not only solution of simultaneous differential equations for each bed with boundary conditions not fully specified but also an overall trial and error solution.

6.7.1 Simplify the Complexity of the Problem

In order not to confront the reader with the complex trial and error solution in simulation of the entire reactor, we shall proceed with the solution in three steps as follows:

Step 1. Solve material and energy balance equations for only one reactor bed where all initial conditions are specified.

Step 2. Assume that we have a two-bed reactor with quench and heat exchanger. Simulate the entire reactor with given temperature $T_2(L)$. The reader can, at that stage, see how multiple runs can be made by specifying different guesses of $T_2(L)$ using the PARAM statement in DSL. In this step, then, various values of temperature $T_2(L)$, which is the guessed temperature at the exit of the second catalyst bed, may be used to see how they affect the performance of the first catalyst bed. The reader is then ready for step 3.

Step 3. In this step, the quench calculations are performed for mixing of the gases exiting the first reactor bed and quench gases, which are portions of the reactor feed gases (see Fig. 6.4). With calculations in step 2, then, the entire two-bed reactor calculations can be performed. Thus, the simulation sequence will be

a. The heat exchanger calculations,
b. The first reactor bed calculations,
c. The quench calculations, and
d. The second bed calculations.

Since step 3 constitutes the simulation of the entire reactor with an assumed value of temperature at the second bed exit, the trial and error (two-point boundary) calculation for the reactor can be repeated by repeating sequence a–d with a newly assumed value of $T(L)$. This last step will be especially useful in illustrating for the reader not only the use of DSL in simulation of such complex physical systems but also the interplay between FORTRAN and DSL as well as the DSL program execution flow.

6.7.2 Single-Bed Calculations with Step 1

The DSL code for step 1 is shown in Table 6.10. To simplify the readability of the program, the variables are named and defined to match the reactor description. Comments are specified along the code to make it self-explanatory.

We see that in step 1 the solution of the reactor equations for a single bed is carried out with all initial conditions specified. For purposes of illustration, two initial temperature runs are made. Note that the heat capacities for the various gaseous components are expressed as functions $H(T)$, $N(T)$, $NH(T, P)$, and $CH(T)$. These may be FORTRAN-callable functions stored on the 1130 disk, or, as an alternative, they may be defined as functions prior to the first executable statement in the UPDAT source deck, which the DSL program punches out as a first step in simulation. This is illustrated in Table 6.11, which shows the UPDAT routine listing for step 1 in ammonia simulation, where the heat capacity functions are inserted.

The results for single-bed simulation shown in Table 6.12 indicate that for low temperatures the reaction progresses slowly and that the temperature profile is always linear near the bed exit. At higher temperatures, the reaction profile is steep at the bed entrance and flattens out near the exit of the reaction bed, as one expects in an exothermic reversible reaction which is not cooled.

TABLE 6.10
DSL Translator Input for Ammonia Reactor
Simulation Step 1

```
                   *** DSL/1800 TRANSLATOR INPUT ***
TITLE AMMONIA REACTOR SIMULATION FOR SINGLE BED
RENAME TIME=Z,FINTI=L,DELT=DELZ
RELERR HYD=0.0001,T=0.01
INTEG MILNE
*      SPECIFY INLET CONDITIONS FOR VARIOUS COMPONENTS
INCON HYDO=2400.,NITO=800.,ARG=160.,MET=400.,AMMO=100.,TTOP=660.,...
      P=200.
*      SPECIFY REACTOR CONSTANTS
PARAM CAREA=31000.,L=500.,GAM1=1.2,THETA=0.
*      SPECIFY STEP SIZE AT INTEGRATION START
CONTRL DELZ=1.,DELMI=0.0001
*      MATERIAL BALANCE ON NITROGEN AND AMMONIA
       NIT=NITO-(HYDO-HYD)/3.
       AMM=AMMO+(HYDO-HYD)*2./3.
*      TOTAL MOLES
       TOTAL=HYD+NIT+AMM+ARG+MET
*      HEAT OF REACTION
       TSQ=T*T
       TCB=T*T*T
       HTRN=-9184.0-7.2949*T+0.34996E-2*TSQ+0.033563E-5*TCB-0.11625E-9...
       *T*TCB-6329.3-3.1619*P+(14.3595+4.4552E-3*P)*T-(8.3395E-3+ ...
       1.928E-6*P)*TSQ-51.21+0.14215*P
*      CALCULATE K FOR FUGACITY CORRECTION
       FUGCO =1.7343-8.143E-4*P+(-2.6714E-3+5.714E-7*P)*T+2.0E-6*TSQ
*       CALCULATE K FOR EQUILLIBRIUM
       EQLTR =(9184.0/T-7.2949*ALOG(T)+0.34996E-2*T+0.016781E-5*TSQ ...
       -.03875E-9*TCB+23.050)/1.9876
       EQLCO =EXP(EQLTR)
*       CALCULATE ENERGY OF ACTIVATION
       ENRAC = -24092.2/T+33.5566
*       CALCULATE CATALYST EFFICIENCY
       CATEF =GAM1-THETA*Z
*       CALCULATE MOLES*HEAT CAPACITY SUM
       SMCPT=HYD*H(T)+NIT*N(T)+AMM*NH(T,P)+MET*CH(T)+ARG*4.9675
*       CALCULATE DN1/DZ
       XTRM1=(EQLCO/FUGCO)**2*P**1.5*NIT*HYD**1.5/(AMM*TOTAL**1.5)
       XTRM2=AMM*SQRT(TOTAL)/(HYD**1.5*SQRT(P))
       XTRM3=(CATEF*3.*(300./P)**.63*EXP(ENRAC)*9.81E-6*CAREA)
       H2DT=XTRM3*(XTRM2-XTRM1)
*       CALCULATE DT/DZ
       TDT=( 2.*HTRN*H2DT/3.)/SMCPT
*       INTEGRATE
       HYD=INTGR(HYDO,H2DT)
       T=INTGR(TTOP,TDT)
PRINT 10.,HYD,T,AMM
END
PARAM TTOP=710.
END
STOP
```

TABLE 6.11
*DSL Translator Output for Ammonia Reactor
Simulation Step 1*

```
PAGE    2                   DSL/1800 (DIGITAL SIMULATION LANGUAGE)

        SUBROUTINE UPDAT
        REAL   HYDO ,NITO ,MET  ,L     ,NIT  ,N    ,NH    ,INTGR
        COMMON NALRM,IZZZ(833) ,Z     ,DELZ ,DELMI,L     ,DELTP,DELTC,HYD   ,
       1T     ,NIT   ,AMM   ,TOTAL,TSQ   ,TCB  ,HTRN ,FUGCO,EQLTR,EQLCO,ENRAC,
       1CATEF,SMCPT,XTRM1,XTRM2,XTRM3,H2DT ,TDT   ,HYDO ,NITO ,ARG   ,MET   ,
       1AMMO ,TTOP ,P     ,CAREA,GAM1 ,THETA
C       BELOW ARE HEAT CAPACITY FUNCTIONS PUT IN AFTER DSL SIMULATION HAS
C       COMPLETED CREATING THE UPDATE SUBROUTINE.
        H(T)=6.952-.0004576*T+.956E-6*T*T-.0279E-9*T*T*T
        N(T)=6.903-.375E-3*T+.193E-5*T*T-.686E-9*T*T*T
        CH(T)=4.75+1.2E-2*T+.303E-5*T*T-2.63E-9*T*T*T
        NH(T,P)=6.5846+.6125E-2*T+.2366E-5*T*T-1.6E-9*T*T*T+96.17-
       1 .06757*P+(1.68474E-4*P-.2225)*T+(1.2890E-4-1.0095E-7*P)*T*T
C       HEAT CAPACITY DEFINITIONS COMPLETE.
        AMM=AMMO+(HYDO-HYD)*2./3.
        NIT=NITO-(HYDO-HYD)/3.
        TOTAL=HYD+NIT+AMM+ARG+MET
        TSQ=T*T
        FUGCO =1.7343-8.143E-4*P+(-2.6714E-3+5.714E-7*P)*T+2.0E-6*TSQ
        TCB=T*T*T
        EQLTR =(9184.0/T-7.2949*ALOG(T)+0.34996E-2*T+0.016781E-5*TSQ
       1-.03875E-9*TCB+23.050)/1.9876
        EQLCO =EXP(EQLTR)
        XTRM1=(EQLCO/FUGCO)**2*P**1.5*NIT*HYD**1.5/(AMM*TOTAL**1.5)
        XTRM2=AMM*SQRT(TOTAL)/(HYD**1.5*SQRT(P))
        ENRAC = -24092.2/T+33.5566
        CATEF =GAM1-THETA*Z
        XTRM3=(CATEF*3.*(300./P)**.63*EXP(ENRAC)*9.81E-6*CAREA)
        H2DT=XTRM3*(XTRM2-XTRM1)
        HYD=INTGR(HYDO,H2DT)
        SMCPT=HYD*H(T)+NIT*N(T)+AMM*NH(T,P)+MET*CH(T)+ARG*4.9675
        HTRN=-9184.0-7.2949*T+0.34996E-2*TSQ+0.033563E-5*TCB-0.11625E-9
       1*T*TCB-6329.3-3.1619*P+(14.3595+4.4552E-3*P)*T-(8.3395E-3+
       11.928E-6*P)*TSQ-51.21+0.14215*P
        TDT=( 2.*HTRN*H2DT/3.)/SMCPT
        T=INTGR(TTOP,TDT)
        RETURN
        END
```

6.7.3 *Simulation for Single-Bed Reactor with a Heat Exchanger*

Next, in step 2, the simulation is to be performed for a single-bed reactor with a heat exchanger. The heat exchanger is where the feed comes in and is heated by the reaction products. It is to be noted that the reaction product composition and temperature must be assumed to be able to calculate the temperature of the feed gas as it leaves the heat exchanger and enters the first

TABLE 6.12
Simulation Results for the Ammonia Reactor Step 1

```
                    *** DSL/1800 SIMULATION DATA ***
TITLE AMMONIA REACTOR SIMULATION FOR SINGLE BED
RELERR HYD=C.0001,T=0.01
INTEG MILNE
INCON HYDC=2400.,NITO=800.,ARG=160.,MET=400.,AMMO=100.,TTOP=660.,...
      P=200.
PARAM CAREA=31000.,L=500.,GAM1=1.2,THETA=0.
CONTRL DELZ=1.,DELMI=0.0001
PRINT 10.,HYD,T,AMM
END
```

AMMONIA REACTOR SIMULATION FOR SINGLE BE

Z	HYD	T	AMM
0.000E 00	C.2400E 04	C.6600E 03	C.1000E 03
0.100E 02	0.2395E 04	0.6612E 03	C.103CE 03
0.200E 02	C.2390E 04	0.6625E 03	0.1060E 03
0.300E 02	C.2386E 04	0.6638E 03	C.109CE 03
0.400E 02	0.2381E 04	0.6651E 03	0.112CE 03
0.500E 02	C.2377E 04	C.6664E 03	C.115CE 03
0.440E 03	0.2169E 04	0.7253E 03	C.2536E 03
C.450E 03	0.2160E 04	0.7279E 03	C.2596E 03
0.460E 03	0.2152E 04	0.7300E 03	C.2648E 03
0.470E 03	C.2141E 04	C.7332E 03	C.2723E 03
C.480E 03	C.2129E 04	0.7365E 03	C.280CE 03
0.490E 03	C.2117E 04	0.7400E 03	C.2884E 03

```
                    *** DSL/1800 SIMULATION DATA ***
PARAM TTOP=710.
END
```

Z	HYD	T	AMM
0.000E 00	C.2400E 04	0.7100E 03	C.1000E 03
0.100E 02	C.2385E 04	0.7142E 03	C.1099E 03
C.200E 02	C.2370E 04	0.7183E 03	0.1198E 03
0.300E 02	0.2355E 04	0.7225E 03	C.1295E 03
0.400E 02	0.2340E 04	0.7266E 03	C.1393E 03
0.500E 02	0.2326E 04	0.7308E 03	C.1491E 03
0.440E 03	C.1976E 04	0.8286E 03	C.3821E 03
0.450E 03	0.1974E 04	0.8291E 03	C.3833E 03
0.460E 03	0.1976E 04	0.8286E 03	C.3823E 03
C.470E 03	0.1975E 04	0.829CE 03	C.3832E 03
C.480E 03	0.1976E 04	0.8287E 03	C.3824E 03
0.490E 03	C.1974E 04	0.8293E 03	C.3839E 03

catalyst bed. In addition, as the diagram of the reactor in Fig. 6.4 shows, not all the feed gas is heated by the heat exchanger. About 80% of the feed gas (DELTA = .8) passes through the exchanger and enters the first bed. The simulation then must proceed in two steps.

1. Calculate the entrance temperature for gases in the first bed.

2. Solve the differential equations for the catalyst bed.

Table 6.13 shows the DSL listing of the second step. The major addition to the single-bed simulation is in a section under NOSORT. Here the feed components entering the heat exchanger are calculated and then the temperatures of the gases leaving the heat exchanger are calculated, based on the temperature and gas composition entering the heat exchanger, by solving Eqs. (6.26) and (6.27). Note again that this involves a trial and error calculation, so that the DSL implicit function IMPL can be readily used. Furthermore, to reduce the number of iterations, the temperature of the product gas leaving the heat exchanger is set to the product temperature entering the heat exchanger, so that an arbitrary value is not used in the first iteration. Once the gas temperature at the top of the first bed TTOP is calculated, integration can begin. Note the use of the flag MGO, which permits bypassing the heat exchanger calculations during integration.

For each new run, where parameter values are changed, the flag MGO must be reset to 1, so that the simulation will resume with heat exchanger calculations first. Again, as in step 1, the heat capacity functions H(T), N(T), etc., are inserted in the UPDAT deck, punched by DSL as a first operational step. Table 6.14 shows the UPDAT deck, whereas Table 6.15 shows a typical simulation run for the ammonia reactor and the heat exchanger.

6.7.4 Simulation of the Entire Reactor

Next we proceed with step 3, the simulation of the entire ammonia reactor, with two catalyst beds. The simulation in this case proceeds as follows:

1. Based on assumed gas composition and temperature leaving the second reactor bed, and given feed gas composition and temperature, perform heat exchanger calculations to evaluate gas temperature entering the first reactor bed. This is the same step as in the previous simulation.

2. Integrate differential equations in the first bed, and calculate exit composition and temperature.

3. With mixing equations (6.30) and (6.31) and knowing quench gas composition as well as the quench gas temperature, evaluate the gas temperature entering the second bed. This step also involves trial and error calculation.

```
                    *** DSL/1800 TRANSLATOR INPUT ***
TITLE AMMONIA SIMULATION STEP 2
RENAME TIME=Z,FINTI=L,DELT=DELZ
RELERR HYD=0.0001,T=0.01
INTEG MILNE
*   SPECIFY INITIAL FEED CONDITIONS TO THE REACTOR
INCON FDH2=3278.,FDN2=1087.,FDAR=206.5,FDCH4=520.0,FDNH3=143.3
INCON TZRO=423.,DELTA=0.8,TAS=800.,P=200.
*    SPECIFY FIRST BED DIMENSIONS
PARAM CAREA=31000.,L=500.,GAM1=1.2,THETA=0.
*      SPECIFY STEP SIZE AT INTEGRATION START
CONTRL DELZ=1.,DELMI=0.0001
*    SPECIFY HEAT EXCHANGER PARAMETERS
PARAM HTXL=185.,UPRIM=.02,ARHTX=24000.,CRCP=8.
*      SPECIFY SECOND BED EXIT CONDITIONS FOR HT XCH CALCULATIONS
PARAM H2AS=2500.,N2AS=850.,NH3AS=600.
*    SET FLAGS AS INTEGERS AND THEIR VALUES
INTGER MGO
PARAM MGO=1
NOSORT
        GO TO (100,200,300),MGO
  100 PP=P-CROP
        FH2=DELTA*FDH2
        FN2=FDN2*DELTA
        FNH3=DELTA*FDNH3
        FCH4=DELTA*FDCH4
        FAR=DELTA*FDAR
        FTOT = FH2+FN2+FNH3+FCH4+FAR
        TL=TAS
*    SET UP LOOP TO CALCULATE TEMPERATURE
        TGO=IMPL(TZRO,.01,AGO)
        T1=.5*(TL+TAS)
        T2=.5*(TZRO+TGO)
        BEE=DELTA*(FDH2*H(T2)+FDN2*N(T2)+FDNH3*NH(T2,P)+FDCH4*CH(T2)+...
        FDAR*4.9675)
        ALF=FDCH4*CH(T1)+FDAR*4.9675+H2AS*H(T1)+NH3AS*NH(T1,PP)+N2AS*...
        N(T1)
        BEALF=BEE/ALF
        UCON=EXP((1.-BEALF)*UPRIM*ARHTX*HTXL/BEE)
        AL=TAS+BEALF*(TZRO-TGO)
        TL=AL
        AGO=(TZRO*(BEALF-1.)+TAS*(1.-UCON))/(BEALF-UCON)
        TTOP=TGO
        HYDO=FH2
        NITO=FN2
        MET=FCH4
        ARG=FAR
        AMMO=FNH3
*    SET TO START INTEGRATION NEXT TIME
        MGO=2
  200 CONTINUE
SORT
*      MATERIAL BALANCE ON NITROGEN AND AMMONIA
        NIT=NITO-(HYDO-HYD)/3.
        AMM=AMMO+(HYDO-HYD)*2./3.
*      TOTAL MOLES
```

TABLE 6.13 (Continued)

```
      TOTAL=HYD+NIT+AMM+ARG+MET
*     HEAT OF REACTION
      TSQ=T*T
      TCB=T*T*T
      HTRN=-9184.0-7.2949*T+0.34996E-2*TSQ+0.033563E-5*TCB-0.11625E-9...
      *T*TCB-6329.3-3.1619*P+(14.3595+4.4552E-3*P)*T-(8.3395E-3+ ...
      1.928E-6*P)*TSQ-51.21+0.14215*P
*     CALCULATE K FOR FUGACITY CORRECTION
      FUGCO =1.7343-8.143E-4*P+(-2.6714E-3+5.714E-7*P)*T+2.0E-6*TSQ
*      CALCULATE K FOR EQUILLIBRIUM
      EQLTR =(9184.0/T-7.2949*ALOG(T)+0.34996E-2*T+0.016781E-5*TSQ ...
      -.03875E-9*TCB+23.050)/1.9876
      EQLCO =EXP(EQLTR)
*      CALCULATE ENERGY OF ACTIVATION
      ENRAC = -24092.2/T+33.5566
*      CALCULATE CATALYST EFFICIENCY
      CATEF =GAM1-THETA*Z
*      CALCULATE MOLES*HEAT CAPACITY SUM
      SMCPT=HYD*H(T)+NIT*N(T)+AMM*NH(T,P)+MET*CH(T)+ARG*4.9675
*      CALCULATE DN1/DZ
      XTRM1=(EQLCO/FUGCO)**2*P**1.5*NIT*HYD**1.5/(AMM*TOTAL**1.5)
      XTRM2=AMM*SQRT(TOTAL)/(HYD**1.5*SQRT(P))
      XTRM3=(CATEF*3.*(300./P)**.63*EXP(ENRAC)*9.81E-6*CAREA)
      H2DT=XTRM3*(XTRM2-XTRM1)
*      CALCULATE DT/DZ
      TDT=( 2.*HTRN*H2DT/3.)/SMCPT
*      INTEGRATE
      HYD=INTGR(HYDO,H2DT)
      T=INTGR(TTOP,TDT)
NOSORT
  300 CONTINUE
PRINT 10.,HYD,T,AMM
END
STOP

OUTPUT VARIABLE SEQUENCE
   1    2    3    4    5    6    7    8    9   10   11   12   13   14   15   16   17   18   19   20
  21   22   23   24   25   26   27   29   28   30   31   34   32   35   36   40   41   37   38   42
  43   45   39   33   44   46   47

STORAGE USED/MAXIMUM
INTGR    2/50,  IN VARS 122/300,  OUT VARS  47/100, PARAMS 26/  ,, SYMBOLS  86/200,
```

4. Integrate differential equations in the second bed.

Table 6.16 shows the DSL listing for step 3 simulation. A block of code is required to perform the calculations of part 3 above. Notice that the values of hydrogen and of the gas temperature, which are both integrator outputs, must be saved for use after the integration is complete. The calculations in step 3 are entered by setting the value of the flag MGO to 3. This value is set after the integration, i.e., after the END statement. We also note that the gas composition exiting from the second bed, which was assumed for part 1 above, is automatically calculated during integration. This is to facilitate itera-

TABLE 6.14
DSL Translator Output for Ammonia Simulation
Step 2

```
        SUBROUTINE UPDAT
        REAL   FDH2 ,L      ,N2AS ,NH3AS,INTGR,IMPL ,N      ,NH   ,NITO ,MET  ,
       1NIT
        INTEGER  MGO
        COMMON NALRM,IZZZ(833) ,Z    ,DELZ ,DELMI,L    ,DELTP,DELTC,HYD  ,
       1T     ,PP    ,FH2  ,FN2  ,FNH3 ,FCH4 ,FAR  ,FTOT ,TL   ,T1   ,T2   ,
       1BEE   ,ALF   ,BEALF,UCON ,AL   ,TGO  ,TTOP ,HYDO ,NITO ,MET  ,ARG  ,
       1AMMO  ,NIT   ,AMM  ,TOTAL,TSQ  ,TCB  ,HTRN ,FUGCO,EQLTR,EQLCO,ENRAC,
       1CATEF,SMCPT,XTRM1,XTRM2,XTRM3,H2DT ,TDT  ,FDH2 ,FDN2 ,FDAR ,FDCH4
        COMMON
       1FDNH3,TZRO ,DELTA,TAS  ,P    ,CAREA,GAM1 ,THETA,HTXL ,UPRIM,ARHTX,
       1DROP ,H2AS ,N2AS ,NH3AS,INZZC,MGO
C       BELOW ARE HEAT CAPACITY FUNCTIONS PUT IN AFTER DSL SIMULATION HAS
C       COMPLETED CREATING THE UPDATE SUBROUTINE.
        H(T)=6.952-.0004576*T+.956E-6*T*T-.0279E-9*T*T*T
        N(T)=6.903-.375E-3*T+.193E-5*T*T-.686E-9*T*T*T
        CH(T)=4.75+1.2E-2*T+.303E-5*T*T-2.63E-9*T*T*T
        NH(T,P)=6.5846+.6125E-2*T+.2366E-5*T*T-1.6E-9*T*T*T*T+96.17-
       1 .06757*P+(1.68474E-4*P-.2225)*T+(1.2890E-4-1.0095E-7*P)*T*T
C       HEAT CAPACITY DEFINITIONS COMPLETE.
        GO TO (100,200,300),MGO
  100   PP=P-DROP
        FH2=DELTA*FDH2
        FN2=FDN2*DELTA
        FNH3=DELTA*FDNH3
        FCH4=DELTA*FDCH4
        FAR=DELTA*FDAR
        FTOT = FH2+FN2+FNH3+FCH4+FAR
        TL=TAS
 9100   TGO=IMPL(899,TZRO,.01,AGO)
        IF(NALRM) 9200, 9200, 9300
 9300   CONTINUE
        T1=.5*(TL+TAS)
        T2=.5*(TZRO+TGO)
        BEE=DELTA*(FDH2*H(T2)+FDN2*N(T2)+FDNH3*NH(T2,P)+FDCH4*CH(T2)+
       1FDAR*4.9675)
        ALF=FDCH4*CH(T1)+FDAR*4.9675+H2AS*H(T1)+NH3AS*NH(T1,PP)+N2AS*
       1N(T1)
        BEALF=BEE/ALF
        UCON=EXP((1.-BEALF)*UPRIM*ARHTX*HTXL/BEE)
        AL=TAS+BEALF*(TZRO-TGO)
        TL=AL
        AGO=(TZRO*(BEALF-1.)+TAS*(1.-UCON))/(BEALF-UCON)
        GO TO 9100
 9200   CONTINUE
        TTOP=TGO
        HYDO=FH2
        NITO=FN2
        MET=FCH4
        ARG=FAR
        AMMO=FNH3
        MGO=2
  200   CONTINUE
        AMM=AMMO+(HYDO-HYD)*2./3.

        NIT=NITO-(HYDO-HYD)/3.
        TOTAL=HYD+NIT+AMM+ARG+MET
        TSQ=T*T
        FUGCO =1.7343-8.143E-4*P+(-2.6714E-3+5.714E-7*P)*T+2.0E-6*TSQ
        TCB=T*T*T
```

TABLE 6.14 (Continued)

```
      EQLTR =(9184.0/T-7.2949*ALOG(T)+0.34996E-2*T+0.016781E-5*TSQ
     1-.03875E-9*TCB+23.050)/1.9876
      EQLCO =EXP(EQLTR)
      XTRM1=(EQLCO/FUGCO)**2*P**1.5*NIT*HYD**1.5/(AMM*TOTAL**1.5)
      XTRM2=AMM*SQRT(TOTAL)/(HYD**1.5*SQRT(P))
      ENRAC = -24092.2/T+33.5566
      CATEF =GAM1-THETA*Z
      XTRM3=(CATEF*3.*(300./P)**.63*EXP(ENRAC)*9.81E-6*CAREA)
      H2DT=XTRM3*(XTRM2-XTRM1)
      HYD=INTGR(HYDC,H2DT)
      SMCPT=HYD*H(T)+NIT*N(T)+AMM*NH(T,P)+MET*CH(T)+ARG*4.9675
      HTRN=-9184.0-7.2949*T+0.34996E-2*TSQ+0.033563E-5*TCB-0.11625E-9
     1*T*TCB-6329.3-3.1619*P+(14.3595+4.4552E-3*P)*T-(8.3395E-3+
     11.928E-6*P)*TSQ-51.21+0.14215*P
      TDT=( 2.*HTRN*H2DT/3.)/SMCPT
      T=INTGR(TTOP,TDT)
  300 CONTINUE
      RETURN
      END
```

tive calculations for the entire reactor. The DSL listing in Table 6.16 indicates that actually two such iterations are performed. Note that only a new value of TAS is required to be specified, since all other assumed values are corrected. The only other variable which needs to be specified is the flag MGO, which is reset to 1.

Finally, Table 6.16 also illustrates that the iterative mixing calculations make use of the implicit function IMPL in DSL.

The UPDAT listing generated by the DSL program and the results are shown in Tables 6.17 through 6.21. Note, again, the inserted statements for heat capacity functions in the UPDAT subroutine.

An interesting exercise for the sophisticated user who has been able to follow this complex example is to extend this problem where part of the quench is also sent to the top of the first reactor. This quench gas may be a different temperature than the feed temperature. In this case the mixing calculations must precede the integrations but follow the heat exchanger calculations for the first bed. For further details on this simulation, the reader should consult Reference (31).

6.8 Solution of Multiple Differential Equations and Use of Subscripted Variables

6.8.1 Simulation of a Cracking Furnace

In this example we shall illustrate the use of the STORAG and MEMORY features of DSL as well as show the use of DSL in the solution of multiple first-order differential equations. The problem is in simulation of process conditions in

TABLE 6.15
Simulation Results for Ammonia Reactor Step 2

*** DSL/1800 SIMULATICN DATA ***
```
TITLE AMMONIA SIMULATION STEP 2
RELERR HYD=0.0001,T=0.C1
INTEG MILNE
INCON FDH2=3278.,FDN2=1087.,FDAR=206.5,FDCH4=520.0,FDNH3=143.3
INCON TZRC=423.,DELTA=0.8,TAS=800.,P=200.
PARAM CAREA=31000.,L=500.,GAM1=1.2,THETA=C.
CONTRL DELZ=1.,DELMI=0.0001
PARAM HTXL=185.,UPRIM=.02,ARHTX=24000.,DRCP=8.
PARAM H2AS=2500.,N2AS=850.,NH3AS=600.
PARAM MGO=1
PRINT 10.,HYD,T,AMM
END
```

AMMONIA SIMULATION STEP 2

Z	HYD	T	AMM
0.000E 00	0.2622E 04	0.7179E 03	0.1146E 03
0.100E 02	0.2605E 04	0.7223E 03	0.1259E 03
0.200E 02	0.2588E 04	0.7267E 03	C.1372E 03
0.300E 02	0.2571E 04	0.7311E 03	0.1483E 03
C.400E 02	0.2555E 04	0.7354E 03	0.1595E 03
0.500E 02	0.2538E 04	0.7398E 03	0.1707E 03
0.600E 02	0.2521E 04	0.7442E 03	C.1819E 03
0.700E 02	0.2504E 04	0.7486E 03	0.1933E 03
0.240E 03	0.2232E 04	0.8188E C3	C.3747E 03
0.250E 03	0.2222E 04	0.8213E 03	0.3812E 03
0.260E 03	0.2214E 04	0.8235E 03	0.3868E 03
0.270E 03	0.2206E 04	0.8254E 03	0.3917E 03
0.280E 03	0.2200E 04	0.827CE C3	0.3958E 03
0.290E 03	0.2195E 04	0.8283E C3	0.3991E 03
0.300E 03	0.2191E 04	0.8293E C3	C.4018E 03
0.310E 03	0.2188E 04	0.8301E C3	C.4038E 03
0.320E 03	C.2185E 04	0.8308E C3	C.4056E 03
0.330E 03	C.2184E 04	0.8312E 03	C.4067E 03
0.340E C3	C.2182E C4	0.8316E 03	C.4078E 03
0.350E 03	0.2181E 04	0.8318E 03	0.4082E 03
0.360E 03	0.2180E 04	0.8321E 03	0.4091E 03
0.370E 03	0.2179E 04	0.8323E 03	0.4095E 03
0.380E 03	0.2180E 04	0.8322E 03	0.4094E 03
0.390E 03	0.2179E 04	0.8325E 03	C.4101E 03
0.400E 03	C.2179E C4	0.8325E 03	C.4101E 03
0.41CE C3	C.2178E 04	0.8326E 03	C.4104E 03
0.420E C3	0.2179E 04	0.8325E 03	C.410CE 03
0.430E 03	0.2178E 04	0.8327E 03	C.4106E 03
0.440E C3	C.2178E 04	0.8325E 03	0.4102E 03
0.450E 03	C.2178E 04	0.8328E C3	C.4108E 03
0.46CE C3	C.2178E 04	0.8326E C3	C.4104E 03
0.470E C3	0.2179E 04	0.8325E C3	C.4101E 03
0.480E C3	C.2177E 04	0.8328E 03	C.4109E 03
0.490E C3	C.2177E 04	0.8328E 03	C.4109E 03

TABLE 6.16
DSL Translator Input for Ammonia Reactor
Simulation Step 3

```
                    *** DSL/1800 TRANSLATOR INPUT ***
TITLE AMMONIA SIMULATION STEP 3
RENAME TIME=Z,FINTI=L,DELT=DELZ
RELERR HYD=0.0001,T=0.01
INTEG MILNE
*   SPECIFY INITIAL FEED CONDITIONS TO THE REACTOR
INCON FDH2=3278.,FDN2=1087.,FDAR=206.5,FDCH4=520.0,FDNH3=143.3
INCON TZRO=423.,DELTA=0.8,TAS=800.,PZRO=200.,OMEGA=.008
*    SPECIFY DIMENSIONS OF THE TWO BEDS
PARAM CAREA=31000.,CATL1=510.,CATL2=520.,GAM1=1.2,THETA=0.
PARAM MXDRP=1.
*       SPECIFY STEP SIZE AT INTEGRATION START
CONTRL DELZ=1.,DELMI=0.0001
*    SPECIFY HEAT EXCHANGER PARAMETERS
PARAM HTXL=185.,UPRIM=.02,ARHTX=24000.,DROP=8.
*     SPECIFY SECOND BED EXIT CONDITIONS FOR HT XCH CALCULATIONS
PARAM H2AS=2500.,N2AS=850.,NH3AS=600.
*    SET FLAGS AS INTEGERS AND THEIR VALUES
INTGER MGO
PARAM MGO=1
NOSORT
        GO TO (100,200,300),MGO
  100 PP=PZRO-DROP
        FH2=DELTA*FDH2
        FN2=FDN2*DELTA
        FNH3=DELTA*FDNH3
        FCH4=DELTA*FDCH4
        FAR=DELTA*FDAR
        FTOT = FH2+FN2+FNH3+FCH4+ FAR
        TL=TAS
*    SET UP LOOP TO CALCULATE TEMPERATURE
        TGO=IMPL(TZRO,.01,AGO)
        T1=.5*(TL+TAS)
        T2=.5*(TZRO+TGO)
        BEE=DELTA*(FDH2*H(T2)+FDN2*N(T2)+FDNH3*NH(T2,PZRO)+FDCH4*...
        CH(T2)+FDAR*4.9675)
        ALF=FDCH4*CH(T1)+FDAR*4.9675+H2AS*H(T1)+NH3AS*NH(T1,PP)+N2AS*..;
        N(T1)
        BEALF=BEE/ALF
        UCON=EXP((1.-BEALF)*UPRIM*ARHTX*HTXL/BEE)
        AL=TAS+BEALF*(TZRO-TGO)
        TL=AL
        AGO=(TZRO*(BEALF-1.)+TAS*(1.-UCON))/(BEALF-UCON)
*   HEAT EXCHANGER CALCULATIONS COMPLETED. WRITE VALUES OF T AROUND IT
        TTOP=TGO
        HYDO=FH2
        NITO=FN2
        MET=FCH4
        ARG=FAR
        AMMO=FNH3
*   SET CATALYST LENGTH OF THE FIRST BED
        L=CATL1
*    SET TO START INTEGRATION NEXT TIME
        MGO=2
  200 CONTINUE
SORT
*     PRESSURE DROP IN BED A FUNCTION OF BED LENGTH
        P=PZRO-OMEGA*Z
*     MATERIAL BALANCE ON NITROGEN AND AMMONIA
        NIT=NITO-(HYDO-HYD)/3.
        AMM=AMMO+(HYDO-HYD)*2./3.
*     TOTAL MOLES
        TOTAL=HYD+NIT+AMM+ARG+MET
```

TABLE 6.16 (Continued)

```
*      HEAT OF REACTION
       TSQ=T*T
       TCB=T*T*T
       HTRN=-9184.0-7.2949*T+0.34996E-2*TSQ+0.033563E-5*TCB-0.11625E-9...
       *T*TCB-6329.3-3.1619*P+(14.3595+4.4552E-3*P)*T-(8.3395E-3+ ...
       1.928E-6*P)*TSQ-51.21+0.14215*P
*      CALCULATE K FOR FUGACITY CORRECTION
       FUGCO =1.7343-8.143E-4*P+(-2.6714E-3+5.714E-7*P)*T+2.0E-6*TSQ
*       CALCULATE K FOR EQUILLIBRIUM
       EQLTR =(9184.0/T-7.2949*ALOG(T)+0.34996E-2*T+0.016781E-5*TSQ ...
       -.03875E-9*TCB+23.050)/1.9876
       EQLCO =EXP(EQLTR)
*       CALCULATE ENERGY OF ACTIVATION
       ENRAC = -24092.2/T+33.5566
*       CALCULATE CATALYST EFFICIENCY
       CATEF =GAM1-THETA*Z
*       CALCULATE MOLES*HEAT CAPACITY SUM
       SMCPT=HYD*H(T)+NIT*N(T)+AMM*NH(T,P)+MET*CH(T)+ARG*4.9675
*       CALCULATE DN1/DZ
       XTRM1=(EQLCO/FUGCO)**2*P**1.5*NIT*HYD**1.5/(AMM*TOTAL**1.5)
       XTRM2=AMM*SQRT(TOTAL)/(HYD**1.5*SQRT(P))
       XTRM3=(CATEF*3.*(300./P)**.63*EXP(ENRAC)*9.81E-6*CAREA)
       H2DT=XTRM3*(XTRM2-XTRM1)
*       CALCULATE DT/DZ
       TDT=( 2.*HTRN*H2DT/3.)/SMCPT
*      INTEGRATE
       HYD=INTGR(HYDO,H2DT)
       T=INTGR(TTOP,TDT)
       H2AS=HYD
       N2AS=NIT
       NH3AS=AMM
       TS=T
NOSORT
       GO TO 600
   300 P=PP+MXDRP
*      RESET INITIAL CONDITIONS,RETRIEVE SAVED VALUE OF HYD REQUIRED TO
*       CALCULATE TEMPERATURE OF FEED TO NEXT BED AFTER QUENCH MIXING.
       MGO=2
       HYD=H2AS
       AMMO=AMM+BETA*FDNH3
       BETA=1.-DELTA
       NITO=NIT+BETA*FDN2
       HYDO=HYD+BETA*FDH2
       ARG=ARG+FAR*BETA
       MET=MET+FCH4*BETA
       L=CATL2
       TTOP=IMPL(TS,.01,FOFT)
       BET=BETA*TZRO
       SMCP=SMCPT*TS
       FOFT=(SMCP +(FDH2*H(TZRO)+FDN2*N(TZRO)+FDNH3*NH(TZRO,PZRO)+...
       CH(TZRO)*FDCH4+FAR*4.9675)*BET )/(AMMO*NH(TTOP,P)+NITO*N(TTOP)...
       1+CH(TTOP)*MET+ARG*4.9675+HYDO*H(TTOP) )
       GO TO 200
   600 CONTINUE
PRINT 10.,HYD,T,AMM
END
PARAM MGO=3
END
PARAM MGO=1,TAS=810.
END
PARAM MGO=3
END
STOP
```

TABLE 6.17
*DSL Translator Output for Ammonia Reactor
Simulation Step 3*

```
      SUBROUTINE UPDAT
      REAL  FDH2 ,MXDRP,N2AS ,NH3AS,INTGR,IMPL ,N      ,NH    ,NITO ,MET  ,
     1L    ,NIT
      INTEGER    MGO
      COMMON NALRM,IZZZ(833) ,Z     ,DELZ ,DELMI,L      ,DELTP,DELTC,HYD  ,
     1T    ,PP    ,FH2   ,FN2   ,FNH3 ,FCH4 ,FAR  ,FTOT ,TL   ,T1   ,T2   ,
     1BEE  ,ALF  ,BEALF,UCON ,AL   ,TGO  ,TTOP ,HYDO ,NITO ,MET  ,ARG  ,
     1AMMO ,P    ,NIT  ,AMM  ,TOTAL,TSQ  ,TCB  ,HTRN ,FUGCO,EQLTR,EQLCO,
     1ENRAC,CATEF,SMCPT,XTRM1,XTRM2,XTRM3,H2DT ,TDT  ,TS   ,BETA ,BET
      COMMON
     1SMCP ,FDH2 ,FDN2 ,FDAR ,.DCH4,FDNH3,TZRO ,DELTA,TAS  ,PZRO ,OMEGA,
     1CAREA,CATL1,CATL2,GAM1 ,THETA,MXDRP,HTXL ,UPRIM,ARHTX,DROP ,H2AS ,
     1N2AS ,NH3AS,INZZO,MGO
C     BELOW ARE HEAT CAPACITY FUNCTIONS PUT IN AFTER DSL SIMULATION HA
C     COMPLETED CREATING THE UPDATE SUBROUTINE.
      H(T)=6.952-.0004576*T+.956E-6*T*T-.0279E-9*T*T*T
      N(T)=6.903-.375E-3*T+.193E-5*T*T-.686E-9*T*T*T
      CH(T)=4.75+1.2E-2*T+.303E-5*T*T-2.63E-9*T*T*T
      NH(T,P)=6.5846+.6125E-2*T+.2366E-5*T*T-1.6E-9*T*T*T+96.17-
     1 .06757*P+(1.68474E-4*P-.2225)*T+(1.2890E-4-1.0095E-7*P)*T*T
C     HEAT CAPACITY DEFINITIONS COMPLETE.
      GO TO (100,200,300),MGO
  100 PP=PZRO-DROP
      FH2=DELTA*FDH2
      FN2=FDN2*DELTA
      FNH3=DELTA*FDNH3
      FCH4=DELTA*FDCH4
      FAR=DELTA*FDAR
      FTOT = FH2+FN2+FNH3+FCH4+ FAR
      TL=TAS
 9100 TGO=IMPL(899,TZRO,.01,AGO)
      IF(NALRM) 9200, 9200, 9300
 9300 CONTINUE
      T1=.5*(TL+TAS)
      T2=.5*(TZRO+TGO)
      BEE=DELTA*(FDH2*H(T2)+FDN2*N(T2)+FDNH3*NH(T2,PZRO)+FDCH4*
     1CH(T2)+FDAR*4.9675)
      ALF=FDCH4*CH(T1)+FDAR*4.9675+H2AS*H(T1)+NH3AS*NH(T1,PP)+N2AS*
     1N(T1)
      BEALF=BEE/ALF
      UCON=EXP((1.-BEALF)*UPRIM*ARHTX*HTXL/BEE)
      AL=TAS+BEALF*(TZRO-TGO)
      TL=AL
      AGO=(TZRO*(BEALF-1.)+TAS*(1.-UCON))/(BEALF-UCON)
      GO TO 9100
 9200 CONTINUE
      TTOP=TGO
      HYDO=FH2
      NITO=FN2
      MET=FCH4
      ARG=FAR
      AMMO=FNH3
```

TABLE 6.17 (Continued)

```
      L=CATL1
      MGO=2
  200 CONTINUE
      AMM=AMMO+(HYDO-HYD)*2./3.
      NIT=NITO-(HYDO-HYD)/3.
      TOTAL=HYD+NIT+AMM+ARG+MET
      P=PZRO-OMEGA*Z
      TSQ=T*T
      FUGCO =1.7343-8.143E-4*P+(-2.6714E-3+5.714E-7*P)*T+2.0E-6*TSQ
      TCB=T*T*T
      EQLTR =(9184.0/T-7.2949*ALOG(T)+0.34996E-2*T+0.016781E-5*TSQ
     1-.03875E-9*TCB+23.050)/1.9876
      EQLCO =EXP(EQLTR)
      XTRM1=(EQLCO/FUGCO)**2*P**1.5*NIT*HYD**1.5/(AMM*TOTAL**1.5)
      XTRM2=AMM*SQRT(TOTAL)/(HYD**1.5*SQRT(P))
      ENRAC = -24092.2/T+33.5566
      CATEF =GAM1-THETA*Z
      XTRM3=(CATEF*3.*(300./P)**.63*EXP(ENRAC)*9.81E-6*CAREA)
      H2DT=XTRM3*(XTRM2-XTRM1)
      HYD=INTGR(HYDO,H2DT)
      SMCPT=HYD*H(T)+NIT*N(T)+AMM*NH(T,P)+MET*CH(T)+ARG*4.9675
      HTRN=-9184.0-7.2949*T+0.34996E-2*TSQ+0.033563E-5*TCB-0.11625E-9
     1*T*TCB-6329.3-3.1619*P+(14.3595+4.4552E-3*P)*T-(8.3395E-3+
     11.928E-6*P)*TSQ-51.21+0.14215*P
      TDT=( 2.*HTRN*H2DT/3.)/SMCPT
      T=INTGR(TTOP,TDT)
      H2AS=HYD
      N2AS=NIT
      NH3AS=AMM
      TS=T
      GO TO 600
  300 P=PP+MXDRP
      MGO=2
      HYD=H2AS
      AMMO=AMM+BETA*FDNH3
      BETA=1.-DELTA
      NITO=NIT+BETA*FDN2
      HYDO=HYD+BETA*FDH2
      ARG=ARG+FAR*BETA
      MET=MET+FCH4*BETA
      L=CATL2
 9101 TTOP=IMPL(897,TS,.01,FOFT)
      IF(NALRM) 9201, 9201, 9301
 9301 CONTINUE
      BET=BETA*TZRO
      SMCP=SMCPT*TS
      FOFT=(SMCP +(FDH2*H(TZRO)+FDN2*N(TZRO)+FDNH3*NH(TZRO,PZRO)+
     1CH(TZRO)*FDCH4+FAR*4.9675)*BET )/(AMMO*NH(TTOP,P)+NITO*N(TTOP)
     1+CH(TTOP)*MET+ARG*4.9675+HYDO*H(TTOP) )
      GO TO 9101
 9201 CONTINUE
      GO TO 200
  600 CONTINUE
      RETURN
      END
```

TABLE 6.18
Simulation Results for Step 3, First Bed Output in Ammonia Reactor

```
TITLE AMMONIA SIMULATION STEP 3
RELERR HYD=0.0001,T=0.01
INTEG MILNE
INCON FDH2=3278.,FDN2=1087.,FDAR=206.5,FDCH4=520.0,FDNH3=143.3
INCON TZRO=423.,DELTA=0.8,TAS=800.,PZRO=200.,OMEGA=.008
PARAM CAREA=31000.,CATL1=510.,CATL2=520.,GAM1=1.2,THETA=0.
PARAM MXDRP=1.
CONTRL DELZ=1.,DELMI=0.0001
PARAM HTXL=185.,UPRIM=.02,ARHTX=24000.,DROP=8.
PARAM H2AS=2500.,N2AS=850.,NH3AS=600.
PARAM MGO=1
PRINT 10.,HYD,T,AMM
END
```

AMMONIA SIMULATION STEP 3

Z	HYD	T	AMM
0.000E 00	0.2622E 04	0.7179E 03	0.1146E 03
0.100E 02	0.2605E 04	0.7223E 03	0.1259E 03
0.200E 02	0.2588E 04	0.7267E 03	0.1371E 03
0.300E 02	0.2571E 04	0.7310E 03	0.1483E 03
0.400E 02	0.2555E 04	0.7354E 03	0.1595E 03
0.500E 02	0.2538E 04	0.7397E 03	0.1706E 03
0.600E 02	0.2521E 04	0.7441E 03	0.1819E 03
0.700E 02	0.2504E 04	0.7485E 03	0.1932E 03
0.800E 02	0.2487E 04	0.7529E 03	0.2045E 03
0.900E 02	0.2470E 04	0.7574E 03	0.2160E 03
0.100E 03	0.2452E 04	0.7619E 03	0.2276E 03
0.110E 03	0.2435E 04	0.7664E 03	0.2392E 03
0.120E 03	0.2417E 04	0.7709E 03	0.2509E 03
0.130E 03	0.2400E 04	0.7755E 03	0.2626E 03
0.140E 03	0.2382E 04	0.7800E 03	0.2744E 03
0.150E 03	0.2365E 04	0.7845E 03	0.2860E 03
0.160E 03	0.2347E 04	0.7890E 03	0.2976E 03
0.170E 03	0.2330E 04	0.7934E 03	0.3089E 03
0.180E 03	0.2314E 04	0.7977E 03	0.3199E 03
0.190E 03	0.2298E 04	0.8018E 03	0.3306E 03
0.200E 03	0.2283E 04	0.8057E 03	0.3406E 03
0.210E 03	0.2269E 04	0.8093E 03	0.3499E 03
0.220E 03	0.2256E 04	0.8126E 03	0.3584E 03
0.230E 03	0.2244E 04	0.8156E 03	0.3662E 03
0.240E 03	0.2234E 04	0.8183E 03	0.3733E 03
0.250E 03	0.2224E 04	0.8207E 03	0.3796E 03

AMMONIA SIMULATION STEP 3

Z	HYD	T	AMM
0.500E 03	0.2183E 04	0.8312E 03	0.4069E 03
0.510E 03	0.2182E 04	0.8316E 03	0.4079E 03
0.510E 03	0.2182E 04	0.8316E 03	0.4079E 03

TABLE 6.19
Ammonia Reactor Simulation Step 3, Second Bed
Output

*** DSL/1800 SIMULATION DATA ***

PARAM MGO=3
END

AMMONIA SIMULATION STEP 3

Z	HYD	T	AMM
0.000E 00	0.2838E 04	0.7666E 03	0.4079E 03
0.100E 02	0.2827E 04	0.7689E 03	0.4150E 03
0.200E 02	0.2816E 04	0.7711E 03	0.4221E 03
0.300E 02	0.2806E 04	0.7734E 03	0.4291E 03
0.400E 02	0.2795E 04	0.7756E 03	0.4362E 03
0.500E 02	0.2785E 04	0.7778E 03	0.4431E 03
0.600E 02	0.2774E 04	0.7800E 03	0.4500E 03
0.700E 02	0.2764E 04	0.7821E 03	0.4569E 03
0.800E 02	0.2754E 04	0.7843E 03	0.4636E 03
0.900E 02	0.2744E 04	0.7863E 03	0.4701E 03
0.100E 03	0.2735E 04	0.7884E 03	0.4766E 03
0.110E 03	0.2725E 04	0.7903E 03	0.4828E 03
0.120E 03	0.2716E 04	0.7923E 03	0.4889E 03
0.130E 03	0.2707E 04	0.7941E 03	0.4947E 03
0.140E 03	0.2699E 04	0.7959E 03	0.5003E 03
0.150E 03	0.2691E 04	0.7976E 03	0.5057E 03
0.160E 03	0.2683E 04	0.7992E 03	0.5108E 03
0.170E 03	0.2676E 04	0.8007E 03	0.5156E 03
0.180E 03	0.2669E 04	0.8021E 03	0.5201E 03
0.190E 03	0.2663E 04	0.8035E 03	0.5244E 03
0.200E 03	0.2657E 04	0.8047E 03	0.5283E 03
0.210E 03	0.2651E 04	0.8059E 03	0.5320E 03
0.220E 03	0.2646E 04	0.8069E 03	0.5353E 03
0.230E 03	0.2642E 04	0.8079E 03	0.5384E 03
0.240E 03	0.2638E 04	0.8088E 03	0.5412E 03
0.250E 03	0.2634E 04	0.8096E 03	0.5438E 03
0.260E 03	0.2630E 04	0.8103E 03	0.5461E 03
0.270E 03	0.2627E 04	0.8110E 03	0.5482E 03
0.280E 03	0.2624E 04	0.8116E 03	0.5501E 03
0.290E 03	0.2622E 04	0.8121E 03	0.5518E 03
0.300E 03	0.2619E 04	0.8126E 03	0.5533E 03
0.310E 03	0.2617E 04	0.8130E 03	0.5546E 03
0.320E 03	0.2616E 04	0.8134E 03	0.5559E 03
Z	HYD	T	AMM
0.500E 03	0.2603E 04	0.8162E 03	0.5646E 03
0.510E 03	0.2608E 04	0.8150E 03	0.5610E 03
0.520E 03	0.2602E 04	0.8163E 03	0.5649E 03
0.520E 03	0.2602E 04	0.8163E 03	0.5649E 03

TABLE 6.20
Ammonia Reactor Step 3 Simulation—Repetition
with New Value of TAS

*** DSL/1800 SIMULATION DATA ***

PARAM MGO=1,TAS=810.
END

AMMONIA SIMULATION STEP 3

Z	HYD	T	AMM
0.000E 00	0.2622E 04	0.7264E 03	0.1146E 03
0.100E 02	0.2602E 04	0.7317E 03	0.1281E 03
0.200E 02	0.2582E 04	0.7369E 03	0.1415E 03
0.300E 02	0.2562E 04	0.7421E 03	0.1547E 03
0.400E 02	0.2542E 04	0.7472E 03	0.1679E 03
0.500E 02	0.2522E 04	0.7523E 03	0.1811E 03
0.600E 02	0.2502E 04	0.7575E 03	0.1943E 03
0.700E 02	0.2482E 04	0.7627E 03	0.2076E 03
0.800E 02	0.2462E 04	0.7678E 03	0.2210E 03
0.900E 02	0.2442E 04	0.7730E 03	0.2344E 03
0.100E 03	0.2422E 04	0.7782E 03	0.2478E 03
0.110E 03	0.2402E 04	0.7834E 03	0.2612E 03
0.120E 03	0.2382E 04	0.7885E 03	0.2745E 03
0.130E 03	0.2362E 04	0.7936E 03	0.2876E 03
0.140E 03	0.2343E 04	0.7985E 03	0.3003E 03
0.150E 03	0.2325E 04	0.8033E 03	0.3126E 03
0.160E 03	0.2307E 04	0.8078E 03	0.3244E 03
0.170E 03	0.2291E 04	0.8121E 03	0.3353E 03
0.180E 03	0.2276E 04	0.8159E 03	0.3454E 03
0.190E 03	0.2262E 04	0.8194E 03	0.3544E 03
0.200E 03	0.2250E 04	0.8225E 03	0.3623E 03
0.210E 03	0.2240E 04	0.8251E 03	0.3691E 03
0.220E 03	0.2231E 04	0.8274E 03	0.3750E 03
0.230E 03	0.2224E 04	0.8293E 03	0.3800E 03
0.240E 03	0.2218E 04	0.8308E 03	0.3840E 03
0.250E 03	0.2213E 04	0.8321E 03	0.3873E 03
0.260E 03	0.2209E 04	0.8331E 03	0.3899E 03
0.270E 03	0.2206E 04	0.8338E 03	0.3918E 03
0.280E 03	0.2204E 04	0.8344E 03	0.3934E 03
0.290E 03	0.2202E 04	0.8348E 03	0.3944E 03
0.300E 03	0.2201E 04	0.8352E 03	0.3953E 03
0.310E 03	0.2200E 04	0.8354E 03	0.3959E 03
0.320E 03	0.2199E 04	0.8355E 03	0.3963E 03
0.330E 03	0.2199E 04	0.8357E 03	0.3967E 03
0.340E 03	0.2199E 04	0.8356E 03	0.3965E 03

Z	HYD	T	AMM
0.500E 03	0.2197E 04	0.8361E 03	0.3976E 03
0.510E 03	0.2199E 04	0.8356E 03	0.3963E 03
0.510E 03	0.2199E 04	0.8356E 03	0.3963E 03

TABLE 6.21
Step 3 Simulation Results for Second Bed with New Value of TAS

*** DSL/1800 SIMULATION DATA ***

PARAM MGO=3
END

AMMONIA SIMULATION STEP 3

Z	HYD	T	AMM
0.000E 00	0.2855E 04	0.7640E 03	0.4250E 03
0.100E 02	0.2845E 04	0.7660E 03	0.4315E 03
0.200E 02	0.2835E 04	0.7681E 03	0.4380E 03
0.300E 02	0.2826E 04	0.7701E 03	0.4444E 03
0.400E 02	0.2816E 04	0.7721E 03	0.4509E 03
0.500E 02	0.2806E 04	0.7741E 03	0.4573E 03
0.600E 02	0.2797E 04	0.7761E 03	0.4637E 03
0.700E 02	0.2787E 04	0.7781E 03	0.4700E 03
0.800E 02	0.2778E 04	0.7801E 03	0.4763E 03
0.900E 02	0.2769E 04	0.7820E 03	0.4824E 03
0.100E 03	0.2760E 04	0.7839E 03	0.4884E 03
0.110E 03	0.2751E 04	0.7857E 03	0.4943E 03
0.120E 03	0.2742E 04	0.7875E 03	0.5001E 03
0.130E 03	0.2734E 04	0.7893E 03	0.5057E 03
0.140E 03	0.2726E 04	0.7910E 03	0.5111E 03
0.150E 03	0.2718E 04	0.7926E 03	0.5164E 03
0.160E 03	0.2710E 04	0.7942E 03	0.5214E 03
0.170E 03	0.2703E 04	0.7957E 03	0.5262E 03
0.180E 03	0.2696E 04	0.7972E 03	0.5308E 03
0.190E 03	0.2690E 04	0.7985E 03	0.5351E 03
0.200E 03	0.2684E 04	0.7998E 03	0.5392E 03
0.210E 03	0.2678E 04	0.8010E 03	0.5431E 03
0.220E 03	0.2672E 04	0.8021E 03	0.5467E 03
0.230E 03	0.2667E 04	0.8032E 03	0.5500E 03
0.240E 03	0.2663E 04	0.8041E 03	0.5531E 03
0.250E 03	0.2658E 04	0.8050E 03	0.5560E 03
0.260E 03	0.2654E 04	0.8059E 03	0.5586E 03
0.270E 03	0.2651E 04	0.8066E 03	0.5611E 03
0.280E 03	0.2647E 04	0.8073E 03	0.5633E 03
0.290E 03	0.2644E 04	0.8080E 03	0.5653E 03
0.300E 03	0.2642E 04	0.8085E 03	0.5671E 03
0.310E 03	0.2639E 04	0.8091E 03	0.5689E 03
0.320E 03	0.2637E 04	0.8095E 03	0.5704E 03
0.330E 03	0.2635E 04	0.8099E 03	0.5716E 03
0.340E 03	0.2633E 04	0.8104E 03	0.5730E 03

Z	HYD	T	AMM
0.500E 03	0.2621E 04	0.8128E 03	0.5809E 03
0.510E 03	0.2621E 04	0.8128E 03	0.5807E 03
0.520E 03	0.2620E 04	0.8130E 03	0.5813E 03
0.520E 03	0.2620E 04	0.8130E 03	0.5813E 03

---

the furnaces which are employed in the production of ethylene and propylene by steam-cracking reactions, where the furnace feedstock is a mixture of ethane and propane. In actual practice, heavier hydrocarbons may also be present in the feedstock; however, for this illustration consideration of ethane and propane will suffice. The following chemical reactions occur in the furnace:

ethane: $C_2H_6 \rightleftharpoons C_2H_4 + H_2$
$2C_2H_6 \rightleftharpoons 2CH_4 + C_2H_4$
propane: $C_3H_8 \rightleftharpoons C_3H_6 + H_2$
$C_3H_8 \rightleftharpoons C_2H_4 + CH_4$

Although other side routines may also occur, for this example we shall assume them to be negligible.

Let us consider the case of a plant which has four furnaces, with each furnace operating slightly differently, in that the amount of heat which can be transferred to the reactor mixture varies because of the fouling conditions of the furnace tubes.

This means that the temperature profile of the tube wall as well as the coefficients for heat transfer from the wall to the gas are different for each furnace. The purpose of the simulation is to examine the effect of various furnace control parameters on each furnace production and exit gas composition as well as on the total production of ethylene and propylene.

The equations for material and energy balance in the ethane furnace are given below. See Reference (30) for details of ethylene furnace operation and derivation of the equations.

Reaction rate for reaction 1:

$$r_1 = \exp\left(A_1 - \frac{E}{RT}\right)[C_2H_6 \cdot P/(RT \cdot \text{TOTML}) - C_2H_4 \cdot H_2$$
$$\cdot P^2/(R^2T^2 \cdot \text{TOTML}^2 \cdot K_1)] \tag{6.30}$$

Reaction rate for reaction 2:

$$r_2 = \exp\left(A_2 - \frac{E_2}{RT}\right)[C_2H_6 \cdot P/(RT \cdot \text{TOTML})] \tag{6.31}$$

Reaction rate for reaction 3:

$$r_3 = \exp\left(A_3 - \frac{E_3}{RT}\right)[C_3H_8 \cdot P/(RT \cdot \text{TOTML}) -$$
$$- C_3H_6 \cdot H_2 \cdot P^2/(R^2T^2 \cdot \text{TOTML}^2 \cdot K_2)] \tag{6.32}$$

Reaction rate for reaction 4:

$$r_4 = \exp\left(A_4 - \frac{E_4}{RT}\right)[C_3H_8 \cdot P/(RT \cdot \text{TOTML})] \tag{6.33}$$

The material balance is given by

for ethane: $\dfrac{dn(\text{ethane})}{dl} = (-r_1 - r_2) \cdot \text{area}$

for propane $\dfrac{dn(\text{propane})}{dl} = (-r_3 - r_4) \cdot \text{area}$

for ethylene: $\dfrac{dn(\text{ethylene})}{dl} = r_1 + \dfrac{r_2}{2} + r_4 \ \cdot \text{area}$

for propylene: $\dfrac{dn(\text{propylene})}{dl} = r_3 \cdot \text{area}$ (6.33a)

for methane: $\dfrac{dn(\text{methane})}{dl} = r_2 \cdot \text{area}$

for hydrogen: $\dfrac{dn(\text{hydrogen})}{dl} = (r_1 + r_3) \cdot \text{area}$

Total moles:

$$\text{TOTML} = C_2H_6 + C_2H_4 + C_3H_8 + C_3H_6 + CH_4 + H_2 + \text{steam}$$

The temperature T is given by

$$\frac{dT}{dl} = \left[\pi \cdot \text{diam} \cdot h(T_w - T) + \text{area} \cdot \left(\sum_{i=1}^{4} -\Delta H_i r_i \right) \right] \bigg/ \left(\sum_{i=1}^{7} n_i C_{p_i} \right) \quad (6.33b)$$

Before we begin the discussion of coding this problem with DSL, the user must be reminded that the DSL language does *not* permit subscripting of the variables which are integrators. Thus, we cannot subscript n (ethylene), n (ethane), etc., although this would be very convenient. The variables which can be subscripted in this case are Arrhenius constants A, activation energies E, and heat capacities and heats of reaction. The last two are actually functions of temperature and can be subscripted only if they are assumed to be constant parameters. For illustration purposes here, we shall use mean values for the heat capacities and heats of reactions. The reader may wish to express them as functions of temperature and supply them as independent FORTRAN functions stored on the 1130 disk (1800, System/3, or System/7), or, as was illustrated in the previous problem of ammonia reactor simulation, they may be expressed as functions and inserted prior to the first executable statement in the UPDAT card deck punched by DSL.

It may also be desirable to subscript values of tube diameter, heat transfer coefficients, and even the coefficients of the tube wall temperature profiles. We shall show, however, in the illustration that this is not necessary because we can handle each furnace separately and sequentially, so that the individual furnace parameters can be specified. This also reduces the amount of coding required and avoids conflicts of double-subscripted variables. The disadvantage is that the user must add up individual furnace yields to get the total production figures.

TABLE 6.22
DSL Translator Input for Ethylene Furnace
Simulation

```
                    *** DSL/1800 TRANSLATOR INPUT ***
TITLE SIMULATION OF STEAM CRACKING FURNACE FOR ETHYLENE PRODUCTION
INTEG MILNE
CONTRL DELMI=0.01,DELT=.1,FINTI=200.
RELERR C2H6=.001,T=.001
ABSERR C2H6=.1,C3H8=.1,T=.01
STORAG A(4),E(4),DELH(4),SPHT(6)
*    SPECIFY VALUES OF ARHENIUS CONSTANTS AND ACTIVATION ENERGIES FOR
*    REACTIONS.SPECIFY HEAT TRANSFER COEFFICIENTS OF AND PRESSURES IN
*    THE FOUR FURNACES. PROVIDE HEATS OF REACTIONS AND MEAN SPECIFIC
*    HEATS. ALSO PROVIDE FURNACE TUBE DIAMETERS .
TABLE A(1-4)=43.45,48.,37.2,40.0,E(1-4)=1.3E5,1.5E5,1.1E5,1.143E5,...
       DELH(1-4)=35.E3,9.E3,30.E3,20.E3,SPHT(1-6)=31.,42.,26.8,42.4,...
       18.9,6.67
*    FURNACE REACTION RATE CALCULATIONS
*    WALL TEMPERATURE
       TW=TWO+ALF*TIME-BET*TIME*TIME
       AREA=3.14*(DIAM/2.)**2
       VALUE=PRES/.7302/TEMP/TOTML
*    EQUILLIBRIUM CONSTANTS
       K1=EXP(14.6-56340./1.987/TEMP)
       K2=EXP(16.64-55512./1.987/TEMP)
*    REACTION RATES FOR THE FOUR REACTIONS
       R1 = EXP(A(1)-E(1)/1.987/TEMP)*VALUE*(C2H6-C2H4*H2*VALUE/K1)
       R2 = EXP(A(2)-E(2)/1.987/TEMP)*VALUE*C2H6
       R3=   EXP(A(3)-E(3)/1.987/TEMP)*VALUE*(C3H8-C3H6*H2*VALUE/K2)
       R4 =  EXP(A(4)-E(4)/1.987/TEMP)*VALUE*C3H8
*    MATERIAL BALANCE ON THE COMPONENTS
       C2H6 =INTGR(C2H6O,(-R1-R2)*AREA)
       C3H8=INTGR(C3H8O,(-R3-R4)*AREA)
       C2H4 =INTGR(C2H4O,(R1+R2/2.+R4)*AREA)
       C3H6=INTGR(C3H6O,R3*AREA)
       H2=INTGR(H2O,(R1+R3)*AREA)
       CH4=INTGR(CH4O,R2*AREA)
       T=INTGR(TO,DT)
*    HEAT BALANCE
       DT=(3.14*DIAM*HTTRC*(TW-T)-(DELH(1)*R1+DELH(2)*R2+DELH(3)*R3+...
       DELH(4)*R4)*AREA)/(C2H6*SPHT(1)+C3H8*SPHT(2)+C2H4*SPHT(3)+C3H6*...
       SPHT(4)+CH4*SPHT(5)+H2*SPHT(6)+STEAM*8.4)
       TEMP=T+460.
*     TOTAL MOLES
       TOTML=C2H6+C3H8+C2H4+C3H6+CH4+H2+STEAM
PRINT 20.,C2H6,C3H8,C2H4,C3H6,CH4,H2,T,TW
PARAM TO=1000.,TWO=1400.,ALF=4.,BET=6.0E-3,STEAM=150.,PRES=2.,DIAM=...
       .34,HTTRC=150.,C2H6O=250.,C3H8O=25.
END
PARAM TO=1100.,TWO=1380.,ALF=4.,BET=6.1E-3,STEAM=160.,PRES=2.1
PARAM HTTRC=145.,C2H6O=230.,C3H8O=65.
END
PARAM TO=1050.,TWO=1410.,ALF=3.95,BET=6.1E-3,STEAM=165.,PRES=2.2
PARAM HTTRC=160.,C2H6O=210.,C3H8O=95.
END
PARAM TO=1000.,TW=1350.,ALF=4.,BET=0.64E-2,STEAM=200.,PRES=2.
PARAM HTTRC=145.,C2H6O=100.,C3H8O=185.
END
STOP
```

TABLE 6.22 (Continued)

```
PARAM NOT INPUT, SET=0  C2H4O
PARAM NOT INPUT, SET=0  C3H6O
PARAM NOT INPUT, SET=0  H2O
PARAM NOT INPUT, SET=0  CH4O

OUTPUT VARIABLE SEQUENCE
  2  19  18  3.  7  4  6  10  9  5  8  11  12  13  14  15  1  17  16
```

The DSL code is shown in Table 6.22. Notice the use of dimensioned variables, A, E, DELH, and SPHT, which are the Arrhenius constants, activation energies, heats of reactions, and heat capacities, respectively. Also note that the values of these parameters are specified using the TABLE function of DSL. The language has the restriction that these parameters can be specified *only* with the TABLE statement and *not* by the PARAM or CONST statement. Also note that we did not specify the values of C_2H_4, C_3H_6, and CH_4 in the furnace feed—the program sets

TABLE 6.23
Ethylene Furnace Simulation Translator Output

PAGE   2   DSL/1800 (DIGITAL SIMULATION LANGUAGE)

```
    SUBROUTINE UPDAT
    REAL  K1  ,K2  ,INTGR
    COMMON NALRM,IZZZ(833) ,TIME ,DELT ,DELMI,FINTI,DELTP,DELTC,C2H6 ,
   1C3H8 ,C2H4 ,C3H6 ,H2  ,CH4  ,T  ,TW  ,AREA ,VALUE,K1  ,K2  ,
   1R1  ,R2  ,R3  ,R4  ,DT  ,TEMP ,TOTML,TO  ,TWO ,ALF  ,BET  ,
   1STEAM,PRES ,DIAM ,HTTRC,C2H6O,C3H8O,C2H4O,C3H6O,H2O  ,CH4O
   1 ,ZZ990(  4) ,A  (  4) ,E  (  4) ,DELH (  4) ,SPHT (  6)
    AREA=3.14*(DIAM/2.)**2
    TOTML=C2H6+C3H8+C2H4+C3H6+CH4+H2+STEAM
    TEMP=T+460.
    VALUE=PRES/.7302/TEMP/TOTML
    R2 = EXP(A(2)-E(2)/1.987/TEMP)*VALUE*C2H6
    K1=EXP(14.6-56340./1.987/TEMP)
    R1 = EXP(A(1)-E(1)/1.987/TEMP)*VALUE*(C2H6-C2H4*H2*VALUE/K1)
    C2H6 =INTGR(C2H6O,(-R1-R2)*AREA)
    R4 =  EXP(A(4)-E(4)/1.987/TEMP)*VALUE*C3H8
    K2=EXP(16.64-55512./1.987/TEMP)
    R3=  EXP(A(3)-E(3)/1.987/TEMP)*VALUE*(C3H8-C3H6*H2*VALUE/K2)
    C3H8=INTGR(C3H8O,(-R3-R4)*AREA)
    C2H4 =INTGR(C2H4O,(R1+R2/2.+R4)*AREA)
    C3H6=INTGR(C3H6O,R3*AREA)
    H2=INTGR(H2O,(R1+R3)*AREA)
    CH4=INTGR(CH4O,R2*AREA)
    TW=TWO+ALF*TIME-BET*TIME*TIME
    DT=(3.14*DIAM*HTTRC*(TW-T)-(DELH(1)*R1+DELH(2)*R2+DELH(3)*R3+
   1DELH(4)*R4)*AREA)/(C2H6*SPHT(1)+C3H8*SPHT(2)+C2H4*SPHT(3)+C3H6*
   1SPHT(4)+CH4*SPHT(5)+H2*SPHT(6)+STEAM*8.4)
    T=INTGR(TO,DT)
    RETURN
    END
```

them to zero, which is correct.

The four furnace parameters are specified by making four different run statements using the END cards. Notice also that for the PARAM statement specified after the END statement continuation cards are not permitted by DSL. To circumvent this, multiple PARAM statements may be specified. Table 6.23 shows the UPDAT routine produced by DSL, and Tables 6.24 and 6.25 show the results of the four furnaces.

TABLE 6.24

Simulation Results for Four Ethylene Furnaces in Sequence—Furnaces 1 and 2

```
                    *** DSL/1800 SIMULATION DATA ***
PARAM    A    = 44 ,E    = 48 ,DELH = 52 ,SPHT = 56
TITLE SIMULATION OF STEAM CRACKING FURNACE FOR ETHYLENE PRODUCTION
INTEG MILNE
CONTRL DELMI=0.01,DELT=.1,FINTI=200.
RELERR C2H6=.001,T=.001
ABSERR C2H6=.1,C3H8=.1,T=.01
TABLE A(1-4)=43.45,48.,37.2,40.0,E(1-4)=1.3E5,1.5E5,1.1E5,1.143E5,...
      DELH(1-4)=35.E3,9.E3,30.E3,20.E3,SPHT(1-6)=31.,42.,26.8,42.4,...
      18.9,6.67
PRINT 20.,C2H6,C3H8,C2H4,C3H6,CH4,H2,T,TW
PARAM TO=1000.,TWO=1400.,ALF=4.,BET=6.0E-3,STEAM=150.,PRES=2.,DIAM=...
      .34,HTTRC=150.,C2H6O=250.,C3H8O=25.
END
```

SIMULATION OF STEAM CRACKING FURNACE FOR

TIME	C2H6	C3H8	C2H4	C3H6	CH4	H2	T	TW
0.000E 00	0.2500E 03	0.2500E 02	0.0000E 00	0.0000E 00	0.0000E 00	0.0000E 00	0.1000E 04	0.1400E 04
0.200E 02	0.2499E 03	0.2499E 02	0.7071E-02	0.5730E-03	0.6341E-03	0.5004E-02	0.1120E 04	0.1477E 04
0.400E 02	0.2499E 03	0.2496E 02	0.1124E 00	0.6590E-02	0.1604E-01	0.8194E-01	0.1227E 04	0.1550E 04
0.600E 02	0.2490E 03	0.2474E 02	0.1022E 01	0.4482E-01	0.2074E 00	0.7521E 00	0.1323E 04	0.1618E 04
0.800E 02	0.2444E 03	0.2381E 02	0.5758E 01	0.1982E 00	0.1500E 01	0.4221E 01	0.1399E 04	0.1681E 04
0.100E 03	0.2305E 03	0.2156E 02	0.1935E 02	0.5594E 00	0.5881E 01	0.1410E 02	0.1448E 04	0.1740E 04
0.120E 03	0.2068E 03	0.1828E 02	0.4189E 02	0.1077E 01	0.1386E 02	0.3039E 02	0.1476E 04	0.1793E 04
0.140E 03	0.1768E 03	0.1465E 02	0.6964E 02	0.1642E 01	0.2433E 02	0.5041E 02	0.1496E 04	0.1842E 04
0.160E 03	0.1432E 03	0.1107E 02	0.1002E 03	0.2196E 01	0.3652E 02	0.7244E 02	0.1517E 04	0.1886E 04
0.180E 03	0.1073E 03	0.7732E 01	0.1322E 03	0.2707E 01	0.4999E 02	0.9541E 02	0.1540E 04	0.1925E 04
0.200E 03	0.7080E 02	0.4787E 01	0.1641E 03	0.3152E 01	0.6420E 02	0.1181E 03	0.1567E 04	0.1960E 04
0.200E 03	0.7080E 02	0.4787E 01	0.1641E 03	0.3152E 01	0.6420E 02	0.1181E 03	0.1567E 04	0.1960E 04

```
                    *** DSL/1800 SIMULATION DATA ***
PARAM TO=1100.,TWO=1380.,ALF=4.,BET=6.1E-3,STEAM=160.,PRES=2.1
PARAM HTTRC=145.,C2H6O=230.,C3H8O=65.
END
```

SIMULATION OF STEAM CRACKING FURNACE FOR

TIME	C2H6	C3H8	C2H4	C3H6	CH4	H2	T	TW
0.000E 00	0.2300E 03	0.6500E 02	0.0000E 00	0.0000E 00	0.0000E 00	0.0000E 00	0.1100E 04	0.1380E 04
0.200E 02	0.2299E 03	0.6496E 02	0.5488E-01	0.6751E-02	0.4353E-02	0.3053E-01	0.1177E 04	0.1457E 04
0.400E 02	0.2297E 03	0.6479E 02	0.3575E 00	0.3766E-01	0.3933E-01	0.2059E 00	0.1253E 04	0.1530E 04
0.600E 02	0.2288E 03	0.6411E 02	0.1737E 01	0.1551E 00	0.2581E 00	0.1029E 01	0.1325E 04	0.1598E 04
0.800E 02	0.2252E 03	0.6200E 02	0.6610E 01	0.5056E 00	0.1252E 01	0.3998E 01	0.1387E 04	0.1660E 04
0.100E 03	0.2156E 03	0.5737E 02	0.1863E 02	0.1254E 01	0.4196E 01	0.1142E 02	0.1432E 04	0.1719E 04
0.120E 03	0.1986E 03	0.5032E 02	0.3879E 02	0.2373E 01	0.9759E 01	0.2399E 02	0.1461E 04	0.1772E 04
0.140E 03	0.1758E 03	0.4202E 02	0.6469E 02	0.3678E 01	0.1753E 02	0.4031E 02	0.1482E 04	0.1820E 04
0.160E 03	0.1491E 03	0.3340E 02	0.9398E 02	0.5020E 01	0.2696E 02	0.5895E 02	0.1501E 04	0.1863E 04
0.180E 03	0.1195E 03	0.2500E 02	0.1252E 03	0.6314E 01	0.3772E 02	0.7901E 02	0.1521E 04	0.1902E 04
0.200E 03	0.8840E 02	0.1724E 02	0.1571E 03	0.7499E 01	0.4947E 02	0.9961E 02	0.1543E 04	0.1936E 04
0.200E 03	0.8840E 02	0.1724E 02	0.1571E 03	0.7499E 01	0.4947E 02	0.9961E 02	0.1543E 04	0.1936E 04

TABLE 6.25

Simulation Results for Ethylene Furnaces 3 and 4

```
                *** DSL/1800 SIMULATION DATA ***
PARAM TO=1050.,TWO=1410.,ALF=3.95,BET=6.1E-3,STEAM=165.,PRES=2.2
PARAM HTTRC=160.,C2H6O=210.,C3H8O=95.
END


SIMULATION OF STEAM CRACKING FURNACE FOR


  TIME       C2H6        C3H8        C2H4        C3H6        CH4         H2          T           TW
0.000E 00  0.2100E 03  0.9500E 02  0.0000E 00  0.0000E 00  0.0000E 00  0.0000E 00  0.1050E 04  0.1410E 04
0.200E 02  0.2099E 03  0.9497E 02  0.2992E-01  0.4768E-02  0.1507E-02  0.1404E-01  0.1149E 04  0.1486E 04
0.400E 02  0.2098E 03  0.9479E 02  0.2840E 00  0.3822E-01  0.2294E-01  0.1403E 00  0.1242E 04  0.1558E 04
0.600E 02  0.2090E 03  0.9382E 02  0.1794E 01  0.2041E 00  0.2142E 00  0.9256E 00  0.1327E 04  0.1625E 04
0.800E 02  0.2052E 03  0.9037E 02  0.7936E 01  0.7768E 00  0.1272E 01  0.4225E 01  0.1396E 04  0.1686E 04
0.100E 03  0.1946E 03  0.8246E 02  0.2354E 02  0.2049E 01  0.4587E 01  0.1280E 02  0.1442E 04  0.1743E 04
0.120E 03  0.1762E 03  0.7073E 02  0.4874E 02  0.3904E 01  0.1070E 02  0.2693E 02  0.1471E 04  0.1796E 04
0.140E 03  0.1523E 03  0.5737E 02  0.7979E 02  0.5994E 01  0.1900E 02  0.4465E 02  0.1492E 04  0.1843E 04
0.160E 03  0.1246E 03  0.4380E 02  0.1139E 03  0.8094E 01  0.2893E 02  0.6446E 02  0.1512E 04  0.1885E 04
0.180E 03  0.9453E 02  0.3089E 02  0.1494E 03  0.1007E 02  0.4016E 02  0.8537E 02  0.1535E 04  0.1923E 04
0.200E 03  0.6324E 02  0.1930E 02  0.1844E 03  0.1183E 02  0.5226E 02  0.1063E 03  0.1562E 04  0.1955E 04
0.200E 03  0.6324E 02  0.1930E 02  0.1844E 03  0.1183E 02  0.5226E 02  0.1063E 03  0.1562E 04  0.1955E 04


                *** DSL/1800 SIMULATION DATA ***
PARAM TO=1000.,TW=1350.,ALF=4.,BET=0.64E-2,STEAM=200.,PRES=2.
PARAM HTTRC=145.,C2H6O=100.,C3H8O=185.
END


SIMULATION OF STEAM CRACKING FURNACE FOR


  TIME       C2H6        C3H8        C2H4        C3H6        CH4         H2          T           TW
0.000E 00  0.1000E 03  0.1850E 03  0.0000E 00  0.0000E 00  0.0000E 00  0.0000E 00  0.1000E 04  0.1410E 04
0.200E 02  0.9999E 02  0.1849E 03  0.1147E-01  0.2610E-02  0.1326E-03  0.3617E-02  0.1098E 04  0.1487E 04
0.400E 02  0.9998E 02  0.1848E 03  0.1134E 00  0.2325E-01  0.2352E-02  0.3585E-01  0.1191E 04  0.1559E 04
0.600E 02  0.9987E 02  0.1842E 03  0.7570E 00  0.1402E 00  0.2605E-01  0.2432E 00  0.1278E 04  0.1626E 04
0.800E 02  0.9921E 02  0.1813E 03  0.3714E 01  0.6268E 00  0.1932E 00  0.1221E 01  0.1356E 04  0.1689E 04
0.100E 03  0.9668E 02  0.1726E 03  0.1314E 02  0.2049E 01  0.9352E 00  0.4430E 01  0.1417E 04  0.1746E 04
0.120E 03  0.9063E 02  0.1553E 03  0.3274E 02  0.4807E 01  0.2882E 01  0.1128E 02  0.1457E 04  0.1797E 04
0.140E 03  0.8098E 02  0.1314E 03  0.6088E 02  0.8560E 01  0.6174E 01  0.2140E 02  0.1483E 04  0.1844E 04
0.160E 03  0.6865E 02  0.1046E 03  0.9366E 02  0.1272E 02  0.1054E 02  0.3352E 02  0.1505E 04  0.1886E 04
0.180E 03  0.5438E 02  0.7748E 02  0.1283E 03  0.1690E 02  0.1578E 02  0.4673E 02  0.1527E 04  0.1922E 04
0.200E 03  0.3878E 02  0.5160E 02  0.1628E 03  0.2084E 02  0.2174E 02  0.6031E 02  0.1552E 04  0.1954E 04
0.200E 03  0.3878E 02  0.5160E 02  0.1628E 03  0.2084E 02  0.2174E 02  0.6031E 02  0.1552E 04  0.1954E 04
```

6.8.2 *Further Exploration in Ethylene Furnace Simulation*

For students interested in further exploring ethylene furnace simulation, the following problem variation is suggested: Use SPHT (i) as a function of temperature as in

$$SPHT(I,TEMP) = C1(I) + C2(I) \cdot TEMP + C3(I) \cdot (TEMP)^2$$

where C1, C2, and C3 are STORAG variables whose values are specified by TABLE and the SPHT (I,T) function statement is placed in the UPDAT card deck prior to the first executable statement. In this case it will be just before the area calculation.

The next variation may be to calculate the total heat required in the furnaces, which is given by

$$\int_0^L dQ = \int_0^L \pi dh (T_w - T) dt \qquad (6.34)$$

This will require the specification of additional integration, parameters for which are already specified. Notice that the initial value of Q at $Z = 0$ is zero. The third variation for the student may be to provide the heat of reaction as a function of temperature, such as

DELH(I,TEMP) = D1(I) + D2(I) + D2(I) · TEMP + D3(I) · TEMP · TEMP.

and include these functions in the UPDAT program in the same manner as the SPHT (I, TEMP) function.

In this chapter as well as in Chapter 5 we have introduced the reader to a large variety of examples with differing degrees of complexities to enable the user to become familiar with the utility of simulation as well as the DSL language for small computers. There are many more examples that exist in the field which could be easily solved with the use of DSL (or CSMP on the IBM 360). The list of references in Appendix D will be helpful to the reader, who we hope by now has become an avid believer in digital simulation.

The next three chapters are of importance to the reader, who must learn the setup mechanism so as to run his simulation with the DSL program, must know the use of data switches, and must make sense of the error messages printed by the program on small computers. Also, the advanced language features of DSL as well as the mathematical bases of the subroutines provided are shown for the more advanced user. Users who will be using CSMP on 360/370 should skip this chapter and instead refer to their appropriate program manuals for the discipline to be followed in program use.

Chapters 10 through 13 are for the computer scientist or systems programmer who will install DSL on his system, be it an IBM 1130, IBM 1800, IBM System/3, IBM System/7, or other equivalent small computer. These chapters will certainly not be fascinating to the user who already has someone to install the DSL program system on the computer he is using.

7 SETUP AND EXECUTION OF THE DSL PROGRAM

7.0 Introduction

In this chapter, we shall discuss the procedures for setting up and executing the DSL simulation program as well as running the simulation program generated by DSL from a user's input to DSL. The rules for the user's input are listed in Chapter 4. We shall assume here that the loading of DSL programs on the system disk storage (1130, 1800, System/7, or System/3) has been performed with the help of a systems programmer, following the guidelines discussed in Chapter 9. For users of the IBM 7090 or the program CSMP 360 (15) this procedure is of no value since the user will be running his simulation job in a *batch* environment where he will submit his job using procedures established by his computer operations department. For the user with a small system other than the four IBM systems listed above, the procedures will depend on the operating characteristics of the system on which the DSL program was installed by the system's programmer.

7.1 Run Setup on the 1800 and 1130

To execute the DSL program beginning with translator input (DSL language statements) two cards appear ahead of the DSL statements. This assumes that if the 1800 system is used it has been cold-started and that the system is ready for job execution. Thus, for the 1800 system the card input will be as shown in Table 7.1.

In most running systems of the 1800, the system monitor will require that the program switch 7 be in the "on" position. After loading the required JOB cards followed by DSL statements and blank cards behind the DSL statements in the card

TABLE 7.1

Card Deck Sequence for Execution of DSL on the IBM 1800

COLD   START CARD 1
COLD   START CARD 2   (for MPX systems)
// JOB   (with input for multiple disks)
// XEQ DSL
(DSL translator input statements)
(blank cards)
// END

---

reader, the user hits the CONSOLE INTERRUPT button. This will then start the batch program processing in the 1800. The user must discuss the proper procedure with the systems programmer who is responsible for the 1800 system to confirm this procedure step.

For the 1130 the cards required are shown in Table 7.2. Notice that a cold start card precedes the // JOB card. Depending on where the DSL program disk is mounted, a specific data switch position will be required to be on. For the 1130 systems with only a single disk or when the DSL program disk is mounted in the main frame of the 1130 system (drive 0), all data switches must be down. When the disk is mounted in the disk unit outside the 1130 main frame, either data switch 15 or data switch 14 must be turned on depending on whether it is in the upper mount (drive 1) or lower mount (drive 2).

Load the cards in the card reader, start the card reader, and press the IMMEDIATE STOP followed by the RESET and PROGRAM LOAD buttons on the 1130 console. If the printer is in READY status, the DSL program execution will begin. If the printer is not READY, bring it in READY status and press the START button on the 1130 console.

The procedures for execution on System/3 are sufficiently different and therefore are treated in Chapter 11, where the system generation as well as execution are described.

Note that although in the DSL input of the user the data and structure statements may be intermixed and in any order for input to DSL, the translator pseudo-operations must appear before any use of the affected variables.

The pseudooperations are described in Chapter 4. Thus, for example, if a variable KX is to be used as an integer rather than a floating-point variable, its

TABLE 7.2

Card Deck Sequence for DSL Execution on the 1130

COLD   START CARD   (DM version 1 or 2)
// JOB
// XEQ DSL
(DSL translator input)
(blank cards)

---

declaration as INTGER in DSL must precede its use. Similarly a variable OUT defined as a function in a procedural block via the use PROCED statement in DSL must come prior to its use in a subsequent DSL statement.

Two cards must appear in sequence: an END card, which identifies the end of each parameter study, and a STOP card, which identifies the end of DSL input. Place blank cards after the STOP card. You will need about 20 more blanks than cards in the original problem input. Structure statements are sequenced and translated to a FORTRAN subroutine called UPDAT, which represents the model to be simulated. UPDAT is punched with appropriate control cards for direct execution of the simulation. The data statements which are used at the execution of simulation after appropriate program load module buildup appear last in the punched output.

7.2 User Subroutines

If the user calls for a subroutine or a function in the DSL statements which is not in the DSL or FORTRAN library of the 1130/1800 system, he must compile and store his SUBROUTINE or FUNCTION prior to execution of the punched card deck by DSL. Table 7.3 shows a typical FORTRAN or assembler program, which must be placed after the first card punched by DSL which is the // JOB card. If the subroutine is only loaded temporarily, remove the WS and UA in *STORE cards for the 1800 and place a T in column 9 of the *STORE card. For the 1130, simply remove the // DUP and *STORE cards. For System/3, follow the procedures described in Chapter 11, which describes equivalent FORTRAN control statements required to store the user subroutines.

TABLE 7.3

Typical FORTRAN Subroutine Inserted During Execution of the Simulation Phase

```
// Job
// FOR NAMEA
*ONE WORD INTEGERS
*LIST ALL
        SUBROUTINE NAMEA(A,B)
        (FORTRAN STATEMENTS)
        RETURN
        END
// DUP
*STORE          WS   UA   NAMEA
```

(Insert here the DSL UPDAT subroutine deck and the control cards punched by the DSL translator.)

7.3 Run Execution on the 1130 and 1800

When the translator comes to a WAIT state in the machine, place the newly punched cards in the reader and hit START on the card reader. For the 1130 system, also press START on the console. If a / / JOB card for a two-disk system must be used, then that card must be substituted for the first card punched by DSL, or, alternatively, the format in the DSL source OUTP *must be* altered. (See card DSL02900.) For the 1130 you may also do a cold start (for restart) at this point by placing the cold start card ahead of the first card punched, / / JOB (or by interrupting for restart at a / / JOB step). This permits you to alter the model or structure description at the FORTRAN level without rerunning the translator.

After the FORTRAN compilation of the UPDAT deck punched by DSL, a final core load is built which includes the model description (UPDAT) and integration routines. This core load is stored in core image format as SIMUL using the main routine INTEG. If there has been a previous use of DSL, the core load SIMUL for that model is deleted. DSL data follow the UPDAT subroutine. Note that for the System/3 the system will halt with a halt code "6E" prior to storing the newly built core load on the user's disk. When the user presses START on the user console the previous core load will be replaced.

In the punched output the simulation data cards follow:

$$/ / \text{XEQ SIMX} \qquad \text{FX}$$

Again, the FX is not used in the 1130 format, and therefore the card must be repunched with FX deleted. Any consecutive run may cold-start (or restart) at this point (i.e., the UPDAT subroutine does not have to be recompiled).

7.4 User Interaction with the 1130 and 1800

As pointed out in earlier chapters, the greatest facility that is available to an engineer and a programmer on a small scientific computer is the man-machine interaction which is missing when one uses a batch run facility of a large scientific computer. Thus, any data errors may be corrected, parameter values may be revised, and runs may be aborted by the engineer using the data switch facilities of the 1130, 1800, or System/3 to perform these functions. If the user follows the directions for use of the digital input facilities of the System/7, the data switches built by him may similarly be used on the System/7.

7.4.1 Modifying the Source Program UPDAT

The DSL translator facilitates, in addition, changing of program data cards and even erroneous FORTRAN statements in the UPDAT deck punched by the translator. If the user will interpret his DSL punched deck, he may also be able to add FUNCTION statements in the UPDAT deck itself, thus eliminating the need for separately compiling a required FUNCTION prior to simulation and storing it on the disk. This facility was demonstrated in Chapter 6, Section 6.7, in the example of simulation of an ammonia synthesis reactor. The FORTRAN rules require that the statement which defines the function must be placed before the first executable statement. This means that it will generally follow the last continued statement describing COMMON variables in the UPDAT deck punched by the DSL translator. One further rule of FORTRAN function statements is that only *one* value of the output or dependent variable may be returned, although multiple input or independent variables may determine the output value.

Thus, a function Y = F(X, W, T, P) may be described by a function statement. However, a function Y, Z = F(X, W, T, P) must always be described using the subroutine facility of FORTRAN. Furthermore, the dependent variable must be defined by a single algebraic statement, and the output or dependent variable must be defined by a single algebraic statement; otherwise, the FUNCTION subroutine facility of FORTRAN must be used with separate compilation, where multiple FORTRAN statements including logic, branches, etc., as shown in Table 7.3, may be used.

Note that although the FORTRAN function statement does not permit a function definition such as

$$Y1, Y2 = F(X, Z, W)$$

the DSL function definition facility will permit such a definition.

For example, an input statement to DSL of the form

$$Y1, Y2 = FCN(X, Z, W)$$

will generate, in the UPDAT subroutine punched by DSL, a statement,

$$CALL\ FCN(X, Z, W, Y1, Y2)$$

Thus, a separate subroutine FCN must be compiled in FORTRAN, which will, based on input values of X, Z, and W, return values of Y1 and Y2.

In addition, the PROCED facility of DSL also permits a function block definition within the user's DSL statements to define functions, thus eliminating the need for separate compilation. This facility is discussed in the procedural blocks section of Chapter 8.

7.4.2 Interaction with the 1800

In certain real-time process control applications of the 1800, the data switches may be used for special real-time functions, such as preparing data logs on operator demand. To avoid use of the data switches on the 1800, then, the DSL program calls a subroutine, REMSW, which in turn calls the FORTRAN subroutine

DATSW. The systems programmer on the 1800 then must be consulted to provide appropriate hardware data switches and the program support to rewrite the subroutine REMSW, which examines the status of these hardware switches (via digital inputs) and provides a value of 0 or 1 to the subroutine. In general, most systems programs on the 1800 will not be using the data switches 1 to 6, which are used by DSL for user-machine interaction.

7.4.3 Interaction with the 1130

No problems exist for the 1130, since the 1130 is a batch machine and the user will have control of the use of the machine during simulation. One exception may be where the 1130 is connected to an IBM System/7 via the SAC channel (12) and operating in a time-shared mode with the system support of 1130 DSP (12). In this case, some program changes in the source of the assembler language portion of DSL will be necessary since the location of the COMMON block is different from the normal COMMON location under the standard 1130 monitors (disk monitor versions 1 and 2: DMV1, DMV2). The user, therefore, must consult with the systems programmer prior to system generation to resolve this situation. Again, in most installations the data switches 1 to 6 will be available for user interaction.

7.4.4 System/3–User Interaction

The data switches on the System/3 are rotary dials and not individual switches. An equivalent table for use of the rotary dials on the System/3 to perform the same function as the 1130 data switches is shown in Chapter 11 describing System/3 operation. These data switches may be used during program execution for DSL. However, these data rotary switches are also used for follow-up action after a system error occurs on System/3. Fortunately, only the last two of the four rotary switches are used for system error actions, and the DSL program uses only the first two, so that during program execution no conflict occurs.

7.4.5 Use of the Data Switches

With the data switches, then, the user may add, delete, or alter data cards for modifying problem parameters. He may modify parameters using typewriter keyboard input, for example, for multiple runs in a two-point boundary-value problem, as shown in some examples in Chapters 5 and 6.

Data switch 1 is used for interrupting a run in the middle of a simulation run; i.e., the simulation will be halted at the completion of an integration interval. When the halt is desired at the end of a parameter study, to transfer the input, for example, from card reader to the keyboard, switch 2 must be used after the simulation has started by reading the first set of parameters from the card reader.

Note also that data switch 2 may be used for starting the entire DSL translation and simulation using the keyboard as the input device rather than the card reader. Data switch 3 permits entry through the keyboard in a special format. Data

TABLE 7.4
Data Switch Control

Data Switch	Function
0	Not used.
1 ON	Interrupts execution of the simulation at the completion of an integration interval. Switches 2 and 3 are used in conjunction with 1 to identify format and unit to be used for input.
2 ON	Data input from typewriter keyboard.
2 OFF	Data input from card reader.

Note: Switch 2 is used both for initial entry of data and structure statements and for reentry after interruption or error.

3 ON	Keyboard data input format is as specified in the subroutine TYPIN as A = N, where A is a 5 = character symbol and N is a 12-digit number in FORTRAN E or F format (2A2, A1, 1X, E12.6). This format may be changed (see System Loading Instructions).

Example:

Column:	1	6	18
	XDOT=		200.
	XDOT =		2.0E02
	XDOT =		200.0000
not XDOT =			200.
not XDOT =			200.

Systems variables TIME, DELT, DELMI, FINTI, and DELTP must be changed in the same format. TIME not equal to zero shows continuation.

3 OFF	Allows all DSL data statements to be entered by keyboard or card in standard format.

Note: The advantage to the simple format is that it may be entered through a subroutine existing in the final link and does not cause reloading another core load. In both cases entries are continued until an END is typed or read from cards in columns 1-3.

4 ON	Loops through drawing of the graph axes and label (for adjustment of scope). Set switch off to get out of loop.
5 ON	Redraws graph axes and label. (To be used any time after the first plot or in switching from scope to 1627 output.) If used after the first 1627 plot, the paper will be rolled forward to 3 in. past the horizontal axis prior to drawing the new axes.
6 ON	Suppresses printing of data and structure statements as they are read.

switches 4 and 5 permit looping through the graph axes, printing and redrawing of graph axes, and reprinting of labels on the graph. Switch 6 is for suppressing data printing.

Table 7.4 shows data switch control and the related functions. The greatest use will be made of data switches 1 and 2 by a normal user. The user may start a run from the last calculated point or restart from time = 0 with graphic output superimposed or redrawn.

7.5 Error Messages, Warning Messages and User Information Messages

Errors may occur during the translation phase of the DSL program where the user may have erroneous input. The error messages appear immediately following the erroneous DSL input. The user must carefully check these messages, some of which are warning messages. The translator will provide a punched deck of UPDAT subroutine, in some cases even with erroneous input. For example, the user may decide not to provide data for some parameters, whose value the translator will default to zero. A warning message will state the default.

Table 7.5 shows the various error messages, together with the DSL routines from which the message is printed. The card sequence number is also shown for those users who will change the maximum number of symbols, outputs, inputs, and integrators, as shown in Chapter 10. These cards will require modification if the user wishes to provide more accurate error messages reflecting his changes.

Messages 8, 9, and 10 are not errors but indicate certain information which may be useful to the user. The symbols XX, YY, XXX, and AAAAA will be replaced by the program with either numbers or user input symbols.

The error messages 11 to 16 occur during execution of the user's simulation program. These errors may result from erroneous input of data cards, punched by the user for runs with modified parameter values, or they may result from erroneous input from the keyboard by the user while making multiple parameter runs.

For error numbers 1, 11, and 12, Table 7.6 gives a further breakdown in the error type, listed under function. Note error 28 (YY = 28), which may occur when either the value of DELMI is specified too large by the user or the function being integrated has sharp changes in derivative values.

Figure 7.1 shows a generalized flow diagram of the execution sequence of the DSL program which will assist systems programmers in pinpointing user's errors, if the user is unable to resolve his errors at execution time, either in the translation phase or simulation (run) phase.

While this chapter provides essential data for the DSL user who is a beginner, whether he be a 1130, 1800, System/3 or System/7 DSL user, Chapter 8 is for the

TABLE 7.5

Error Messages from DSL Program

Message Number	Message Translation Phase	Routine	DSL Card Sequence Number
1	ERR IN COL XX, DSL ERR YY (see detail description in Table 7.6 for YY)	ERR1	00090
2	CHECK MAX vs. COMMON VARS XXX	OUTP	02864
3	XXX SYMBOLS, REDUCE TO 140	OUTP	04940
4	PARAM NOT INPUT, SET=0 AAAAA	SORT	08150
5	MORE THAN 120 OUTPUTS IN SORT	SORT	08540
6	MORE THAN 30 OUTPUTS IN SINGLE LOOP	SORT	08840
7	UNDEFINED IMPLICIT LOOP	SORT	09720
8	OUTPUT VARIABLE SEQUENCE		
9	STORAGE USED/MAXIMUM		
10	INTGR XX/50, etc.	SORT	09770
	Simulation Phase		
11	DSL ERR XX	ERR	31002
12	DSL ERR XX, COL. YY (see detail description in Table 7.6 for YY)	ERR	31010
13	DATA TYPE NOT A RESERVE WORD AAAAAA	INTRA	33870
14	UNDEFINED VARIABLE AAAAA	INTRA	34120
15	DATA ERR, START TO EXECUTE	INTRA	35740
16	SYMBOL NOT IN TABLE AAAAA	TYPIN	49030

user who wishes to become familiar with the system and who wants to utilize some of the advanced language features of the system. Even for the beginner, some of the examples in Chapters 5 and 6 showed use of the advanced features, which are described in detail in Chapter 8.

TABLE 7.6

Detailed Description of Error Types

Error Number (YY)	Subroutine	Card Sequence Number	Function
Translator			
4	NUMRC	DSL00320	Illegal character in numeric field
1	SCAN	06070	Illegal character in alphameric field
10		07270	More than 200 symbols More than 20 real variables
10	SORT	08170 09702	More than 60 parameters More than 60 parameters
6	TRANS	12150 13400	Data card error—skip card Illegal character in structure statement—skip card
8		13701	Multiple output block with less than 3 characters before '='
10		15060	Data words more than 1000 Parameters more than 60 Memory blocks more than 15 Storage variables more than 10 Integer variables more than 10 Integrals (special) more than 10 FORTRAN words more than 5000 FORTRAN words per statement more than 200 Output variables more than 100 Integrators more than 50 Input variables more than 300

TABLE 7.6 (Continued)

Error Number (YY)	Subroutine	Card Sequence Number	Function
Simulator			
15	ALPHA	30340	Illegal character in alphameric name—skip card
16	INTRA	34310	More than 25 PRINT variables (skip all more than 25)
17		34630	More than 15 RANGE variables (skip all more than 15)
18		34740	More than 5 FINISH variables (skip all more than 5)
19		36090	Error in TABLE data card
21		36690	No output specifications; continue
28	MILNE	38930 or 39764	Minimum integration interval exceeded; interval halved and run continued
20	NUMER	42570	Illegal character in numeric field—skip card
28	RKS	46614	Minimum interval in integration exceeded; halved and run continued
33	AFGEN	60180	Below minimum argument in table
34		60290	Above maximum argument in table
31	DELAY	62340	DELAY block error; check input parameters
30	IMPL	63750	IMPL block error
32	LOOK	65060	Symbol not in table
33	NLFGN	65490	Below minimum argument in table
34		65600	Above maximum argument in table

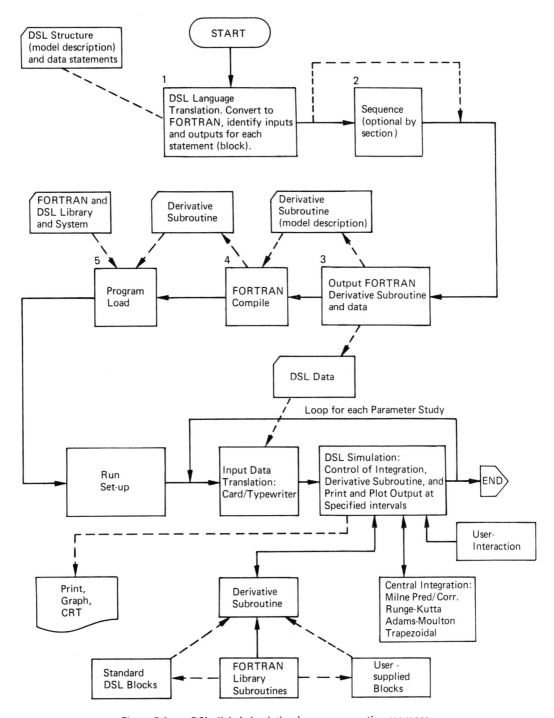

Figure 7.1.　DSL digital simulation language execution sequence.

8 DSL ADVANCED LANGUAGE FEATURES

8.0 Introduction

In this chapter we shall describe certain language features in DSL with a more detailed explanation. Descriptions are given below for the following: centralized integration, statement sequencing, procedural blocks, and optional user variations. The last heading includes additional information on data storage, memory blocks, implicit functions, and function generators. The overall program for diagrams in DSL was shown in Fig. 7.1 (Chapter 7) and may be referred to again. The program linkage between the various DSL program modules is shown in Fig. 8.1.

8.1 Centralized Integration

The block name INTGR is used to specify that centralized integration is specified. During execution of the simulation, all inputs to the integrators are computed and stored consecutively by INTGR. However, computation of outputs is bypassed until the end of the iteration cycle. At this time, all integrator outputs are updated simultaneously. That is, the derivative and integration routines are separated and then executed independently to compute all integrator inputs or outputs consecutively for each iteration cycle. A choice may be made between the fifth-order Milne predictor-corrector, fourth-order Runge-Kutta, trapezoidal, or Adams-Moulton integration methods. If none of the above methods is specified, the Runge-Kutta scheme with interval control is used. The first two methods allow the integration interval to be adjusted by the system to meet a specified error criterion. This

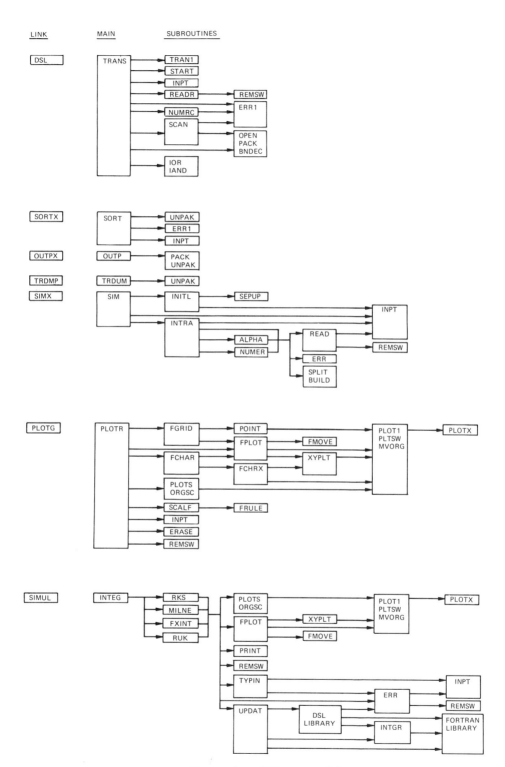

Figure 8.1. DSL program linkage.

factor allows the integration routine to take large or small steps, depending on the rate of change of one or more variables. A different error bound may be specified for each integral. If this specification overflows the statement beyond column 72 in a card, the continuation feature of DSL should be used. The error criteria used to vary the integration interval are as follows: For the Milne fifth-order predictor-corrector integration method,

$$|y^c - y^p| \leqslant R|y^c| \quad \text{for } |y^c| \geqslant 1$$
$$|y^c - y^p| \leqslant A \quad \text{for } |y^c| < 1 \tag{8.1}$$

where y^p and y^c are the values of the Milne predictor and corrector formulas and R and A are the input values of relative and absolute errors. For fourth-order Runge-Kutta integration,

$$\frac{|y - y^s|}{A + R|y|} \geqslant 1 \tag{8.2}$$

where y and y^s are the values of the integral as computed, respectively, by the Runge-Kutta and Simpson's rule integration methods and A and R are the input values of absolute and relative errors.

The use of centralized integration is coordinated closely with the DSL sort technique. By placing the integrator last in each loop and by merging the functions of integration and prediction (advancing an output one time step ahead) blocks within the integration loop are computed at the same time step. This permits a larger integration interval with a corresponding decrease in computer run time, and, in addition, it removes the user's responsibility of deciding which blocks should be advanced in each closed loop. If a closed loop exists which does not contain an integrator, the user may define this as an implicit function, described later in this chapter, to eliminate time delays.

A user may try procedural integration by not using the systems name INTGR for the integration subroutine. In this case, the UPDAT subroutine is called twice by the simulator with the independent variable equal to zero to set initial conditions; then the simulation is executed with only the UPDAT subroutine. If there is integration, a subroutine for performing the integration must be supplied if INTGR is not used.

8.2 Statement Sequencing (Sorting)

A nonprocedural input language such as DSL transfers the responsibility for establishing the execution sequence from the user to the program. To accomplish this process, DSL arranges the sequence of input statements according to the following rule. An operational element (or statement) is properly sequenced if all its inputs are available either as input parameters or initial conditions or as previously computed values in the current iteration cycle. Since integration takes place prior to each iteration, all integrals are considered as known values. Unspecified algebraic loops are identified, and if there are any, the run is halted. The result of this sequencing operation is a properly organized FORTRAN subprogram.

Each loop involving integrators is sorted independently of all others. The loops appear within the program in the same sequence as the original input. For example, the first loop ends with the first integrator and includes all statements involving the computed variables necessary to calculate the integrator input value (except other integrator outputs, which may appear earlier or later in the program).

8.2.1 Configuration Switching

By using the SORT and NOSORT options, the user may sort only sections of his program. For example, a NOSORT section in the first part of the user's DSL statements may be used for run initialization, with the parameter values computed only once; a central NOSORT section may be used to test the run response for the purpose of switching portions of the configuration into or out of the simulation to decrease the run time or alter the information flow. For a particular case, it may be desirable to switch specific elements out of the simulation model so that any problem structure variations may be included in a single run. The FORTRAN "computed GO TO" statement may be used for branching to various sequences of statements within NOSORT sections. The control variable may be computed or input on a PARAM card but must be declared as an integer as shown below. All calculations other than integrals (integrator outputs) may be switched out. Note that the integrals must be computed in every iteration cycle.

An example of FORTRAN branching in a NOSORT block is given below:

```
INTGER  KEY
        GO TO (1, 2, 3), KEY
```

where the value of KEY is specified by a PARAM card in a given run.

Examples of the use of NOSORT and the computed GO TO statement of FORTRAN are shown in Chapters 5 and 6.

8.3 Procedural Blocks

DSL is a nonprocedural language because execution does not depend on input statement order. Structure or connection statements are separated from data and sequenced by the DSL translator (except where a NOSORT option is exercised). In contrast, FORTRAN is a procedural language, where the execution of statements by the computer is influenced by the statement position. To gain the advantage in instances of procedurality, DSL allows a pseudooperation to the translator which sets aside a block of code as procedural. These pseudooperations, PROCED and ENDPRO, punched in columns 1-6, designate the beginning and end of the procedural block of code. Input and output parameters *must* be specified in the PROCED card. This specification enables the procedural block of code to be sorted as a whole relative to the remaining DSL statements. Any DSL or FORTRAN statement may be used within PROCED. Continuation within PROCED is as in FORTRAN with column 6 nonblank. A procedural block may be defined as follows:

1-6	7-72
PROCED	VALUE=BLOCK (TRIG, IN)
	IF (TRIG) 1, 1, 2
1	VALUE=LIMIT (PAR1, PAR2, IN)
	GO TO 3
2	VALUE=IN+TRIG
3	CONTINUE
ENDPRO	

During the sequencing of DSL statements, the procedural statements shown above are treated as a single functional block with the output VALUE and the inputs TRIG and IN. Thus, this block of statements will follow evaluation of TRIG and IN and precede the statements using VALUE. The order of statements within the procedural block remains unchanged. In this example, LIMIT is a standard DSL library block. If an entire program is procedural, as in FORTRAN, or if a section is not to be sorted in relation to other blocks, the NOSORT option should be used instead of PROCED. This option ensures proper sequencing whether statements are DSL or FORTRAN and whether or not centralized integration is used.

8.4 User Variations

Because most of the DSL programs are written in FORTRAN and organized into functional subroutines, a user may easily be able to modify any portion of the system. In addition, at execution time certain areas of the system may be modified by simple changes of the DSL input. Areas permitting this flexibility include (1) location and alteration of simulation variables from a user-supplied program (which may be during the course of a run), (2) control of the simulation from a user-supplied MAIN program, (3) addition of functional blocks to become a temporary or permanent part of the DSL language, and (4) alteration of the implicit function convergence criterion.

8.4.1 Data Storage

Data in DSL are stored primarily in a single vector and are allocated dynamically (i.e., at execution time) according to what portions of the simulator are used and how many integrators, tables, and structure variables are in the simulation model. The user must know where certain data are stored if he intends to use the storage area for nonstandard purposes. The variables in COMMON are shown as

COMMON NALRM, NINTG, NSYMB, I1(8), KPOIN, I2(3), KEEP, I3(818), C(1)

COMMON "C" contains the current value of all DSL simulation and systems variables and function table values. The appearance of COMMON N(834), X(1) in any subroutine makes X equivalent to C in the DSL UPDAT subroutine. This gives the user access to the current value of all his structure variables. Access to a specific variable is described in Section 8.4.2.

COMMON contains, respectively, a switch, NALRM, the number of integrators, and the number of simulation variable names (symbols). NALRM is set negative if the implicit function fails to converge, zero to signify convergence, and positive to cause looping to compute the implicit function. It is also printed as the error number in the ERR1 routine. The twelfth and sixteenth variables in COMMON are KPOIN, which contains the index of available storage in COMMON "C," and KEEP, which is set equal to 1, -1, or 0 by the variable step integration routines if the computation is for the next, past, or present value of the derivatives. KEEP is used by routines such as DERIV and DELAY to signify that the value is to be retained, disregarded, or used to replace the value at the last iteration. A user may advance KPOIN to give himself work storage.

8.4.2 Locating DSL Variables

The variables of the user are stored in a COMMON area by DSL starting at a particular location. This is the COMMON "C" area. To locate a particular variable for use in a subroutine of the user, a LOOK function is provided in DSL. The user's subroutine can make use of the LOOK function as follows: Three words are required for each variable symbol which occupies up to five characters. The symbol may be identified by using a REAL (first letter not I-N) or INTEGER (first letter I-N) variable.

Since in FORTRAN each REAL variable occupies two words, if the symbol name to locate the variable is used as a REAL variable, a dimension of 2 is required for the symbol variable. If the symbol variable is an integer variable, a dimension of 3 is required.

8.4.2.1 An Example Using the LOOK Function

Let us take, for example, two variables XDOT and FCN which are to be located in the user's DSL code. If the symbol XDOT is to be used for variable XDOT and the symbol NAME for variable FCN, then the following dimension statements are required:

$$\text{DIMENSION XDOT(2), NAME(3)}$$

followed by a COMMON statement:

$$\text{COMMON I(834), X(1)}$$

where X(1) is the first user variable location.

Now follow this by the DATA statement:

$$\text{DATA XDOT/'XDOT'/, NAME/'FCN'/}$$

Note that, as is the case in XDOT, it is not required that the variable name itself be used as the symbol name. However, the dimension of 2 or 3 is required depending on whether REAL or INTEGER symbols are used to locate the DSL variable. To locate the index of X in the COMMON location where the XDOT and FCN are located, the statements

K = LOOK (XDOT)
N = LOOK (NAME)

are used.

Thus, X(K) is the reference to the variable XDOT, and X(N) will refer to the variable FCN. Note that in the above DATA statement, 'FCN' and 'FCNbbb' will generate the same DATA values for NAME, since DSL refers to six alphanumeric characters, and blanks are inserted to the right by the DATA statement.

Now suppose that in the example above FCN is a function generated or described by an AFGEN or NLFGEN data card. The LOOK facility of DSL can be used by the user to actually alter the function data via a user's subroutine. The value of FCN appears in COMMON starting at some location J, consecutively as pairs of independent and dependent variable values. Let the function be described by the card

AFGEN FCN = 0., 5.0, 0, 2., 8.0, 3.0, 9.5 . . . . .

In the above example N = LOOK (NAME) will provide X(N) referring to the first value of the independent variable in the function FCN, which in this case is equal to zero. In addition, X(N-1) points to the index of the last independent variable in the table. Thus, if the user provided 20 pairs of values in the AFGEN card, X(N-1) will contain that number. Note, however, that to extract the value of the index, an additional operation of converting the REAL value of X(N-1) to INTEGER is required, as, for example,

$$KX = X(N-1)$$

The following example shows the use of a user subroutine which doubles the dependent variable values of all points in the AFGEN function:

```
              SUBROUTINE DOBLE
              DIMENSION NAME (3)
              COMMON I(834), X(1)
              DATA NAME/'FCN'/
C FIND        INDEX FOR FIRST INDEPENDENT VALUE
              N = LOOK (NAME)
C FIND        INDEX FOR LAST INDEPENDENT VARIABLE IN FCN
              M = X(N-1)
              DO 1 I = 1, M
C CALCULATE INDEXES OF SEQUENTIAL DEPENDENT VARIABLES
              L = (I-1) * 2+N+1
    1         X(L) = 2. *X(L)
              RETURN
              END
```

The data on AFGEN or NLFGEN statements in DSL can also be modified by using a TABLE data card, which is shown in the following section on tabular data. Note in the above discussion that the first variable X(1) in COMMON is TIME followed by DELT, DELMI, FINTI, DELTP, and DELTC. If the INTGR function of DSL is used, these system variables are followed by all integration outputs consecutively in the order of computations. To address or use TIME in the user's subroutine for modifying a parameter dynamically, the COMMON variable X(1) may be used as shown below:

```
COMMON I(834), X(1)
EQUIVALENCE (X(1), TIME)
```

8.5 Tabular Data

In some cases the user may find that data may be handled more conveniently in tabular form. This applies particularly to lengthy subroutine argument strings, function input, history data, and DSL block input parameters. The method of using these tabular data is illustrated in the example shown below:

```
AFGEN FCN = 3.5, 4., 5., 12., 8.0, 10.0
STORAG IC(3), PAR(8)
       Y = FUNCT (IC(1), PAR(1), PAR(5), X)
TABLE IC(1) = 5.0, IC(2) = 0., IC(3) = 0., . . .
       PAR(1-8) = 4.0, 5*6.2, 8.4E-6, -2.3
*CHANGE THIRD VALUE IN AFGEN TABLE
TABLE FCN(3) =5.2
```

In this example we see that

1. Three initial conditions IC and eight parameters PAR are required. These dimensions are declared in the STORAG statement.

2. The values of IC and PAR are supplied by the TABLE data card. Note that repeated identical values of the parameter are indicated by the '*' character. The TABLE and STORAG cards go together.

3. Any value in the AFGEN or NLFGEN card may be altered by the TABLE card as shown above. Note that the value changed will be the *third* entry in the table, which in this case will be the *second* value of the independent variable.

It might be of interest to the reader that in the equation for Y the two arguments of the FUNCT, namely, PAR(1) and IC(1), could be simply written as IC and PAR since 1130/1800 FORTRAN will automatically interpret them as the first entry in the dimensioned variables. The use of tabular data was shown in the example in simulation of the ethylene furnace in Chapter 6.

The following restrictions apply to use of tabular data:

1. The maximum size of the table for a variable must be shown in a STORAG data statement before the name is used in a TABLE statement. The names, in the example above IC and PAR, may be used as subroutine arguments or subroutine inputs, either subscripted to denote a particular element in the table or without subscripts to denote the first element in the array. The facility of FORTRAN also allows the use of all elements in the tabular array if the user in his code for the function FUNCT declares PAR and IC as with DIMENSION PAR(8), IC(3).

2. In the DSL statements, the dimensioned variable *cannot* appear subscripted to the *left* of an equals sign, *except* within a procedural block (enclosed

by PROCED and ENDPRO statements or by NOSORT and SORT statements). The PROCED arguments cannot be dimensioned.

3. When the subscripted variable as declared by using the STORAGE statement is used unsubscripted as a block input or output, for example, as PAR in the example above, the sequencing of the DSL statements referring to the subscripted variable will be done in the same manner as the sequencing for an undimensioned variable. If the user wishes to use a single STORAG variable as output, it must be treated as a multiple output block and used as shown in Section 8.6 on FUNCTION blocks.

8.6   Use of Function Blocks in DSL

Over 50 individual simulation blocks are provided, either by special subroutines or by existing FORTRAN routines. If the user encounters a special problem requiring a particular block that is unavailable in the library, he may define his own block by following the FORTRAN or Assembly language specifications and add it to the library. He may then use this block just as any other DSL block. If this particular block contains multiple outputs, the FORTRAN subroutine form must be used. In this case, the output arguments, as called out in the DSL connection statements, appear in the same sequence but on the right-hand side of the subroutine argument string. The following statement is an example:

Y1, Y2=FLIP (RESET,TRIG)

The user must program a FORTRAN subroutine with an input argument string as follows

SUBROUTINE FLIP (RESET,TRIG,Y1,Y2)

for the DSL translator to communicate with the user's subroutine.

The DSL simulator generates the following form from the function definition block:

CALL FLIP (RESET,TRIG,Y1,Y2)

Examples of the function block usage were shown in Chapters 5 and 6. If the user-supplied subroutine or function contains integration, then the integrals (integrator output variables) must appear as output arguments of the subprogram and must have appeared on an INTGRL data card prior to their use as

INTGRL   Y, YDOT
         Y, YDOT=CMPXP (IC1, IC2, P1, P2, X)

A user's block may be stored as a subroutine on the disk library to become a permanent part of his DSL language, using the standard FORTRAN/Assembler store features of the system monitor.

8.7   Memory Blocks

If the user wishes to reserve a block of memory for initial conditions or variable history, special considerations are required. With each use of the block, a unique storage area must be assigned for history, and the initial values of history at time = 0 must be set by the user (the derivative subroutine UPDAT is executed twice at time = 0). Since a memory block may be the starting point of a closed loop, as in an integrator, this block must be identified to the DSL SORT routine. Note that in a feedback loop a time lag is created regardless of the block sequence. Either a predictor formula may be inserted within one block of the loop, as is the case with INTGR, or a special block may be added to advance an output. The method of handling such blocks is best illustrated by the following example: Assume that a block FUNCT with one input and one output history for each use is to be coded. The DSL statements for this block are shown below:

MEMORY   FUNCT(2)
         Y = FUNCT (IC, PARAM, X)

The MEMORY data card will reserve two words, which will be used for storing history of one input and one output. The MEMORY card tells the DSL translator how many storage locations for history are to be reserved. The translator assigns the required storage in COMMON and inserts the *index* of the assigned location for the first storage location for that block. Thus, the above DSL statement for Y may generate the following translated statement in the UPDAT routine:

$$Y = FUNCT (299, IC, PARAM, X)$$

where 299 may be the relative location in COMMON in "C," which appears after the DSL system variables in COMMON.

Now, if the user will code another function which uses the MEMORY block of N words, another block of words will be assigned at a location 299+N.

The function subprogram FUNCT can then be written as follows:

```
FUNCTION FUNCT (K, YIC, PARAM, X)
COMMON IDUMY (834), C(1)
```

Note that IC has been replaced by YIC since in the UPDAT subroutine the translator will declare IC as REAL. The user must assign initial values for the variable, in this case for YIC, to the location in the MEMORY blocks at TIME = 0. Thus, he may code the subprogram statements as follows:

```
      EQUIVALENCE (C(1), TIME)
      IF (TIME) 4, 4, 5
4     C(K+1)=YIC
      Go to 6
5     C(K+1)=FCN (X, C(K), C(K+1), PARAM, TIME)
6     C(K)=X
      FUNCT=C(K+1)
      RETURN
      END
```

Note that K has the value 299 in the above example and FCN is another function, which may be a user subprogram or an available function in the DSL or FORTRAN library.

8.8 Implicit Function

DSL provides an implicit function block called IMPL for the solution of an equation $f(y) = 0$ expressed in the form $y = g(y)$. Clearly, some iterative technique must be employed. The iterations are performed by DSL within each integration interval until, ideally, some convergence criterion is satisfied. The method used within IMPL, as developed by Wegstein (41), accelerates the rate of convergence if the iteration converges and often induces convergence if the iteration diverges. To use the implicit function block, the user simply writes the following DSL statement:

$$Y = IMPL (YO, ERROR, FOFY)$$

where YO is the initial value of Y, FOFY is the function, and ERROR is the error tolerance required for iteration.

This statement is followed by a linked set of DSL or FORTRAN statements evaluating FOFY. The DSL translator system then sets up the necessary iterative loop, which includes the proper sequencing for information flow. This is illustrated by solving the implicit equation

$$y = .75 + .2 \sin y$$

The DSL statements are as follows:

Y = IMPL (YO, ERROR, FOFY)
FOFY = 0.75+0.2*SIN(Y)

The DSL translator automatically generates the following statements in UPDAT:

9001   Y = IMPL (YO, ERROR, FOFY)
        IF (NALRM) 9002, 9002, 9003
9003   FOFY=0.75+0.2*SIN(Y)
        GO TO 9001
9002   CONTINUE

Only three statements are added to the ones written by the user. The first time the IMPL routine is entered, NALRM is set to 1, and Y is calculated from the initial guess YO. After each calculation of FOFY, the program flow returns to the IMPL subroutine where the convergence criterion is tested. If the criterion is satisfied, NALRM is set equal to zero and Y assumes the most recently calculated value of FOFY. If the convergence criterion is not satisfied, the iteration continues. NALRM is set negative if convergence fails. The error criterion is as follows:

$$\left| \frac{(y_{n+1} - y_n)}{y_{n+1}} \right| \leqslant error \qquad for \; |y_{n+1}| > 1 \tag{8.3}$$

$$|y_{n+1} - y_n| \leqslant error \qquad for \; |y_{n+1}| \leqslant 1$$

For computing an implicit function, the user may substitute his own block with the name IMPLC. The translator handles this replacement block exactly as it handles IMPL. The user is expected to set NALRM based on his own convergence criterion. Implicit blocks are treated as MEMORY blocks with five available history elements (as described above in Section 8.7). The following statements appear in IMPLC:

*ONE WORD INTEGERS
        REAL FUNCTION IMPLC (N, FIRST, ERROR, FUNCT)
        COMMON NALRM, IDUMY (833), C(1)
        EQUIVALENCE (C(1), TIME)

8.9 Arbitrary Functions

DSL provides two functional blocks, AFGEN and NLFGN, for handling arbitrary functions of one variable, The X- and Y-coordinates of the function points are entered sequentially following an identifying label and the symbolic name of the function, e.g.,

```
1-6      7-72
AFGEN    FC1 = -10.2, 2.3, -5.6, 6.4, 1.0, 5.9, etc.
```

Although the total number of data storage locations is necessarily fixed by machine size, there is no restriction on the number of points one may use to define any function. The only requirement is that the X-coordinates in the sequence X1, Y1, X2, Y2, ... are monotonically increasing. Any number of arbitrary functions may be defined, identified only by their symbolic names assigned by the user. As an example, the DSL statement Y3 = AFGEN(FC1,XIN) will refer to the function called FC1. AFGEN provides linear interpolation between consecutive points, while NLFGN uses a second-order Lagrange interpolation formula.

Data to function generators may be altered in several ways. One may substitute a complete new table by using an AFGEN or NLFGN card with the same function name in the next data set. The new table may overlay the old to conserve space. Individual points may also be altered by using the TABLE data card as shown in Section 8.5.

8.10 Data Output

Output from DSL may be on a printer or typewriter or in the case of 1130 and 1800 systems on an IBM 1627 plotter at preselected intervals of the independent variable. A CRT storage scope may also be used with the 1800. Scope and 1627 output are identical, as are print and typed output. The formats are fixed by DSL, although the size and labels of graphic output can be controlled through data statements. There are two print formats. A column format with variable-name headings is selected if there are 10 variables or less to output. Otherwise data are printed in equation form with four variables per line. A maximum of 25 variables may be printed in a single run. All variables may be printed by the use of DEBUG (see Section 8.11). Only one dependent variable is plotted per run.

In plotting, the independent variable is placed along the horizontal axis and must be the first variable specified on a GRAPH or SCOPE card. Dependent variables may appear on both axes. If, on successive runs, a different dependent variable is named, a new vertical axis will be drawn 1 inch to the left of the origin. If a new independent variable is named (or if switch 5 is on), the graph will be moved to a position 3 inches beyond the former graph and new axes will be drawn. Data switch 5 will cause erasing and redrawing of axes on the scope. Switch 4 may be placed on if the drawing of the axes on the scope is unsatisfactory and must be redrawn. Switch 4 may also be used to force drawing of the axis on the 1627. Leave switch 4 off to get out of loop.

8.11 Problem Checkout
Facilities (DEBUG)

The subroutine DEBUG, a standard DSL block, may have considerable value to the user in the early stages of problem checkout. DEBUG may be used to print the current value and associated variable name of all simulation model variables and parameters including systems variables. The printout may begin at some prespecified value of the independent variable and continue for as many consecutive iterations as desired. The calling sequence is Z=DEBUG(N,T), where N is an integer specifying the desired number of consecutive iterations to print, T the value of the independent variable (TIME) at which printout is to begin, and Z a dummy variable not used.

Within the DSL translator a subroutine has been included to dump data tables for DSL program debugging. This routine may be activated by using the DUMP data card (pseudooperation). A dump using this feature would assist the user in further error analysis if there is difficulty in translating or sorting a user's problem.

Program diagnostic messages were described earlier in Chapter 7.

An example of a typical DEBUG output is shown in Table 8.1.

8.12 Modification of the
DSL Main Program

If the user wishes to take over the execution from the DSL MAIN program, once the translated UPDAT check is produced and the INTEG core load is built by the DSL system, the DSL program permits this if the user will write his own MAIN program in FORTRAN and compile it at execution time. This feature enables the user to test conditions for multiple boundary-value problem solutions or to use a DSL simulated program in optimization techniques such as the "steepest descent." He may wish to use convergence techniques in a trial-error solution of his problem, such as the Regula-Falsi, Newton-Rhapson, and other convergence formulas. The user, of course, will have to provide the program and logic in his MAIN program to perform the various functions he requires, entirely through FORTRAN, and thus he can

TABLE 8.1

A Typical DEBUG Output

TIME	0.1000E-02	DELT	0.2000E-02	DELMI	0.1000E-09	FINTI	0.6
DELTG	0.0	D2250	0.0	DELTS	0.0	DELTC	0.0
DELMX	0.0	XDOT	0.2900E 03	X	0.2900E 00	Y3DOT	0.1
ZZ005	0.2095E-04	Y2	0.5111E-08	ZZ001	-0.4832E-05	ZZ002	0.2
ZZ004	0.0	ZZ006	0.6439E-01	ZZ007	0.1533E-04	SQR1	0.1
FK1	0.1458E 01	Y23	0.7666E-08	FK2	0.1940E-03	FD	0.0
K2	0.2530E 05	M1	0.1400E 04	M2	0.4528F 02	M3	0.2
H	0.1250E 03	XDOTO	0.2900E 03				

TABLE 8.2

INTEG Routine in DSL

```
*ONE        WORD INTEGERS
*IOCS       (CARD, KEYBOARD, TYPEWRITER, PLOTTER, 1443 PRINTER)
            COMMON NALARM, NINTG, NSYMB, KPT, KP(4), KREL,
            KABS, KMEM, KPOIN, KLOCK, ID (821), C(300)
            K1=KPOIN
            K2=KPOIN+NINTG
            CALL RKS (C (KMEM), C (KABS), C (KREL), C (K1), C(K2))
            CALL LINK (SIMX)
            END
```

eliminate multiple stacking of data cards and sitting at the console and making multiple runs.

The user can also control the data INPUT/OUTPUT and graph plotting by modifying the DSL MAIN program, and he may modify parameter values or control variable values between parameter studies. As shown in Section 8.4.2, the system function LOOK may be used to locate the simulation variables in COMMON.

To use this feature, the user must substitute his own routine as a FORTRAN 'MAIN' program INTEG and call the DSL integration subroutine from his own statements. The INTEG routine supplied from the PID library appears as shown in Table 8.2.

KPT is the logical unit number for the printer. COMMON should be specified as $C(n)$ to set the maximum dimensions, where n is selected prior to loading the system. (See Chapter 10, System Generation.)

The user must provide some method (as in subroutine INTRA) of entering data to control plotting and printing and the simulation run. Insert FORTRAN statements after CALL RKS which test run response, establish initial conditions for the next run, and determine whether acceptable performance is achieved. Set KLOCK=1 to continue a run, and use CALL

LINK(SIMX) to take advantage of the DSL data translator for keyboard or card input.

8.13 Program Restrictions

The DSL program has limitations on the number of certain components for a given problem. These restrictions were made in an attempt to balance the computer storage capacity available on small systems with the components found in the largest number of simulation problems. Recognizing that this is

TABLE 8.3
DSL Program Restrictions

Item	Maximum
1. Integrators (first-order differential equations)	50
2. Total number of characters in data statements (1000 words of 2 characters)	2,000
3. Total number of characters in structure statements (measured from col. 1 to the last character punched)	10,000
4. Number of DSL structure statements	100
5. Cards per structure statement (set by FORTRAN)	6
6. Characters per structure statement	400
7. Parameters (variables, not literals)	60
8. Block or statement input variables	300
9. Block or statement output variables	100
Note: Statements between PROCED and ENDPRO are not counted.	
10. Total unique symbols	140
11. Variables in single closed loop under SORT	30
12. Total variables under SORT	120
13. Real variables with first character I to N	20
14. MEMORY blocks	15
15. STORAG variables	10
16. INTGER variables	10
17. INTGRL variables (integrators in user subroutines)	10
18. Special cards (D in col. 1)	1
19. *LOCAL cards	1
20. PRINT variables	25
21. RANGE variables	15
22. FINISH variables	5
23. SORT–NOSORT sections	10

an arbitrary choice that may not be suitable for a particular problem, the routines have been set up for some flexibility in storage allocation. For example, in less than an hour the first four maximums could be doubled by following directions in Chapter 10. The restrictions are listed in Table 8.3.

In this chapter we have discussed in more detail some advanced language features which were not covered in Chapter 4, although some of the features were used in illustrations in Chapters 5 and 6. This chapter could very well have followed Chapter 4; however, since the examples are what attract an engineer, we interspersed the example chapters among the chapters which provide the "rules" of a language or a programming system.

Chapter 9 is for the reader who wishes to know the mathematical details of the subroutine programs which are provided as a part of DSL. They are primarily provided to point out to the user any limitations due to their mathematical formulations—which may result in instabilities in simulation in certain cases.

9 MATHEMATICAL FORMULATION FOR SOME DSL ROUTINES

9.0 Introduction

In this chapter we shall discuss mathematical formulation for some of the DSL routines which perform, among other things, centralized integration with various integration techniques and provide some functional block capability for simulation. The formula descriptions are of necessity brief, and the reader should consult Appendix D to find references for more detailed descriptions and derivation of the formulas.

9.1 Centralized Integration Schemes

9.1.1 Milne Fifth-Order Predictor-Corrector Integration Technique

To integrate $y = x\, dt$, y is evaluated at various values of $t + \Delta t$, with the increment Δt determined from the error criterion specified by the user in the RELERR and ABSERR cards of DSL.

$$y_{t+\Delta t} = .96119\, y^c_{t+\Delta t} + .03884 y^p_{t+\Delta t} \tag{9.1}$$

where $y^p_{t+\Delta t}$ is the predicted value evaluated by

$$y^p_{t+\Delta t} = y_{t-\Delta t} + \frac{\Delta t}{3}(8x_t - 5x_{t-\Delta t} + 4x_{t\ -2\Delta t}) \tag{9.2}$$

and $y^c_{t+\Delta t}$ is the corrected value based on the formula

$$y^c_{t+\Delta t} = \frac{1}{8}(y_t + 7y_{t-\Delta t}) + \frac{\Delta t}{192}(65x_{t+\Delta t} + 243x_t + 51x_{t-\Delta t} + x_{t-2\Delta t}) \quad (9.3)$$

The integration interval control is based on the criteria

$$\begin{aligned}|y^c - y^p| &\leqslant R|y^c| &\text{for } |y^c| > 1\\|y^c - y^p| &\leqslant A &\text{for } |y^c| \leqslant 1\end{aligned} \quad (9.4)$$

Please note that the Milne routine in DSL uses the fourth-order Runge-Kutta routine as a starter to calculate the first four intervals.

9.1.2 Runge-Kutta (Fourth-Order) Integration

To evaluate $y = \int f\,dt$, the formula for y is

$$y_{t+\Delta t} = y_t + \frac{1}{6}(k_1 + k_2 \cdot 2 + k_3 \times 2 + k_4) \quad (9.5)$$

where

$$\begin{aligned}k_1 &= \Delta t \cdot f(t, y_t)\\k_2 &= \Delta t \cdot f\left(t + \frac{\Delta t}{2}, y_t + \frac{k_1}{2}\right)\\k_3 &= \Delta t \cdot f\left(t + \frac{\Delta t}{2}, y_t + \frac{k_2}{2}\right)\\k_4 &= \Delta t \cdot f(t + \Delta t, y_t + k_3)\end{aligned} \quad (9.6)$$

If variable step integration is used, then the interval is reduced to satisfy the criteria

$$\frac{|y_{t+\Delta t} - y^s|}{(A + R|y_{t+\Delta t}|)} \leqslant 1 \quad (9.7)$$

where y^s is $y_{t+\Delta t}$, calculated by Simpson's rule, and A and R are the absolute and relative errors corresponding to the particular integrator value.

9.1.3 Simpson's Rule Integration

Simpson's rule integration is achieved using the formula

$$y_{t+\Delta t} = y_t + \frac{\Delta t}{3}(x_t + 4x_{t+\Delta t/2} + x_{t+\Delta t}) \quad (9.8)$$

where $y = \int x\,dt$.

9.1.4 Adams-Moulton Integration

This technique uses the fixed step method wherein the trapezoidal integration technique is used to evaluate the first value of $y_{t+\Delta t}$:

$$y_{t+\Delta t} = y_t + \frac{\Delta t}{2}(3x_{t+\Delta t} - x_t) \tag{9.9}$$

9.1.5 Trapezoidal Integration

Using a rectangular predictor to advance integrator output values,

$$y_{t+\Delta t} = y_t + \Delta t \cdot x_t \tag{9.10}$$

the following trapezoidal formula is used for integration:

$$y_{t+\Delta t} = y_t + \frac{\Delta t}{2}(x_t + x_{t+\Delta t}) \tag{9.11}$$

9.2 Formulation for Some Standard DSL Blocks

9.2.1 Derivative

$$Y = \text{DERIV}(IC, X)$$
$$Y = \frac{dx}{dt}, \qquad Y_{t=0} = IC \tag{9.12}$$

A second-order Lagrange interpolation formula is used to approximate the function. The derivative of this polynomial evaluated at X_j is

$$Y_j = \prod_{k=0}^{2}(t_j - t_k)\sum_{k=0}^{2}\frac{X_k}{(t_j - t_k)\prod_{n=0}^{2}(t_k - t_n)} + \sum_{k=0}^{2}\frac{X_j}{t_j - t_k} \qquad \text{for } k \neq j, n \tag{9.13}$$

For example,

$$Y_2 = (t_2 - t_0)(t_2 - t_1)\left\{ \frac{X_0}{[(t_2 - t_0)(t_0 - t_1)(t_0 - t_2)]} \right.$$

$$+ \left. \frac{X_1}{[(t_2 - t_1)(t_1 - t_0)(t_1 - t_2)]} \right\} + \frac{X_2}{(t_2 - t_0)} + \frac{X_2}{(t_2 - t_1)} \tag{9.14}$$

9.2.2 Nonlinear Function Generator

$$Y = \text{NLFGEN (F, X)} \qquad (9.15)$$
$$Y = F(X)$$

A second-order Lagrange interpolation formula is used to approximate the function $F(X)$, where X is not an input value in the table but lies between two points specified in the table. Let Y_0, Y_1, and Y_2 be specified at three consecutive points of X. The function Y is to be evaluated at a point X in between those three points. The approximated function $F(X)$ is given by

$$F(X) = Y_0 L_0^2(X) + Y_1 L_1^2(X) + Y_2 L_2^2(X) \qquad (9.16)$$

where L_0, L_1, and L_2 are given by

$$L_j^2(X) = \frac{\displaystyle\prod_{k=0}^{2}(X - X_k)}{(X - X_j)\displaystyle\prod_{k=0}^{2}(X_j - X_k)} \qquad \text{for } k \neq j \qquad (9.17)$$

For example, L_1^2 is evaluated as

$$
\begin{aligned}
L_1^2(X) &= \frac{(X - X_0)(X - X_1)(X - X_2)}{(X - X_1)(X_1 - X_0)(X_1 - X_2)} \\
&= \frac{(X - X_0)(X - X_2)}{(X_1 - X_0)(X_1 - X_2)}
\end{aligned} \qquad (9.18)
$$

9.2.3 The Implicit Function IMPL

The logic for evaluation of the implicit function in DSL is shown in Fig. 9.1 by means of a flow diagram. Both formulations and the program flow are self-explanatory in the figure.

The evaluation of the complex pole, real pole, and lead-lag functions are described using self-explanatory block diagrams in Figs. 9.2 through 9.4.

9.2.4 Complex Pole

$$Y, \dot{Y} = \text{CMPXP } (IC_1, IC_2, P_1, P_2, X)$$
$$\ddot{Y} + 2 P_1 P_2 \dot{Y} + P_2^2 Y = X$$

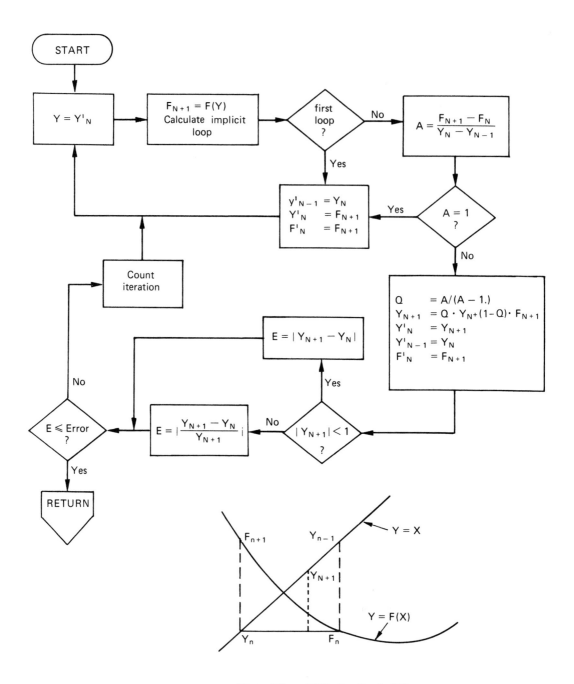

Figure 9.1.   IMPL function in DSL.

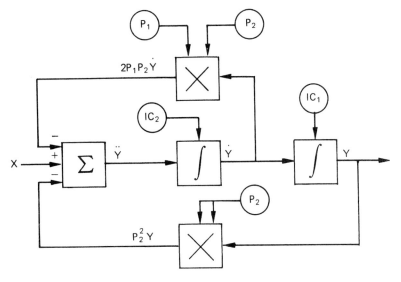

Figure 9.2.

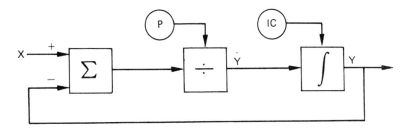

Figure 9.3.

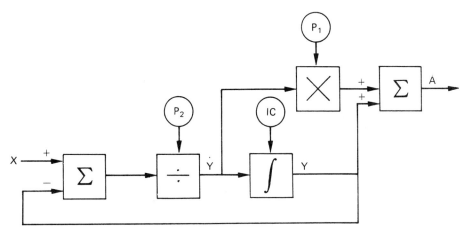

Figure 9.4.

Equivalent Laplace transform:

$$\frac{1}{S^2 + 2\,P_1\,P_2\,S + P_2^2}$$

9.2.5 Real Pole

$$Y = \text{REALP (IC, P, X)}$$
$$P\dot{Y} + Y = X$$

Equivalent Laplace transform:

$$\frac{1}{PS + 1}$$

9.2.6 Lead-Lag

$$A, Y = \text{LEDLG (IC, P1, P2, X)}$$
$$P_2\,\dot{A} + A = P_1\,\dot{X} + X$$

Equivalent Laplace transform:

$$\frac{P_1\,S + 1}{P_2\,S + 1}$$

$$Y_0 = IC$$

$$A_0 = (X_0 - IC)\,\frac{P_1}{P_2} + IC$$

(9.19)

9.2.7 Higher-Order Transfer Functions

To obtain the response of a higher-order transfer function for a linear system with constant coefficients, the equivalent system of differential equations is integrated numerically with the DSL centralized integration method selected by the user.

Transfer function:

$$\frac{Y}{X} = \frac{b_{n-1}p^{n-1} + b_{n-2}p^{n-2} + \cdots + b_1 p + b_0}{p^n + a_{n-1}p^{n-1} + a_{n-2}p^{n-2} + \cdots + a_1 p + a_0}$$

(9.20)

Differential equation:

$$D^n Y_1 + D^{n-1}(a_{n-1} Y_1 - b_{n-1} X) + D^{n-2}(a_{n-2} Y_1 - b_{n-2} X)$$

$$+ \cdots + D(a_1 Y_1 - b_1 X) + (a_0 Y_1 - b_0 X) = 0 \qquad (9.21)$$

Equipment system of n differential equations (first-order):

$$DY_1 = b_{n-1} X - a_{n-1} Y_1 + Y_2$$
$$DY_2 = b_{n-2} X - a_{n-2} Y_1 + Y_3 \qquad (9.22)$$
$$\cdot$$
$$\cdot$$
$$DY_{n-1} = b_1 X - a_1 Y_1 + Y_n$$
$$DY_n = b_0 X - a_0 Y_1$$

In this chapter we have provided a brief review of the mathematical bases of some of the DSL subroutines included in the DSL program. Table 4.2 may provide some insight into the remaining subroutines/functions. Needless to say, all the subroutines included in DSL are not mathematically defined here, since for a majority of them conventional methods have been used for their mathematical descriptions.

The remaining chapters are for the systems programmer or the brave novice user who is interested in installing the program on his system, be it IBM 1130, 1800, System/7, or System/3. It will certainly be worthwhile to read the next two chapters if the user plans to run on any other small system that is equivalent to the above listed systems.

10 SYSTEM GENERATION AND LOADING INSTRUCTIONS

10.0 Introduction

In this chapter we shall describe only the system generation and setup for the systems IBM 1800 and IBM 1130. For users of DSL on System/3, and System/7, the procedures are described in Chapter 11.

10.1 Preload Setup

Eight routines are distributed by the IBM Program Information Department in source language as part of the type III basic material. The routines and corresponding card sequences are INPT (DSL00123-00156), ERASE (DSL30610-30900), INTEG (DSL33210-33360), ORGSC (DSL42700-42948), PLOTI (DSL42950-43625), PLOTR (DSL43850-44780), RKS (DSL45500-47740), and SIM (DSL48170-48360).

Object programs for the above are also included in case no changes are necessary. The user may also choose to order the entire source for this program from the IBM Program Information Department (Hawthorne, New York), optional material. The program source listing is included in Appendix A.

The following system and program variations may be made at the source level prior to compilation and loading.

10.1.1 Choice of Integration Package

Card number DSL33320 may be altered to force inclusion of one of four possible sets of integration routines for the execution of the simulation run. The word RKS may be changed as follows:

RKS: Includes fourth-order Runge-Kutta fixed and variable step methods and trapezoidal and Adams-Moulton fixed step methods of integration. May be used with 8K 1130 or 16K TSX 1800.

MILNE: Adds fifth-order Milne predictor-corrector method to RKS. Suitable only for 32K 1800 at present. Uses 2042 words of storage over RKS.

FXINT: Includes only fixed step methods of integration, viz., Runge-Kutta, trapezoidal, and Adams-Moulton. Saves about 656 words of storage over RKS, all of which may be used for the model description.

RUK: Second-order Runge-Kutta variable step scheme adapted for problems of wider range in time constants. Routine also includes Adams-Moulton and trapezoidal methods.

10.1.2 Source Program Variations

1. Basic material release.

   a. The FORTRAN FUNCTION INPT(K) is called to determine logical unit numbers for I/O as follows: K=1 (typewriter output), 2 (card punch), 3 (primary output device), 6 (keyboard input), 7 (plotter), and 8 (card input). K=4 is used to set the number of COMMON locations for DSL simulation data (see part b below). In the released object deck the logical units returned by the function INPT are set as follows: INPT(1) = 1, INPT(2) = 2, INPT(3) = 3, INPT(6) = 6, INPT(7) = 7, and INPT(8) = 2.

   b. Four cards must be altered to set the number of COMMON storage locations to be used with the model simulation data. Data include current values of variables, tables for input to function generators, and working storage assigned by the translator such as in DELAY. COMMON is currently set at 300 on cards DSL00139, DSL33290, DSL43980, and DSL48260. The cards are in the subroutine INPT, INTEG, PLOTR, and SIM, respectively. For the 16K 1130 or 32K 1800 this number may be set at the maximum, 998. The three-digit restriction exists within the subroutine BNDEC only.

c. Analog and pulse output addresses (1800 only). If the type III scope support package program no. 1800-08.7.001 from PID is used; then the routines ORGSC, ERASE, and PLOTI may be discarded. They are included in simplified form for those users who do not have other scope applications:

(1) A pulse output address or ECO may be set at DSL30870 for the scope erase function.

(2) Analog output addresses for the horizontal and vertical axes on the scope are set at DSL43340 and DSL43350, respectively.

(3) The basic scope increment is set at DSL42895 and DSL43445. Movement with input N (integer) is 10N/32000 centimeters.

The object programs for these routines in the released system are merely dummy routines to return to the calling program. Source decks must be compiled and inserted to get scope output. The user is cautioned in using these routines without considering other outputs affected by their use. For example, the pulse output may be ORed into the output word.

2. Secondary material (source) variations.

a. *IOCS cards are at DSL01880, (OUTP), 07530 (SORT), 11585 (TRANS), 15250 (TRDUM), 33250 (INTEG), 43910 (PLOTR), and 48220 (SIM).

b. DEFINE FILE cards are at DSL01900 (OUTP) and 11587 (TRANS). References to these files are at DSL04560 and 05340 (OUTP) and 11670 and 12030 (TRANS).

c. Keyboard input format in TYPIN is set at DSL48930. This format is used when switch 3 is on following a simulation interrupt.

d. All switch control is through REMSW (DSL05610-DSL05680), which is set up to merely CALL DATSW. REMSW may be substituted by the user.

e. All print output is from the subroutine PRINT (DSL44790-45280). Formats are at DSL45060 and DSL45100-10.

f. The main routine OUTP writes the model subroutine UPDAT. It may be desirable to alter job control output before and after UPDAT. These formats are at DSL02900, 04700-10, and 04760.

g. If no plotter is available, the core load PLOTG may be eliminated from the system. The only CALL to PLOTG is in the main routine SIM at DSL 48310-20.

10.2 DSL/1130 Variations
(from DSL/1800 Source)

Although the 1130 source program at DSL is included in Appendix A, for a 16K 1130, the following variations must be noted by the user:

1. All *STORE card formats must be altered. (UA is replaced by WS UA in columns 13–18). A // DUP card must precede each *STORE card.

2. Remove cards DSL37350-DSL41200 (for 8K 1130), DSL30610-DSL30900, and DSL42700-DSL43625, the MILNE and scope subroutines. Delete DSL45882-86 in RKS and DSL44015, 44040, and 44260 in PLOTR to eliminate CALLs to scope routines.

3. Cards DSL04700-DSL04760 must be exchanged for

```
252  FORMAT (12H      RETURN/9H      END/6H//
     DUP/25H*DELETE                  UPDAT/25H*STORE
     WS   UA   UPDAT/25H*DELETE              SIMUL/
     25H*DUMP      UA    WS    INTEG/25H*STORECI
     WS   UA  S   IMUL/15H// XEQ SIMX   1)
     IF  (NLOC)    167, 167, 166
166  WRITE  (L, 251)  (LOCAL (K), K=1, 36)
     GO TO 168
167  WRITE (L, 256)
256  FORMAT    (35H*LOCALSIMUL, PRINT, XYPLT, TYPIN, UPDAT)
168  CONTINUE
```

Note: The *LOCAL is unnecessary for the 16K 1130; remove the last six cards above and the 1 in the XEQ SIMX card.

4. The following DSL blocks are stored as SOCALs, subtype 8 for the 8K 1130 system only (column 11 of *STORE card): AFGEN, CMPXP, DELAY, DERIV, HSTRS, IMPUL, INTGR, NLFGN, PULSE, RST, TRNFR, and TRNFX. Any block not using I/O may be subtype 8 (I/O in DEBUG and LOOK).

5. The 1130 system's subroutine TYPEZ is replaced to eliminate the character inserted in core when the keyboard EOF key is used (DM V1 only).

6. In the 1130 disk monitor Mod II or III, three subroutines must be dumped and restored but not as type 8 SOCALs. These are EBPRT, HOLEB, and PAPEB (DM V1 only).

7. If core is not available for 8K DSL/1130, cut down the vector storage for outputs, inputs, and integrators (see Chapter 13).

8. Hardware variations may cause *IOCS card changes as follows:

    a. Card reader: TRANS, SIM
    b. Card punch: OUTP
    c. Printer: SORT, TRANS, TRDUM, INTEG, SIM

9. The pseudoaccumulator for the 1130 differs from the 1800. Alter cards DSL 68170 and 68200 to use register 3 and locations 126 and 125, respectively.

10. The FORTRAN function subprograms IOR and IAND used in TRANS are not available in the 1130 system. Object decks only are included in the DSL release.

10.3   Loading the DSL System

10.3.1   Basic Material, DSL/1800

1. Load deck setup to store into the disk relocatable library.

2. Follow with deck to store temporarily routines to be used for core load building.

3. Cards DSL80730-DSL80850 will cause six core loads to be built and stored in core image format.

4. Cards DSL90130-DSL90610 may be run as a sample problem. Place blank cards after the STOP card. You will need about 20 more blanks than cards in the original input. The DSL translator will convert the original language to a card FORTRAN subroutine UPDAT with appropriate control cards for direct execution of the simulation. When the translator comes to a WAIT, place the newly punched cards in the reader and START. You may also do a cold start (or restart) at this point by placing the cold start cards ahead of the first card punched (// JOB) (or restarting). This permits you to alter the model or structure description at the FORTRAN level without rerunning the translator. You may need to substitute a // JOB X X card for a two-disk system. After the FORTRAN compilation of UPDAT a final core load is built which includes the model description (UPDAT). The integration routines included depend on the choice in Section 10.1.1. This core load is stored in core image format as SIMUL. If there has been a previous use of DSL, that model, SIMUL, is deleted. DSL data follow the UPDAT subroutine. One may restart at this point if data only are to be changed. The UPDAT subroutine does not have to be recompiled.

10.3.2 Secondary Material, DSL/1800 (Source)

1. All routines are set up to compile and load into the disk relocatable library.

2. Repeat steps 3 and 4 in Section 10.3.1.

3. DSL82000-82700 and DSL80000-80720 are included to allow setup of the basic decks from the secondary material.

10.3.3 Basic Material, DSL/1130

1. List object program decks to locate identification cards for *IOCS assignments. If routines are not included corresponding to I/O at your installation, you must use secondary material for loading. All released decks are set up to load directly to the disk relocatable library.

2. Follow the basic material loading with the control cards for building core loads (DSL95000-95140). For a 16K 1130, remove the *LOCAL card and the 1 in the preceding card.

3. Use the 1800 sample problem (part 4, Section 10.3.1) with FX removed from the // XEQ card.

10.3.4 Secondary Material, DSL/1130 (Source)

1. Use DSL/1800 source with the variations described in Section 10.2. Important changes are in *IOCS cards and in INPT.

2. Load the entire DSL system from source to relocatable library. *DUMP and *STORE cards may be generated from the 1800 cards (Section 10.3.1).

3. Repeat steps 2 and 3 from Section 10.3.3.

10.4 Variations for Other Computer Systems

The 1130 and 1800 are similar systems, at least from the FORTRAN compiler and Assembler standpoint as well as in the use of the control statements. While the FORTRAN is becoming a more universally accepted language for all computer systems in the United States, the Assembler language as well as control statements are certainly quite different for different computer systems. The user effort of converting a substantially large program such as DSL to run on other machines can be reduced substantially if all the source programs are converted to FORTRAN. The only task for running the DSL program on another machine (once the converted source is compiled and stored on a disk of the user's system) is to learn the operation control language and commands for system operation of that system.

In Chapter 11, we shall discuss not only the conversion of DSL to almost all FORTRAN but also the operation control language changes to run on IBM System/3 and System/7. This exercise should give sufficient information to the user so that he will be able to achieve conversion to other small computer systems if he needs to. Admittedly, this conversion will not be possible for an engineer who is a novice in computer systems and/or programming. However, a systems programmer's help will go a long way in installing the system as a first step, essential to operation of the engineer's simulation system.

To repeat what was said earlier, the user who has access to IBM System 360/370 in a batch environment has no such worries since the CSMP 360/370 (15) is the program he will use and program installation is not a problem for him.

11 SYSTEM/3 AND SYSTEM/7 DSL PROGRAM PREPARATION AND OPERATION

11.0 Introduction

In this chapter we shall discuss the steps necessary for DSL system generation for users who wish to execute the DSL program on the IBM System/3 and System/7. As we shall see during this chapter, once the user becomes familiar with this procedure, the DSL program (system) can be used with very minor changes on any small scientific computer with sufficient storage (generally about 16,000 to 24,000 16-bit words), a disk memory for bulk storage, and standard FORTRAN IV language capability with compilation facility on the operating system of the computer.

11.1 System/3 Conversion

The description of System/3 was briefly covered in Chapter 2. The System/7 conversion will be covered in Section 11.2. To generate the digital simulation language program (DSL) on System/3, the following requirements are essential for the System/3 hardware:

1. The control processing unit must have 32,000 bytes of memory (16,000 words).

2. The system must have at least one fixed and one removable drive (standard on all models of System/3).

3. The system must have a card reader and punch.

4. The system must have a printer (1403 or 7431).

All System/3 Models 10 and 15 have a keyboard/console printer, which is used by the DSL program as an additional input/output device or as a sole input/output device.

Furthermore, the IBM program product FORTRAN IV (Program Number 5703-F01) for System/3 must be installed on the System/3 to be able to generate and operate the DSL system. The user must consult with either a systems programmer in his computer center or an IBM field engineer servicing his System/3 to ensure that the FORTRAN system is installed on his system. The following discussion will assume that the System/3 FORTRAN program obtained from the IBM Program Information Department resides on the fixed drive number 1, i.e., F1. We shall also assume that the procedural library for calling the FORTRAN compiler and the FORTRAN linkage editor are installed on the fixed drive F1.

11.1.1 Source Card Preparation
for System/3 Card Reader—5471

The first hurdle to overcome is that the 1130/1800 source cards for DSL are of a different type for System/3 if the user has a System/3 multifunction card reader MFCU1/MFCU2 installed as the card input device. Figure 11.1 shows the two card types. If the user has received a tape from the IBM Program Information Department for the optional source material and he has a tape drive attached to his System/3, then his first order of business is to get the assistance of a systems programmer to write either a program in System/3 RPG or FORTRAN to read the tape and punch the records from the tape in cards using only the 80 columns of the 96-column card for System/3. Remember that both FORTRAN and DSL will use only 80 columns and ignore the rest of the columns (81-96) on the card.

If the user's System/3 installation has a 2501/1442 card reader with standard 80-column cards, then he can obtain the program source cards punched from a tape on an IBM 360/370 or equivalent machine to get the 80-column source cards for input to the 2501/1442 card reader on the System/3. No tape drive is required on the System/3 in that case. The user will need to work with the source cards in both instances (80 columns or 96 columns) for the discussion on program conversion which follows.

11.1.2 Program Modifications

The discussion in Chapter 12 on the description of the program modules in DSL for the IBM 1130 and 1800 clearly indicates that several of these modules are in 1130/1800 Assembler language. In addition, at several places

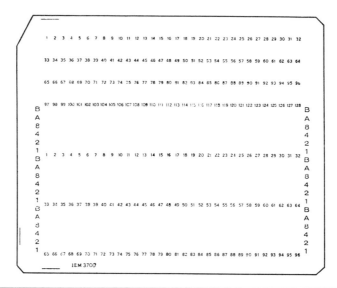

Figure 11.1.   Data cards for the IBM System/3 and IBM 1130.

within the DSL programs written in FORTRAN, five character words [in KA(1) to KA(5)] are packed into two words, with each character using 6 bits. This practice is difficult to transpose from one computer machine to another. The ideal program from the standpoint of transferability from one machine to another is the one that is all written in standard FORTRAN and has no bit manipulation or Assembler coding in the program. The first step in modifying DSL for use on System/3 then requires conversion of the Assembler program in DSL to FORTRAN.

In Appendix A, the program source listings as obtained from PID for DSL 1800 and 1130 are shown. In Appendix B, listings of System/3 modules are shown where changes are required in the source to run on System/3. Most assembly language modules of 1130/1800 which are converted to FORTRAN on System/3 can be seen by comparing the two listings. Reference to these listings will assist in understanding the following discussion on program conversion for System/3 and other computer systems.

11.1.3 Removal of Plot Routines

First, there is no plotting facility available on System/3, and therefore the entire set of program modules PLOTR and its associated plotting subroutines have to be eliminated. Since some of these subroutines are also called from integration routines such as MILNE, RKS, etc., dummy subroutines must be substituted which will print messages indicating no plotting facility available on System/3. In the event the user on System/3 calls for GRAPH, the translator will accept his input cards. However, at execution time, he will obtain a series of messages indicating no plotting facility on System/3, intermixed with his printed output.

Next, all Assembly language routines in DSL for the IBM 1800 are converted to FORTRAN. These include START, SETUP, PACK, UNPAK, SPLIT, BUILD, OPEN, and BNDEC. To perform minimum bit manipulation, which is essential, one small assembly language program needs to be written for System/3. This program, named SHIFT, is also listed in Appendix B and basically performs shifting of bytes (half-words) in order to perform packing/unpacking functions required in the DSL translation phase. This is one subroutine which must be rewritten for other scientific computers when converting the DSL program from System/3 to run on other systems. The functional descriptions of the modules START, SETUP, BUILD, BNDEC, and SPLIT remain the same as discussed in Chapter 12. Note, however, that the PACK subroutine converts five characters in KA(1) to KA(5) with one character per word to three words KA(1) to KA(3) using two characters per word and not to two words as in DSL 1130/1800. The impact of this change is felt in some other modules in the DSL translation phase written in FORTRAN because of this difference. These changes will be pointed out in Section 11.1.5. Notice that the subroutine UNPAK is not required and is only a dummy CALL module.

11.1.4 Modification of Control Cards

For those program modules retained, the control cards must all first be modified to be compatible with the control card requirements for System/3. The same requirement will apply for conversion to other hardware systems, since the job control language is bound to be different for each machine.

Assuming that the FORTRAN is installed on the fixed disk F1 of System/3, Fig. 11.2 shows the replacement cards required for converting 1130/1800 system control functions to System/3 functions.

```
// FOR                                    // CALL FORTRN,F1
                                          // RUN

*ONE WORD INTEGERS                              IMPLICIT INTEGER*2(I-N)
// DUP                                    *PROCESS OBJECT(R,LIB(R1))
*STORE          WS  UA  XXXXX

*IOCS(CARD,KEYBOARD,...)                  // READ DEVICE-'MFCU1,5471'
                                          // PRINT DEVICE-'1403,5471'
                                          // PUNCH DEVICE-'MFCU2'

// DUP                                    // CALL FORTL,F1
*DUMP           UA  WS  XXXXX             // RUN
*STORECI        WS  UA  YYYYY             // PHASE NAME-'YYYYY',UNIT-R1,RETAIN-P
                                          // OPTIONS UPACK-R1
                                          // INCLUDE NAME-XXXXX,UNIT-R1
                                          // END
```

Figure 11.2.   Comparison of FORTRAN control cards for 1130 and
System/3.

The // FOR card which calls the FORTRAN compiler on the 1130 will
be replaced by the two FORTRAN CALL cards for System/3. The two cards
shown actually generate four catalogued procedure statements, and depending
on how the procedure is set up by the systems programmer, the compiler will
set up the two required files $WORK and $SOURCE on a removable disk,
most probably R1. Appropriate file space allocation therefore must be available
on the disk on R1 drive before the compiler can execute. The one-word
integer and name declaration is one-to-one replacement. The *LIST ALL card
for the 1130 is removed and is replaced by the *PROCESS card, which, in
addition to performing the LIST and DECK options, provides the storing of
the compiled object code of a subroutine or of a main program on the
specified disk as a replaceable module (LIB(R), R1) in an object library. This
function is performed on the 1130 by the // DUP function, and therefore
the // DUP cards and *STORE cards are removed for System/3 operation.
However, each program module must end with a /* card. The card describing
I/O units used by the program (*IOCS in 1130/1800) is replaced by individual
DEVICE cards for System/3, as shown in Fig. 11.2.

11.1.5 Alteration of COMMON
and Other Source Changes

In System/3 the various core loads communicate using GLOBAL rather
than COMMON. Thus, all COMMON statements must be modified to
GLOBAL. In addition, in System/3 (as indicated earlier in Section 11.3), the
DSL variables, after reading in the user's DSL statements, are packed into
three words rather than two words (as in DSL 1130/1800), so that the size of
the DSL symbol array KSYMB in GLOBAL must be increased. With the
increase in this array size, one must also alter several statements in translator
program modules to reflect the use of three words for each DSL variable

symbol in System/3 rather than two words in 1130/1800. Actually, this is accomplished easily when one examines all statements using information from KSYMB. The listings in Appendix B show where additional statements are added whenever symbols are equated to values in KSYMBL(J+200), etc. The DSL/1800 UNPAK function becomes unnecessary, and calls to it may be removed. In any case the UNPAK subroutine itself is converted to a dummy subroutine which performs no function in System/3.

System/3 FORTRAN does not permit disk file unit numbers which conflict with logical unit numbers assigned to the input/output devices such as card reader, line printer, or console printer. All disk file unit numbers in DSL translator programs must therefore be altered to avoid this conflict, since logical unit numbers for System/3 normal input/output devices are fixed, as shown in Table 2.3 (Chapter 2). To accomplish this, all DEFINE FILE statements as well as file READ/WRITE statements must be altered accordingly. The disk READ/WRITE is performed only in the translation phase of DSL. Source statements of the following format therefore must be altered:

$$\text{DEFINE FILE M(I, J, U, K)}$$
$$\text{READ (M'N) (-----)}$$
$$\text{WRITE (M'N) (------)}$$

where M is the file unit number to be altered and N is a specified record number in the DSL code. These changes are required in program modules DSL and OUTPX.

Furthermore, all logical unit numbers in READ/WRITE statements, if specified as integer variables, must be of length 4 bytes (two words), and not 2 bytes (one word). This is a peculiarity of System/3 FORTRAN READ/WRITE rules. This then requires that in every module where READ/WRITE statements are encountered, two to three more statements must be added to equate the unit numbers properly and/or declare the variables as 4-byte integers.

Finally, all CALL LINK (XXXXX) statements in the six main programs, namely, DSL, SORTX, OUTPX, TRDMP, SIMX, and SIMUL, must be altered to INVOKE XXXXX to be compatible with System/3 FORTRAN.

11.1.6 Generated Control Cards

The OUTPX program punches control cards for 1130/1800 for a problem run in the simulation phase. These source statements must be changed to be able to punch control cards for simulation on the System/3. Basically, this requires changing of the FORMAT statements at two places in OUTPX, as is seen in Appendix B. These statements punch out the FORTRAN compiler control cards and the linkage editor cards required for the System/3 to build the load module for simulation, namely SIMUL, using the latest punched FORTRAN subroutine UPDAT. Again, these control cards assume that the

FORTRAN linkage editor resides on F1 and that the program load module is to be stored on R1.

The listing in Appendix B, indicating a modified source for DSL on System/3, may be used without any changes by the user who has the following input/output hardware installed on the System/3:

1. 1403 printer.

2. MFCU1/MFCU2 card reader.

3. 5741 console keyboard/printer.

4. User disk pack with DSL program named R1R1R1 and mounted on drive R1.

5. FORTRAN program and link editor installed on fixed drive disk F1.

6. Standard catalogued FORTRAN procedure installed on F1.

Appropriate changes in the OUTPX program will be required if the disk pack name is different and/or other conditions listed above are not satisfied.

Notice that this particular change in OUTPX source statements will apply for program conversion to other systems such as the IBM System/7. The FORMAT change should be made to correspond to the job control language for the compiler and linkage editor for the machine to be used.

Finally, the subroutine INPT must be modified to correspond to the input/output devices of the user on System/3. Refer to Table 2.3 for the logical unit numbers and the discussion on the DSL 1800/1130 variations in Chapter 10. The appropriate unit numbers should be filled in. Remember, again, that corresponding DEVICE OPTIONS cards in the source of the main programs DSL, TRDMP, SORTX, OUTPX, SIMX, and SIMUL must match the devices used. The user whose System/3 peripheral devices are different from those specified in Appendix B should consult the FORTRAN IV Program Reference Manual for System/3. This manual will also be useful in understanding the operations control language (OCL) for the System/3 FORTRAN compiler, link editor, and the various control statements on System/3.

Once the source is created as discussed above and shown in Appendix B, the user is ready to begin compiling his DSL source. Depending on the printer used, this procedure should take approximately 2 hours. The next step is to create executable storage loads from the compiler objects stored on the disk pack R1.

11.1.7 Creating Storage Loads

Again, if the linkage editor cards as shown in Appendix B are used, with user's disk named R1R1R1 mounted on drive R1, and if the FORTRAN link editor is installed on F1, the user will be able to create his required storage loads for DSL directly on System/3. Note especially the // CORE card during the linkage editing step of SIMX. This card is required because, depending on the size of the simulation problem, the SIMUL core load may be bigger than SIMX. Since during simulation SIMX is executed first, the System/3 monitor will allocate only the smaller storage area of SIMX, thus creating program errors when SIMUL is called into execution. The CORE options will permit the use of maximum storage on the 32K-byte System/3. For a System/3 with larger memory size the user may wish to increase this size provided that his SIMUL storage load will require more than 28K bytes.

Once all the storage load modules are built and successfully stored on the disk mounted on R1, the user is ready to start executing the DSL translator on System/3. To execute the translator he will read the execution control cards in the card reader as shown in Fig. 11.3.

The user must make sure that all data switches (rotary dials) are set to 0. Notice that the files used by the DSL translator must also be identified during execution of the program with the // FILE cards.

The DSL data cards may then follow in the card reader with at least one blank card behind the STOP card. For System/3 Model 10 and Model 15, if the card punch device is declared as MFCU2 (in the INPT subroutine), blank cards for punching the UPDAT subroutine must be placed in hopper 2. If the punch unit is declared the same unit as the card reader, blank cards must follow the DSL data cards. The proper selection of logical unit numbers in the INPT subroutine as shown in Appendix B is important, and the reader must choose his appropriate values for his hardware based on the logical unit numbers discussed for System/3 in Chapter 2.

Once the translator creates and punches out the proper UPDAT subroutine with its control cards for compilation and link-editing storage load, the punched cards, just as in the 1130/1800, may be read back into the card reader to perform the simulation run. Remember that only printed output can be obtained in System/3—no plot facilities should be requested by GRAPH.

The data switch usage must be performed using the rotary switches on the System/3 console. Figure 11.4 indicates in which position the dials must be to correspond to the 1130/1800 data switches being "on." Do not use the last rotary dial, since it is used for control purposes by the System/3 monitor. The corresponding data switches (12-15) are not used by DSL, and therefore that dial is not of any consequence either during the translation or simulation phase of the DSL program.

```
// DATE 010174
// LOAD DSL,R1
// FILE UNIT-R1,PACK-R1R1R1,NAME-FT00012,LABEL-FILE12,RETAIN-T,TRACKS-20
// FILE UNIT-R1,PACK-R1R1R1,NAME-FT00011,LABEL-FILE11,RETAIN-T,TRACKS-20
// RUN
*      USER DSL SOURCE CARDS  ***
*      FOLLOWED BY BLANK CARDS
```

Figure 11.3.   Control cards to run DSL on System/3.

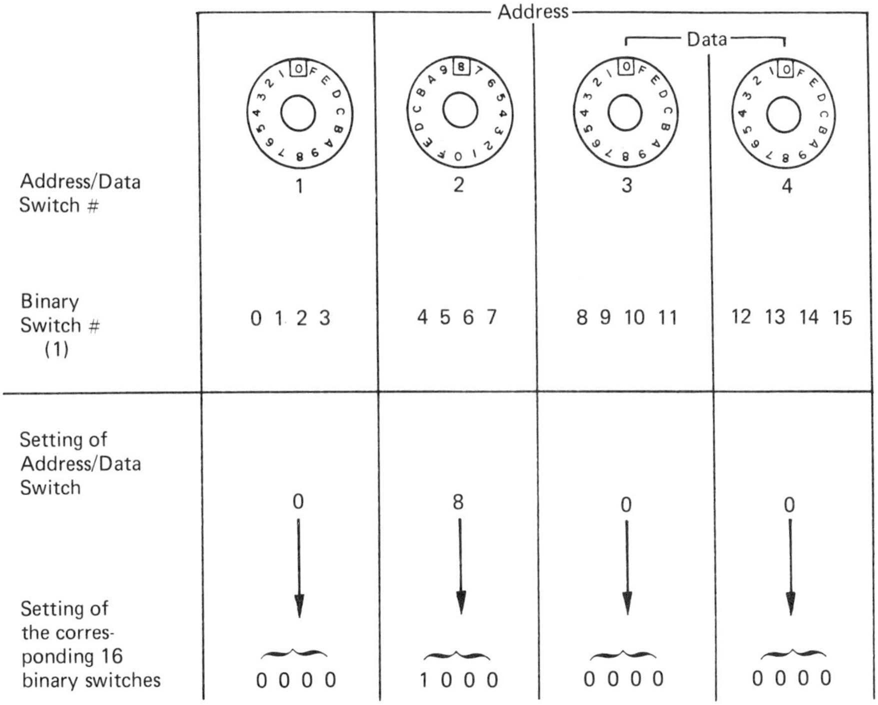

Figure 11.4.   System/3 data switch addressing.

Finally, during the linkage editing of SIMUL with the newly created UPDAT deck of the user, the system will halt with a halt code of 6E. This warns the user that if he wishes to replace the existing SIMUL load modules on the disk he must turn the fourth rotary dial to zero position and press the start button. This will then provide him with the capability of simulation of his new problem. Ensure that the dial is in position zero before proceeding.

11.1.8 Summary of Conversion Steps for System/3

To summarize, then, the conversion of DSL for operation on System/3, the user must perform the following steps:

1. Convert the DSL/1800 source obtained from the IBM Program Information Department to the System/3 cards using a tape to card program (record length = 80 bytes).

2. Remove all 1130/1800 control cards for FORTRAN compiler and supervisor DUP function (// DUP and *STORE) cards and replace them with the System/3 compiler and storage call cards as in Fig. 11.2.

3. Remove all 1800 Assembly language program modules and replace them with the FORTRAN modules shown in Appendix B. Also add the SHIFT assembly language routine for System/3.

4. Replace the PLOTR module and its associated routines with the dummy modules shown in Appendix B.

5. Make the required source changes in COMMON statements in the translator portions of program modules for the array KSYMB and change COMMON to GLOBAL.

6. Make appropriate source changes to reflect changes in source for KSYMB array size change in DSL, SORTX, and OUPTX as shown in Appendix B. Modify FORMAT statements in OUTPX for machine differences.

7. Make changes in the DEFINE FILE and READ/WRITE (file records) for alteration of file unit numbers. Make logical unit numbers in other READ/WRITE statements as 4-byte integers (INTEGER*4).

8. Prepare the control cards for linkage editing after compilation as shown in Appendix B.

9. Compile all the prepared source decks after alteration (approximately 2 hours) and then create storage loads using the linkage editor control cards.

10. Execute the DSL program as shown in Fig. 11.3 with the sample problem shown in Appendix B.

One of the sample problems shown in Appendix B may be used to ensure that the system created by the user is properly operational. In Appendix B, the UPDAT subroutine together with the control cards and all input data required for execution of the simulation phase are shown. The user can compare his output with the output this author obtained as shown in Appendix B.

11.2 Conversion of DSL to System/7

The IBM System/7 is a smaller sensor-based computer of the category of the IBM 1800, except that its operating system is more limited than an 1800 and also has a more limited set of peripheral I/O devices available for purposes of batch programs such as DSL. However, since the FORTRANs for System/3 and System/7 are quite similar, once the user has understood the mechanism of conversion of the DSL/1800 source to System/3, there is, but for the differences in compiler and linkage editor control cards, very little additional source modification required.

11.2.1 *Modification of DSL Source*

A large portion of the effort will be expended in building a suitable system monitor nucleus, which has very little to do with DSL source changes. If the user will follow the steps outlined below and use the recommendations in these steps, he should have the least difficulty in generating a suitable DSL program for his System/7. For his system monitor nucleus, he should consult a systems programmer in his organization to ensure obtaining proper System/7 hardware parameters required in the nucleus generation.

The steps required to modify a DSL source obtained in connecting the DSL/1800 to DSL for System/3 are outlined below:

1. Rewrite the SHIFT program which is in System/3 Assembler language using the System/7 FORTRAN ISHIFT function. The SHIFT program can now be in FORTRAN, as shown in Appendix C.

2. The OUTPX source must be modified at the FORMAT statement to punch out the System/7 compiler/link editor control cards required for the simulation phase. As a matter of fact, the System/7 allows a user to store his own special catalogued procedure so that he can simply reduce the FORMAT to one or two statements. In Appendix C, then, the required catalogued procedure to reflect the use of the two statements is shown. This procedure is to be stored in the system prior to start of the simulation phase once and for all.

3. All // CALL FORTRAN, F1 cards, // RUN cards, *PROCESS cards, and /* cards should be removed. All DSL translator source cards can be compiled in one sequence using only one set of compiler control cards as shown in Appendix C.

4. All INVOKE XXXXX cards must be replaced by CALL $OVLY (XXXXX, K111, K222). You may recall that these cards originally were CALL LINK (XXXXX) cards in DSL/1800. If the user has a System/7 with a storage size of 26K words, he does not need to use overlays; however, the source code in DSL, SORTX, TRDMP, OUTPX, SIMX, and SIMUL must be modified prior to END statements as shown in Appendix C. If you use overlays, the called program XXXXX must be declared external using the statements EXTERNAL XXXXX right after the PROGRAM statement. The system nucleus monitor must also reflect additional changes to indicate usage of overlays.

5. You are ready to compile the DSL source after the changes to the source using either the System/7 native FORTRAN compiler (21) or, on a 360/370 host, using the System/7 host FORTRAN COMPILER (21).

Because of the large number of FORTRAN statements involved, it is suggested that the user find access to an IBM 360/370 with a System/7 FORTRAN host compiler (21) and create the DSL translator storage loads for transfer to the System/7 using object cards. Attempting to compile the 3000 cards on the System/7 will be a rather formidable job. Follow Appendix C in the generation of the system nucleus, compilation, and building of storage loads to obtain the object cards. You are, of course, required to fill in a couple of places your system configuration numbers which must be obtained from the systems programmer for your System/7. These are the module numbers for the printer (7431) and card reader (129) units attached to the System/7.

11.2.2 Compilation of Simulation Programs on System/7

At this point the user must at least become familiar with the use of the System/7 (native) FORTRAN compiler, since all simulation runs will be made using this compiler program. Appendix C again shows how all the DSL library subroutines and the main program SIMX and SIMUL are to be compiled and stored on the System/7 disk. A special disk data set is created first with the name DSLS7, which is used both by the translator and the simulator. Punch out the cards as shown in Appendix C and follow the instructions to generate your DSL program library.

Notice that the control cards are quite different from the 360/370, 1130/1800, and System/3. No // cards are used to define the control card, but the first letter in column 1 will indicate what function is to be performed.

11.2.3 Operation of the System/7 for Simulation

The user must first be warned that, unlike the IBM 1800, the DSL program created for the System/7 cannot be used in the real-time environment. It is to be used only in a batch environment where the user has complete control of the System/7 and it is not performing any other function or controlling another process. In fact, the simulation phase uses the standard DSS/7 nucleus provided with the System/7 program support by IBM.

The source program which was used as a test problem with the System/3 can be used to operate on the System/7 and to check out the DSL system.

It is the experience of this author that the System/7 will require considerably more assistance from a programmer, knowledgeable in System/7 FORTRAN, than does a System/3 to be able to prepare and install the system. However, once installed it will most likely perform considerably faster than System/3, 1130, or 1800.

Last, the user *must* install a hardware consisting of digital inputs with switches (a small power supply or a battery of about 10 volts providing voltage through the switches) to be able to use DATSW. Otherwise he must modify the REMSW routine to return without performing any function. This will, however, eliminate his capability of using data switches for the interactive mode of operation in problem solution.

In this chapter we have attempted to provide information to a user who is more of a computer scientist and is interested in modifying the program to run on IBM System/3, IBM System/7, or other small scientific computers by modifying the DSL/1800 source. Although a complete set of instructions cannot be within the scope of this book, the computer scientist should consult the publications for his system to get more details on their operation, whether for compilation and system building or for actual execution/operation of the DSL system program prepared by the user.

12 DSL PROGRAM DESCRIPTION

12.0 Introduction

In this chapter we shall describe the present DSL translator, simulator, and plot subroutines. A summary chart is given in Fig. 8.1 to show how all programs within the system are linked together (i.e., to show which subroutines are used at each level and to identify the calling program). Main programs and subroutines will be described by core load. The system exists as seven complete core load overlays. The core load names and corresponding main program names are DSL/TRANS, OUTPX/OUTP, SORTX/SORT, TRDMP/TRDUM, PLOTG/PLOTR, SIMX/SIM, and SIMUL/INTEG. The core load SIMUL contains the model description or derivative subroutine for the simulation phase and is built at execution time. DSL library blocks are not described here. Some of the library routine description was given in Chapter 9. Flow charts are included only for the routines which are generally modified for special applications or hardware such as integration and plot routines. For those users who find difficulty in obtaining the program source from the IBM Program Information Department, a source listing is given in Appendix A which may be referred to for further understanding of the program descriptions elaborated on here.

12.1 Core Load DSL

1. Main program: TRANS.
   Function:
   To convert the higher-level problem-oriented input language statements

into an equivalent set of FORTRAN statements. In the process a series of internal tables is created containing each variable name, selected according to the variable's function in the statement and stored as input, output, or parameter variables. In addition, key tables are developed containing macros, integrators, literals, vectors, and other pertinent data for later communication with the SORT and OUTP routines. An example of the use of these tables is demonstrated at the end of this section. Key variables are also described in the example.

2. Subroutine: ERR1.
   Function:
   To print error messages and error number.
   COMMON NALRM, IERR, ICONT, JPT, KEY, ISTRT
   NALRM: Error number identifying error as shown in Chapter 7.
   JPT: Output unit number.
   ISTRT: Column of DSL statement where error occurred.

3. Subroutine: INPT(K).
   Function:
   To set I/O system unit numbers (logical units). See Chapter 10 for a complete description.

4. Subroutine: NUMRC.
   Function:
   Conversion of BCD or EBCDIC number to binary.

   COMMON NALRM, IERR, ICONT, XPT, KEY, ISTRT, KA(6), LETRS(72)
   ISTRT: Column of statement to begin conversion.
   LETRS: DSL statement, one 2-byte word per character.
   NMBR: KA(6), where result is stored.

5. Subroutine: OPEN.
   Function:
   To separate the DSL input statement into 72 words with one character per word and leading zeros. Other entries reverse the process for single words and convert binary to EBCDIC.
   COMMON
   ID(6), KA(6), LETRS(72), IDUMY(1608), KFORT(2)
   ENTRY OPEN
   Separates 72 characters in two-byte words in LETRS(1) to LETRS(36) into 72 two-byte words with leading zeros in LETRS(1) to LETRS(72).

ENTRY PACK

Packs five characters from five 2-byte words with leading zeros in KA(1) to KA(5) into two 2-byte words, KA(1) and KA(2), truncating each letter to 6 bits.

ENTRY BNDEC

Converts a binary number in NMBR, less than 1000, [KA(6)], and stores it as an EBCDIC number in KFORT starting at KFORT(LOC) where LOC is KA(5)+1.

6. Subroutine: READR.
Function:
Reading and printing DSL input statement.

COMMON NALRM, IERR, ICONT, JPT, KEY, ISTRT, KA(6), LETRS(72), ID(1810), KEYB, KARD

LETRS: Input/output area.
JPT: Output systems unit.
KEYB: Input systems unit for keyboard (interrupt).
KARD: Input systems unit for card.

7. Subroutine: REMSW(J,K).
Function:
To recognize console key interrupt. At present REMSW merely calls DATSW, but it may be altered to test remote switches. Switch J is tested; k is set to 1 if interrupt has occurred and 2 if not.

8. Subroutine: SCAN.
Function:
To scan input DSL statement for next name. When a first usage of name appears, then it is stored in the symbol table and in subsequent usages only the symbol table location is returned to the calling program for that name.

COMMON NALRM, IERR, ICONT, JPT, KEY, ISTRT, KA(6), LETRS(72)

ISTRT: Starting location of scan.
LETRS: Input area to be scanned; one character in each 2-byte word.
KA: Five characters are packed in the first two 2-byte words of KA (Note that for System/3 as well as System/7 they are packed into three words without truncation) with leading bits per character truncated.
KEY: KEY is used to identify the name returned. KEY is set as follows: 1 (numeric), 2 (+,*,-,/,), ** or a comma or end of statement), 3 (=), 4 (left paren), 5 (greater than five characters), 6 (end of record, no alpha) and 7 (continuation).

9. Subroutine: START.
Function:
Assembly language routine which places symbol tables and other data in COMMON. Can be replaced by BLOCK DATA if available.

10. Subroutine: TRAN1.

    Function:

    Initialization of constant data areas. This subroutine could be replaced by a BLOCK DATA routine if more space is desired.

12.1.1 Translator Data Table

The following is an example showing the translator table contents built for the DSL input statements which follow the table:

KOUT	23	25	27	31	30	34	38	39	36	35	29
LFORT	1	16	27	40	62	73	131	146	154	171	183
NFORT	15	11	13	22	11	-58	15	8	17	12	22
LIN	1	3	6	8	13	14	17	19	21	24	25
NIN	2	3	2	5	1	3	2	2	3	1	3

INPUT	25	26	27	28	29	30	28	32	23	30	27
	33	31	35	36	37	39	40	29	35	38	34
	41	36	23	38	42						

KSYMB TIME   DELT   (systems vars. & blocks –20 locs)

	FK1	INSW		Y12	K1	SQR1	H	Y2		X
	XDOT	XDOT0		M1	FD	Y3	Y3DOT	FOFY3		KF2
	Y23	K2		M3	M2					

LINT	4	5		9	10	0	11			
KPAR	1	2.....		6	37	28		26	40	33	42	41
	32									

DSL input statements:

```
              FK1=INSW(Y12,0.,K1*Y12)
              Y12=SQR1-H-2.*Y2
              SQR1=SQRT(X**2+H**2)
              XDOT=INTGR (XDOT0,-2.*FK1*(X/SQR1)/M1)
              X=INTGR(0.,XDOT)
PROCED        FD=FCYN3(Y3,Y3DOT,FOFY3)
              IF(Y3-324.)4,4,5
4             FD=AFGEN(FOFY3,Y3)*Y3DOT**2
              GO TO 6
5             FD=90.*Y3DOT**2
6             CONTINUE
ENDPRO
              FD2=INSW(Y23,0.,K2*Y23)
              Y23=Y2-Y3
              Y3DOT=INTGR(0.,(FK2-FD)/M3)
              Y3=INTGR(0.,Y3DOT)
```

$$Y2=INTGR(0.,INTGR(0.,(2.*FK1-FK2)/M2))$$

AFGEN FOFY3= 0., 8.33, 30., 4., 60., 1.6,. . . .

CONST H=125., K1=4550., K2=25300., M1=1400.,. . .

   M2=45.28, M3=20.

INCON XDOT0=290

SCAN is used to find all variable names within DSL language statements and place all unique names in KSYMB. The five characters are packed into two words using the lower 6 bits of each character; the first word is stored starting at KSYMB(1) and the second word at KSYMB(200). If DSL names are used as input or output to a block, the relative location (index) of the name in KSYMB is placed in the corresponding table. Table names and functions are as follows:

KOUT: Index of output variable name or names for each DSL statement. PROCED blocks are considered as a single statement.

NFORT: Number of 2-byte words in statement.

LIN: Location (index) in INPUT of input variable indices identifying block inputs.

NIN: Number of inputs to block identified by corresponding entry in KOUT.

INPUT: Indices of input variables in order of inputs to DSL statements. Indices show location of variable in KSYMB.

LINT: Statement number of integrals. The number is zero for nested integrals and negative if integral is not a direct output of the block.

KPAR: KSYMB index of variables which appear on PARAM, CONST, and INCON data statements.

The total number of variables or indices stored in the above tables can be found in locations NOUT, NSYMB, NINP, NINT, and NPAR. The first five tables use NOUT.

12.2 Core Load SORTX

1. Main program: SORT.
   Function:
   To sequence DSL language statements to provide proper information flow in the system or network. The tables KOUT and INPUT are used to identify input and output to each block. Translated FORTRAN statements are sequenced such that values of variables used in each statement have been computed in the current iteration cycle or are constant. Integrator values are predicted prior to each iteration and therefore are current. Each sequence is begun with an integrator output and traced via input and output tables to a link with another integrator loop or a constant. The

sequence of variables is then reversed to establish the FORTRAN statement sequence. The final sequence is transferred to the output routine by a table in COMMON KSEQ, which contains the correct sequence in terms of output statement sequence.

2. Subroutine: OPEN.

    ENTRY UNPAK

    Function:
    To convert truncated five-character names packed in two words by PACK to EBCDIC five-character names in three words of two characters each into KA(1) to KA(3). (See Section 12.1, part 5, for storage alignment.) This function is not required in System/3 and System/7 DSL program operation.

3. Subroutine: ERR1. (See Section 12.1, part 5, for description.)

4. Subroutine: INPT(K). (See Chapter 10 for description.)

12.3 Core Load OUTPX

1. Main program: OUTP.
    Function:
    To write a FORTRAN subroutine UPDAT in the sequence established by SORT. Control cards, COMMON statements, etc., are also written. FORTRAN statements are transferred from the random disk data set 1 to the systems unit set by INPT(2). DSL data are transferred from random disk data set 2 to the systems unit set by INPT(2). INPT(2) identifies the card punch if FORTRAN requires card input. DSL data are in the same sequence as input, but a symbol table and header card are added. If STORAG variables are used, a PARAM statement is generated. All DSL variables appear in a COMMON statement generated in UPDAT.

2. Subroutine: OPEN.
    ENTRY PACK. (See Section 12.1, part 5, for description.)
    ENTRY UNPAK. (See Section 12.2, part 2, for description.)

12.4 Core Load TRDMP

1. Main Program: TRDUM.
    Function:
    Intended as a DSL debugging aid, TRDUM can be used to assist the systems programmer to determine if DSL translation, sort, or output routines have an error. TRDUM is executed only if the pseudooperation

DUMP has been used. Data areas that are dumped are shown in the example in Section 12.1.1.

12.5 Core Load SIMX

1. Main program: SIM.
   Function:
   To initialize data areas and input data for the simulation run. PLOTG and SIMUL are linked from SIM.

2. Subroutine: INITL.
   Function:
   Similar to that of TRAN1 in the first phase. Data areas are initialized, constants are set, etc. A BLOCK DATA subroutine could be used in its place. COMMON variables used in INITL are identified at the end of this section.

3. Subroutine: INTRA.
   Function:
   INTRA is the major data language translator. DSL variable names are separated and identified as to function. Numeric data are converted to binary and stored by variable name and function. Control for the simulation is set up as well as input and output functions and variables. Variables used in INTRA, usually COMMON variables, are identified in Section 12.8.

4. Subroutine: ALPHA.
   Function:
   To select the next alphanumeric name from LETRS beginning with LETRS(ISTRT) and store it left-justified with blanks to the right in KA(1) to KA(3) as 2-byte words.

5. Subroutine: INPT(K).
   Function:
   To establish logical units for FORTRAN I/O statements. (See Chapter 10, Section 10.1.5, for description.)

6. Subroutine: ERR.
   Function:
   To print error messages and error number.

   COMMON NALRM,ID(220), ISTRT
   NALRM: Number of error as explained in Chapter 7, Table 7.5.
   ISTRT: Column of card where error occurred.

7. Subroutine: NUMER.
Function:
To convert BCD or EBCDIC numeric data in FORTRAN E, F, or I format to binary. Scan is begun in LETRS(ISTRT), and the result is returned in DWORD. LETRS in INTRA is equivalent to NUMBR in this routine.

COMMON
NALRM,ID(220), ISTRT, DWORD, KA(7), NUMBR(72)
KA and NUMBR are 2-byte integer variables.

8. Subroutine: READ.
Function:
Reading and printing DSL data statements.

COMMON
NALRM, NINTG, NSYMB, JPT, ID(227), LETRS(72)

Data are read into the first 36 words of the 2-byte word area LETRS. If data switch number 2 is up, input is from the keyboard; otherwise it is from cards.

9. Subroutine: REMSW(J,K). (See Section 12.1, part 7, for description.)

10. Subroutine: SETUP.
Function:
Assembly language routine used to store data in COMMON such as constants, symbol tables, etc.

11. Subroutine: SPLIT.

ENTRY SPLIT

Function:
To convert DSL input from 36 2-byte words to 72 2-byte words with leading 1-byte zeros. Input is assumed in LETRS(1) to LETRS(36) (Assembly language).

ENTRY BUILD

Function:
To convert six 2-byte words in KA which have zeros in alternating bytes to three packed 2-byte words in KA(1) to KA(3).

12.6 Core Load PLOTG

1. Mainprogram: PLOTR
   Function:
   PLOTR is called from SIM to draw and label the axes for plots either on the 1627 or scope. If switch 4 is on, the picture is redrawn to allow adjustment of the scope. If switch 5 is on, the paper for the 1627 is advanced to 3 inches past the last graph.
   Note that plotting facilities are not possible in System/3. Core load PLOTG is therefore not usable for System/3 and System/7.
   Variables used in PLOTR:

   > INTEGER TITLE(6)), SYMB(420),
   > COMMON NALRM, NINTG, NSYMB, ID(11), INDX1,
   > ID2(21), TITLE, ID3(310), SCAL1, SCAL2, YPOS,
   > INDX2, ISCOP, SYMB

NSYMB: Total number of symbols in symbol table SYMB.

SYMB: DSL variable and system variable names. SYMB is in 2-byte words with the first 2 bytes of each name in SYMB(1) to SYMB(140), the next 2 starting at SYMB(141), and the last two starting at SYMB(281).

INDX1: Index in SYMB of the variable to be graphed on the horizontal axis.

INDX2: Index of the variable to be graphed on the vertical axis. INDX2 is set negative to identify the first usage as a dependent variable.

SCAL1, SCAL2: Not used by PLOTR; independent and dependent variable scale factors for graphic output.

YPOS: Identifies position from origin to draw vertical axis.

ISCOP: Set = 0 for 1627 graph output; set = 1 for scope output (by subroutine PLOTS).

TITLE: TITLE(41) to TITLE(60) contains the graph label as two characters per word.

All plot subroutines interface through the subroutine PLOTI. The facilities available in the language used by PLOTR can be converted to any device, on- or off-line, by simply altering the output of PLOTI to data equivalent to the eight directional codes generated by all plot routines.

2. Subroutine: FGRID, FPLOT, FCHAR, FCHRX, FMOVE, FRULE, SCALF, POINT, XYPLT, PLOTX, and an adaptation of PLOTI.
Subroutines from the IBM 1130/1800 plotter subroutine package are used for grid and character generation on both the 1627 and scope (see *IBM Reference Manual C26-3755*).

3. Subroutine: ORGSC, ERASE, PLOTI.
Subroutines from the type III library (order file number 1800-08.7.001) are used for output on the scope. Adaptations of these are included with the basic DSL materials in source and object form. Chapter 10 gives card sequence numbers of cards to be altered for variations of analog and pulse output addresses, plot increment, etc.

4. Subroutine: INPT(K). (See Chapter 10 for description.)

5. Subroutine: REMSW(J,K). (See Section 12.1, part 7, for explanation.)

12.7   Core Load SIMUL

1. Main program: INTEG.
INTEG is the DSL simulation phase main control routine. It is short to facilitate replacement by the user. RKS is called to execute the simulation. LINK (SIMX) returns for more data.

2. Subroutine: ERR. (See Section 12.5, part 6, for description.)

3. FUNCTION INPT(K). (See Chapter 10 for description.) Sets FORTRAN logical units.

4. REAL FUNCTION INTGR.
Function:
This is the interface routine to the DSL system integration. INTGR merely collects all derivatives in the sequence in which the function is entered and outputs the previously predicted value of the integral.

COMMON
NALRM, NINTG, ID(8), KMIM, ID2(291), KOUNT, ID3(531), C(1)

NINTG: Number of integrators (first-order equations).

KOUNT: Count of integrators per iteration. The first integrator output is stored in C(7). The current integrator output is in C(KOUNT+7) with the corresponding input in C(KMIM+KOUNT).

5. Subroutine: PRINT.

Function:
All on-line printing is done within PRINT. Two formats are used: column printing if 10 output variables or less, and otherwise equation form. The format is E12.4. COMMON variables are explained in Section 12.8.

6. Subroutine: REMSW(J,K). (See Section 12.1, part 7, for explanation.)

7. Subroutine: UPDAT.
Function:
UPDAT is the user's model description and is called by the system to compute the derivatives. The derivatives are stored sequentially and serve as input to the centralized integration routine. This subroutine is generated from DSL input and compiled at execution time.

8. Subroutine: RKS(P0,XABS,R,YST,DYST).
Function:
The main simulation program. It contains all the systems integration routines, writes intermediate off-line plot output, and controls on-line plot output and printing according to the interval specified by the user.

P0: Vector containing derivatives stored consecutively in order of computation.

XABS: Vector of absolute errors in the same sequence as the derivatives according to integrator output.

R: Vector of relative errors to be used for integration interval control.

YST,DYST: Work storage areas, each at same dimension as number of integrators if used by the integration method specified. Depending on the integration subroutine used, up to 15 vectors are dimensioned at the maximum number of integrators. All 15 areas are used in MILNE; RUK and RKS use P0, XABS, R, and six work areas.

Flow charts are included in Figs. 12.1 through 12.5 so that users may alter integration methods or add new ones. Note that the subroutine RKS represents a part of MILNE. RUK may be substituted for RKS. By changing RKS in INTEG to FXINT, MILNE, or RUK, different integration packages are compiled into the system.

9. Subroutine: TYPIN.
Function:
Called from the simulation routine RKS when data switch 3 is on. This routine is included in the basic DSL merely to allow parameter variations from the keyboard using only the format A = N, where A is a systems or

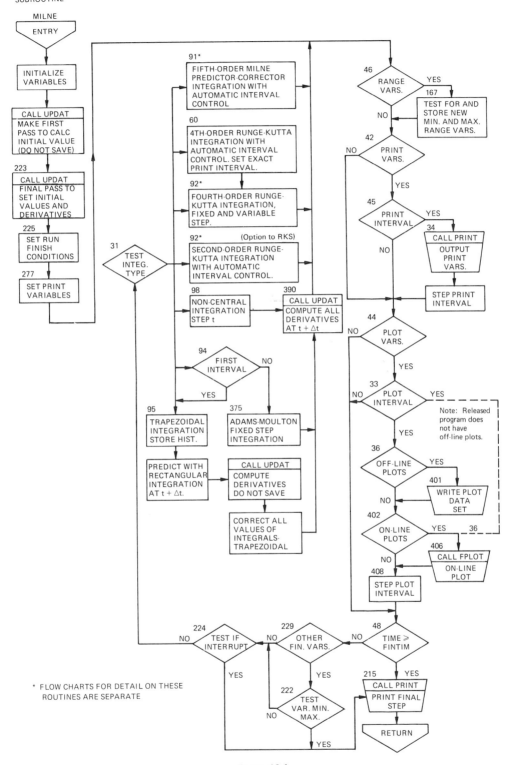

* FLOW CHARTS FOR DETAIL ON THESE
  ROUTINES ARE SEPARATE

Figure 12.1.

MILNE 5TH ORDER
VARIABLE STEP, PREDICTOR/
CORRECTOR INTEGRATION.
DSL38070-DSL39610

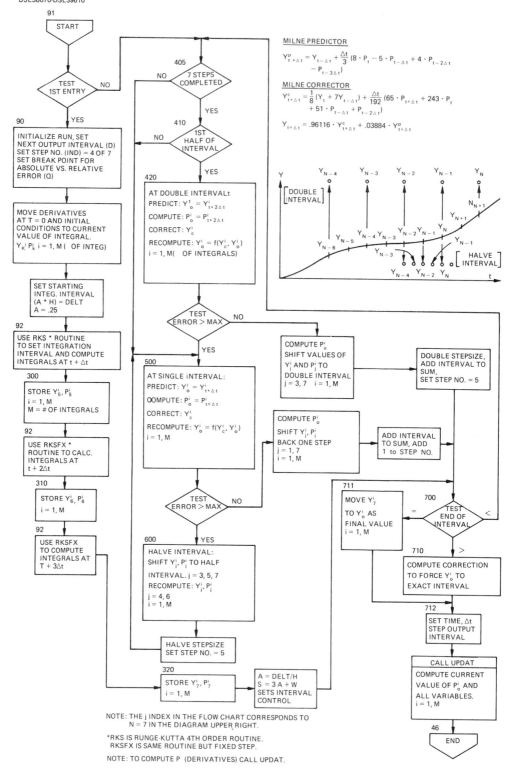

NOTE: THE j INDEX IN THE FLOW CHART CORRESPONDS TO
N = 7 IN THE DIAGRAM UPPER RIGHT.

*RKS IS RUNGE-KUTTA 4TH ORDER ROUTINE.
RKSFX IS SAME ROUTINE BUT FIXED STEP.

NOTE: TO COMPUTE P (DERIVATIVES) CALL UPDAT.

Figure 12.2.

RUNGE-KUTTA VARIABLE STEP
4TH ORDER INTEGRATION
DSL 39660 - DSL 40403 OR
DSL46510 - DSL 47260

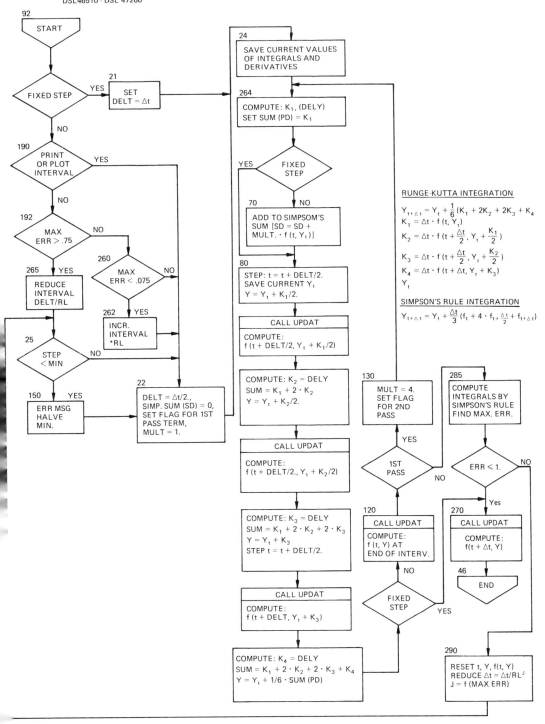

Figure 12.3.

RUNGE-KUTTA 2ND ORDER
VARIABLE STEP INTEGRATION
DSL 51090 - DSL 52510

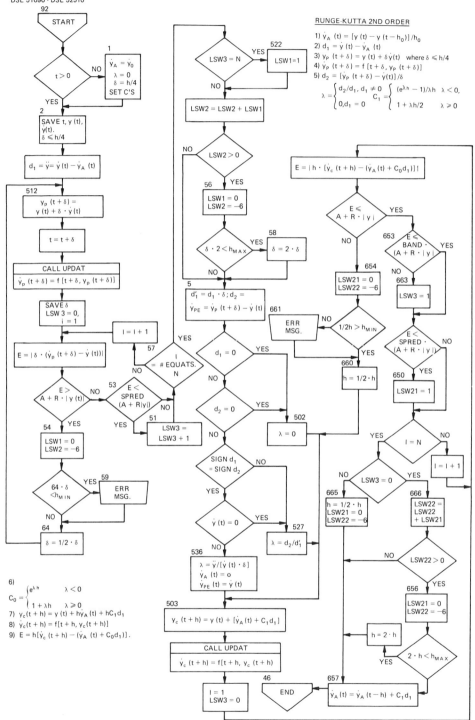

Figure 12.4.

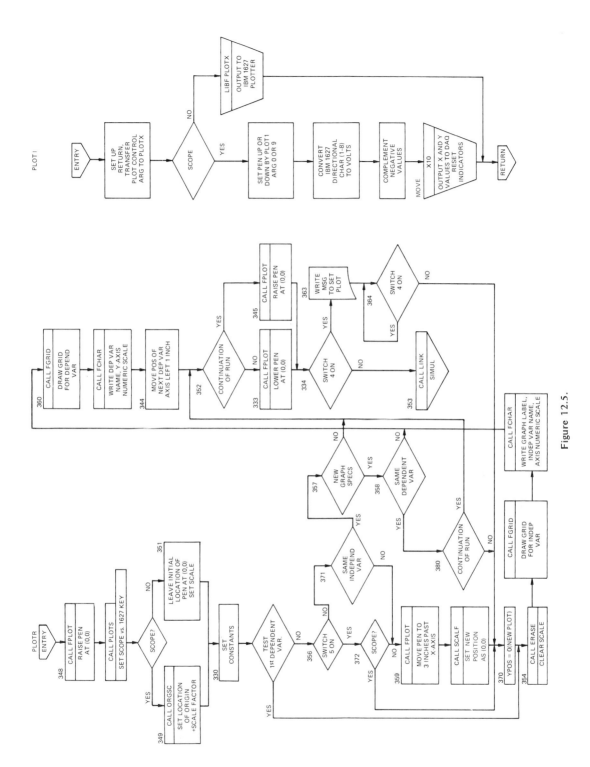

Figure 12.5.

245

model structure variable name and N is a FORTRAN constant. The format used is 2A2, A1, 1X, E12.6. TYPIN could be used for any interrupt function. The variable KLOCK is set to zero to cause initialization of the simulation if TIME = 0 is one of the changes made when this subroutine is called.

10. The plot routines

FPLOT,PLOTS,PLOTI,PLOTX,FMOVE, and XYPLT

are referenced in Section 12.6.

12.8 Variables in COMMON

COMMON C(n) at the end of blank COMMON contains simulation variables and work storage areas. Six systems variables starting with TIME and DELT are first, followed by all integrator outputs in order of computation. All other statement output variables follow in order of input, and then parameter variables and inputs not specified. Data input to function generators, STORAG variables, special tables, etc., are continued following model variables. Inputs to integrators (derivatives) are stored beginning at C(KMEM).

12.8.1 Blank COMMON Variables

NALRM: Set = 1 when systems error occurs. Also used by implicit function to signal when convergence criteria are satisfied. Error routine prints NALRM on each call to identify error with documentation (Chapter 7, Table 7.5).

NINTG: Number of integrators (first-order equations).

NSYMB: Number of symbols in SYMB (symbol table).

JPT: Primary output unit number (printer).

KPRNT: Number of print variables indexed in INDXP.

KRANG: Number of RANGE variables indexed in IRANG.

KFINI: Number of FINISH variables indexed in IFINL.

KTITL: Number of words stored for page titles in TITLE stored.

KREL: Index in COMMON C(n) where relative errors are stored.

KABS: COMMON C(n) index for absolute error table.

KMEM: Starting location of integrator inputs in COMMON C(n).

KPOIN: Index in COMMON C(n) where work storage area is available to user.

KLOCK: Set = 1 to signify continuation of run from last calculated values.

INTYP: Integration type by sequence number.

KEEP: Set to –1, 0, or +1 by integration routines to identify computation with TIME set back, held constant for recomputation or testing, or advanced for new values. KEEP is used by DELAY block to store new table values.

DELTA: Maximum integration interval. Set to minimum of print or graph interval if not given.

KAFGN: Key used to identify number of entries to function generator tables outside range of table. Now set 10 to allow only 10 tries.

LINE, NLINS, etc.: Used for print page spacing.

KODE: Used to communicate to print routine.

KOUNT: Count of integrators active in current iteration, zero for none active, etc. KOUNT is set by INTGR routine on each call.

IRES: Contains all systems names and values.

SYMB: DSL system symbol table; contains systems variables and all model simulation variables. Variable names are stored with the first two bytes at SYMB(K), the next two at SYMB(K+140), and the final two at SYMB (K+240) for the Kth variable.

13 AIDS TO PROGRAM MODIFICATION

This chapter is intended for the advanced DSL user who wishes to modify the program at the source language level. Flow charts are described in Figs. 12.1 through 12.5 in Chapter 12 for two routines: the main simulation subroutine, which includes all the integration methods, and the main plot control subroutine. These are primarily the routines altered at various installatons.

For flexibility in altering DSL program restrictions a list identifies all cards affecting each restriction. Program vector storage has been reduced to allow for a 16K 1800 or 8K 1130 system, forcing many users to consider increasing vector capacities.

To alter	*Change cards*
1. Total characters for structure cards.	DSL01900, 09775, 11587, 11685
1b. Cards per structure statement	Governed by FORTRAN system in use
2. Total characters for data cards	DSL01900, 09755, 11587, 12040
3. Maximum number of integrators (equations)	DSL01760, 01950, 05460, 05790, 07590, 09773, 11010, 11135, 11180 11470, 11593, 13664, 14330, 15310 31140, 37430–40, 45580, 50080
4. Number of DSL structure statements (output variables)	DSL01760, 01940, 05460, 05780, 07580, 09774, 11010, 11135, 11170, 11470, 11592, 13715, 15300
5. Number of DSL input variables	DSL01760, 01940, 05460, 05780, 07580, 09774, 11010, 11135, 11170, 11470, 11592, 14000, 15300

6.	Number of parameters	DSL01760, 01950, 05460, 05790, 07590, 11010, 08160, 09774, 11135, 11180, 11470, 11593, 12200, 15310
7.	Total unique symbols	
	a. Translator	DSL01760, 01920, 02540, 03390, 03904, 04320, 04920, 05010, 05460, 05760, 07180, 07260, 07350, 07560, 07662, 08120, 09775, 10390, 11010, 11150, 11590, 12680, 15280, 15470
	b. Simulator	DSL30970, 31130, 32810, 33050, 33290, 33610, 34070–80, 34250–60, 37400, 43930, 44340, 44520, 44850, 45140, 45230, 45560, 48260, 48870, 48960–70, 50060, 60070, 60440, 61320, 61770, 61880, 62030, 62650, 63220, 63480, 63940, 64420, 64630, 64960, 65010–20, 65380, 66200, 66640, 66810, 66940, 67170, 67330, 68560
8.	Number of STORAG variables	DSL01760, 01950–60, 04350–60, 05460, 05790–800, 07590–600, 11010, 11135, 11180–90, 11470, 11593–4, 12490, 15310–20
9.	Number of REAL variables with first character I–N.	DSL01760, 01960, 05460, 05800, 07400, 07600, 11135, 11190, 11470, 11594, 15320
10.	MEMORY blocks	DSL01760, 01950, 05460, 05790, 07590, 11010, 11135, 11180, 11470, 11593, 12360, 15310
11.	INTGER variables	DSL01760, 01950, 05460, 05790, 07590, 11010, 11135, 11190, 11470, 11594, 12570, 15320
12.	INTGRL variables	DSL01760, 01960, 05460, 05800, 07600, 11135, 11190, 11470, 11594, 12740, 15320
13.	Characters per structure statement	DSL11595, 11657, 13280, 13686, 13699
14.	Random access file, logical unit, etc.	DSL01900, 04560, 05240, 11587, 11670, 12030
15.	Logical unit numbers (not disk)	DSL00132–45
16.	Maximum variables in SORT section	DSL01970, 07610, 08520, 08620
17.	Maximum variables within single closed sort loop	DSL07602, 08820
18.	SORT-NOSORT sections	DSL01760, 01960, 05460, 05800, 07600, 11135, 11190, 11470, 11594, 15320
19.	Maximum outputs per statement	DSL07602, 08490

*20.	Maximum PRINT variables	DSL32810, 33610, 34250–60, 34300, 34370–80, 37400, 43930, 44850, 44860, 44990, 45090, 45560, 50060
*21.	RANGE variables	DSL31180, 33700, 34620, 37480, 43970, 44900, 45620, 50130
*22.	FINISH variables	DSL34730
23.	Maximum data storage for simulator tables, current values, etc.	DSL00139
24.	Maximum COMMON variables in UPDATE subroutine (model)	DSL01980, 02860

Note: PRINT, RANGE, and FINISH also require changes in DSL30070, 30970, 31170, 32840, 33290, 33650, 37470, 41270, 43960, 44890, 45350, 45610, 48260, 48760–70, 48880, 50120, 60070, 61320, 61790, 62030, 62650, 63220, 63480, 63940, 64420, 64630, 64970, 65380, 65400, 66200, 66640, 66810, 66940, 67170, 67330, and 68560.

Note: See Chapter 10 for variations of source decks for scope routines reflecting analog and pulse output addresses. Card sequence numbers are also given in Chapter 12 for *IOCS cards. The program source listing is included in Appendix A.

A
LISTING FOR DSL
1130/1800 PROGRAM

```
// FOR ERR1                                                              DSL00010
*LIST SYMBOL TABLE                                                       DSL00020
*LIST SUBPROGRAM NAMES                                                   DSL00030
*LIST ALL
*ONE WORD INTEGERS                                                       DSL00040
**ERROR PRINT ROUTINE FOR DSL/1800 TRANSLATOR                           DSL00050
      SUBROUTINE ERR1                                                    DSL00060
      COMMON NALRM,IERR,ICONT,JPT,KEY,ISTRT,KA(6),LETRS(72)             DSL00070
      L=JPT                                                              DSL00075
      WRITE(L,8) ISTRT,NALRM                                             DSL00080
    8 FORMAT(12H ERR IN COL I2,9H   DSL ERR I4)                          DSL00090
      NALRM=0                                                            DSL00100
      RETURN                                                             DSL00110
      END                                                                DSL00120
*STORE        UA       ERR1                                              DSL00122
// FOR INPT                                                              DSL00123
*LIST SYMBOL TABLE                                                       DSL00124
*LIST ALL
*ONE WORD INTEGERS                                                       DSL00126
**ROUTINE TO SET LOGICAL UNIT NUMBERS OF I/O                            DSL00127
      FUNCTION INPT(K)                                                   DSL00128
      GO TO(1,2,3,4,20,6,7,8,20),K                                       DSL00129
C...  SET LOGICAL UNIT FOR TYPEWRITER(1),CARD PUNCH(2),PRIMARY OUTPUT   DSL00130
C...  DEVICE(3), KEYBOARD(6), PLOTTER(7), CARD READER(8)                DSL00131
    1 N=1                                                                DSL00132
      GO TO 20                                                           DSL00133
    2 N=2                                                                DSL00134
      GO TO 25                                                           DSL00135
    3 N=3                                                                DSL00136
      GO TO 20                                                           DSL00137
C...  SET NUMBER OF COMMON STORAGE LOCS. TO BE USED IN SIMULATION MODEL DSL00138
    4 N=500                                                              DSL00139
      GO TO 20                                                           DSL00140
    6 N=6                                                                DSL00141
      GO TO 20                                                           DSL00142
    7 N=7                                                                DSL00143
      GO TO 20                                                           DSL00144
    8 N=2                                                                DSL00145
   20 INPT=N                                                             DSL00151
      RETURN                                                             DSL00152
      END                                                                DSL00153
// DUP
*DELETE            INPT
*STORE        WS  UA  INPT
// FOR NUMRC                                                             DSL00158
*LIST SYMBOL TABLE                                                       DSL00160
*LIST SUBPROGRAM NAMES                                                   DSL00170
*LIST ALL
*ONE WORD INTEGERS                                                       DSL00180
**DSL/1800 ROUTINE TO CONVERT INTEGER EBCDIC TO BINARY                  DSL00190
      SUBROUTINE NUMRC                                                   DSL00200
      COMMON NALRM,IERR,ICONT,JPT,KEY,ISTRT,KA(6),LETRS(72)             DSL00210
      EQUIVALENCE (KA(6),NMBR)                                           DSL00220
      NMBR=0                                                             DSL00230
    3 IF(LETRS(ISTRT)-64)10,4,5                                          DSL00240
    5 IF(LETRS(ISTRT)-240)7,6,6                                          DSL00250
    6 NMBR=10*NMBR+LETRS(ISTRT)-240                                      DSL00260
    4 ISTRT=ISTRT+1                                                      DSL00270
      IF(ISTRT-72)3,3,10                                                 DSL00280
    7 IF(LETRS(ISTRT)-107)8,15,10                                        DSL00290
```

251

```
      8 IF(LETRS(ISTRT)-93)9,15,10                              DSL00300
      9 IF(LETRS(ISTRT)-76)10,15,10                             DSL00310
     10 NALRM=4                                                 DSL00320
        CALL ERR1                                               DSL00330
     15 ISTRT=ISTRT+1                                           DSL00340
        END                                                     DSL00360
*STORE        UA      NUMRC                                     DSL00380
// ASM                                                          DSL00390
*LIST                                                           DSL00400
                        ENT     OPEN                            DSL00410
                        ENT     PACK                            DSL00420
                        ENT     UNPAK                           DSL00430
                        ENT     BNDEC                           DSL00440
                OPEN    DC      0                               DSL00450
                        STX   1 SAVE+1                          DSL00460
                        STX   2 SAVE+3                          DSL00470
                        LDX  L1 LETRS-35                        DSL00480
                        LDX   2 -72                             DSL00490
                LOOP    SLT     8                               DSL00500
                        LD    1 0                               DSL00510
                        RTE     8                               DSL00520
                        STO  L2 LETRS+2                         DSL00530
                        SRA     8                               DSL00540
                        SLT     8                               DSL00550
                        STO  L2 LETRS+1                         DSL00560
                        MDX   1 1                               DSL00570
                        MDX   2 2                               DSL00580
                        MDX     LOOP                            DSL00590
                SAVE    LDX  L1 *-*                             DSL00600
                        LDX  L2 *-*                             DSL00610
                        BSC   I OPEN                            DSL00620
                PACK    DC      0                               DSL00630
                        LD    L KA-2                            DSL00640
                        SRT     6                               DSL00650
                        LD    L KA-1                            DSL00660
                        SRT     6                               DSL00670
                        LD    L KA                             DSL00680
                        SLT     9                               DSL00690
                        AND     MASK                            DSL00700
                        STO   L KA                             DSL00710
                        LD    L KA-4                            DSL00720
                        SRT     6                               DSL00730
                        LD    L KA-3                            DSL00740
                        SRT     6                               DSL00750
                        LD    L KA-2                            DSL00760
                        SLT     12                              DSL00770
                        AND     MASK                            DSL00780
                        STO   L KA-1                            DSL00790
                        BSC   I PACK                            DSL00800
                UNPAK   DC      0                               DSL00810
                        STX   1 SAVE2+1                         DSL00820
                        LDX  L1 KA-5                            DSL00830
                        LD      BLANK                           DSL00840
                        STO   1 0                               DSL00850
                        LD    1 4                               DSL00860
                        AND     MASK1                           DSL00870
                        BSC     +-                              DSL00880
                        OR      ZONE1                           DSL00890
                        EOR     ZONE                            DSL00900
                        STO   1 1                               DSL00910
                        LD    1 4                               DSL00920
                        SRA     6                               DSL00930
                        AND     MASK1                           DSL00940
                        BSC     +-                              DSL00950
                        OR      ZONE1                           DSL00960
                        EOR     ZONE                            DSL00970
                        STO   1 2                               DSL00980
                        LD    1 4                               DSL00990
                        SRT     15                              DSL01000
                        LD    1 5                               DSL01010
                        SLT     3                               DSL01020
                        AND     MASK1                           DSL01030
                        BSC     +-                              DSL01040
                        OR      ZONE1                           DSL01050
                        EOR     ZONE                            DSL01060
                        STO   1 3                               DSL01070
                        LD    1 5    .                          DSL01080
                        SRA     3                               DSL01090
                        AND     MASK1                           DSL01100
                        BSC     +-                              DSL01110
                        OR      ZONE1                           DSL01120
                        EOR     ZONE                            DSL01130
```

252

```
                    STO    1   4                                       DSL01140
                    LD     1   5                                       DSL01150
                    SRA        9                                       DSL01160
                    AND        MASK1                                   DSL01170
                    BSC        +-                                      DSL01180
                    OR         ZONE1                                   DSL01190
                    EOR        ZONE                                    DSL01200
                    SLA        8                                       DSL01210
                    OR     1   4                                       DSL01220
                    STO    1   5                                       DSL01230
                    LD     1   3                                       DSL01240
                    SLA        8                                       DSL01250
                    OR     1   2                                       DSL01260
                    STO    1   4                                       DSL01270
                    LD     1   1                                       DSL01280
                    SLA        8                                       DSL01290
                    OR     1   0                                       DSL01300
                    STO    1   3                                       DSL01310
        SAVE2   LDX    L1  *-*                                         DSL01320
                BSC    I   UNPAK                                       DSL01330
        MASK    DC         /7FFF                                      DSL01340
        MASK1   DC         /003F                                      DSL01350
        BLANK   DC         /0040                                      DSL01360
        ZONE    DC         /00C0                                      DSL01370
        ZONE1   DC         /0080                                      DSL01380
        BNDEC   DC         0                                          DSL01390
                LD     L   KA-5                                       DSL01400
                LIBF       BINDC                                      DSL01410
                DC         DCML                                       DSL01420
                LD         LEFTP                                      DSL01430
                STO        DCML+2                                     DSL01440
                LD     L   KA-4                                       DSL01450
                BSC    L   OVER,+                                     DSL01460
                LD         DCML+3                                     DSL01470
                STO        DCML+2                                     DSL01480
                LD         DCML+4                                     DSL01490
                STO        DCML+3                                     DSL01500
                LD         DCML+5                                     DSL01510
                STO        DCML+4                                     DSL01520
                LD         DCML+6                                     DSL01530
                STO        DCML+5                                     DSL01540
                SRA        16                                         DSL01550
                S      L   KA-4                                       DSL01560
        OVER    A          KFOR1                                      DSL01570
                STO        AX                                         DSL01580
                S          ONE                                        DSL01590
                STO        AY                                         DSL01600
                LIBF       HOLEB                                      DSL01610
                DC         0                                          DSL01620
                DC         DCML+2                                     DSL01630
        AX      DC         *-*                                        DSL01640
                DC         2                                          DSL01650
                LIBF       HOLEB                                      DSL01660
                DC         0                                          DSL01670
                DC         DCML+4                                     DSL01680
        AY      DC         *-*                                        DSL01690
                DC         2                                          DSL01700
                BSC    I   BNDEC                                      DSL01710
        DCML    BSS        6                                          DSL01720
                DC         /2420       COMMA                          DSL01730
        LEFTP   DC         /8120       LEFT PAREN                     DSL01740
        ONE     DC         /1                                         DSL01750
        KFOR1   DC         /7963                                      DSL01760
        KA      EQU        /7FF9                                      DSL01770
        LETRS   EQU        /7FF3                                      DSL01780
                END                                                   DSL01790
*STORE       UA      OPEN                                             DSL01810
// FOR OUTP                                                           DSL01820
*LIST SYMBOL TABLE                                                    DSL01830
*LIST SUBPROGRAM NAMES                                                DSL01840
*LIST ALL
*ONE WORD INTEGERS                                                    DSL01850
*NONPROCESS PROGRAM                                                   DSL01860
*NAME OUTP                                                            DSL01870
*IOCS(KEYBOARD,TYPEWRITER,CARD,1132 PRINTER,DISK)                     DSL01880
**ROUTINE TO GENERATE FORTRAN PROGRAM                                 DSL01890
      DEFINE FILE 1(5000,1,U,MFOR),2(1000,1,U,NDATA)                  DSL01900
      COMMON NALRM,IERR,ICONT,JPT,KEY,ISTRT,KA(6),LETRS(73),ICHAR(12),DSL01910
    1 IRES(158),KSYMB(400),NSYMB,ITYPE,NOUT,NINP,NPAR,NINT,NDATA,NFOR,DSL01920
    2 NSORT,ISORT,NMEM,NING,NINGL,NSTOR,NTSEQ,NREAL,MFCR,              DSL01930
    3 KOUT(100),LFORT(100),NFORT(100),LIN(100),NIN(100),INPUT(300),   DSL01940
    4 KPAR( 60),KSTOR(10),LINT(50),INGER(10),MEMRY(15),KMEM(15),      DSL01950
```

253

```
    5 LSTOR(10),INGRL(10),KREAL(20),KSORT(20)                         DSL01960
      COMMON KFORT(128),KSPEC( 36),LOCAL(36),NLOC,NSPEC,KSEQ(120),      DSL01970
    1 KOMN(110),KBUFR(36)                                              DSL01980
      EQUIVALENCE (IRES(106),IDOLR),(IRES(107),IBLNK),(IRES(158),IONE) DSL01990
      KPT=JPT                                                          DSL02000
      L=ISTRT                                                          DSL02005
      IF(NSORT)101,101,103                                            DSL02020
  101 NTSEQ=NOUT                                                       DSL02030
      DO 102  I=1,NOUT                                                 DSL02040
  102 KSEQ(I)=I                                                        DSL02050
C...  TEST IF ALL PARAMETERS INPUT                                     DSL02060
      DO 104  I=1,NINP                                                 DSL02070
      KVAR=INPUT(I)                                                    DSL02080
      DO 105  J=1,NOUT                                                 DSL02090
      IF(KOUT(J)-KVAR)105,104,105                                     DSL02100
  105 CONTINUE                                                         DSL02110
C...  TEST IF PARAMETER OR STORAG VARIABLE                             DSL02120
      DO 106  J=1,NPAR                                                 DSL02130
      IF(KPAR(J)-KVAR)106,104,106                                     DSL02140
  106 CONTINUE                                                         DSL02150
      IF(NSTOR)117,117,108                                            DSL02160
  108 DO 109  J=1,NSTOR                                                DSL02170
      IF(KSTOR(J)-KVAR)109,104,109                                    DSL02180
  109 CONTINUE                                                         DSL02190
  117 IF(KVAR)104,104,118                                             DSL02200
  118 KA(1)=KSYMB(KVAR)                                                DSL02210
      KA(2)=KSYMB(KVAR+200)                                            DSL02220
      CALL UNPAK                                                       DSL02230
      WRITE(KPT,200) KA(1),KA(2),KA(3)                                 DSL02240
  200 FORMAT(25H PARAM NOT INPUT, SET=0   3A2)                         DSL02250
      IF(NPAR-60)107,300,300                                           DSL02260
  300 WRITE(KPT,210) NPAR,NCOM                                         DSL02270
  210 FORMAT(33H CHECK MAX VS. PARAM, COMMON VARS 2I4)                 DSL02280
      GO TO 150                                                        DSL02290
  107 NPAR=NPAR+1                                                      DSL02300
      KPAR(NPAR)=KVAR                                                  DSL02310
  104 CONTINUE                                                         DSL02320
C...  FORM COMMON VARIABLE LIST (ALSO SYMBOL TABLE)                    DSL02330
C...  SIX SYSTEMS VARS, INTEGRALS, OUTPUT VARS, PARAMS, STORAG VARS    DSL02340
  103 KA(3)=240                                                        DSL02350
      KA(4)=240                                                        DSL02360
      KA(5)=240                                                        DSL02370
      DO 5  I=1,6                                                      DSL02380
    5 KCMN(I)=I                                                        DSL02390
      IF(NINT)15,15,6                                                  DSL02400
    6 NKS=1                                                            DSL02402
      M=1                                                              DSL02404
  222 NEXT=KSEQ(NKS)                                                   DSL02406
C...  USE OUTPUT SEQUENCE TO SET UP INTEGRATORS IN COMMON             DSL02408
      DO 10  I=1,NINT                                                  DSL02410
      J=LINT(I)                                                        DSL02420
      IF(J)203,10,223                                                  DSL02430
  223 IF(J-NEXT)10,8,10                                                DSL02432
  203 IF(J+NEXT)10,7,10                                                DSL02434
    7 KA(1)=233                                                        DSL02440
      KA(2)=233                                                        DSL02450
      IF(KA(5)-249)3,2,2                                               DSL02460
    2 KA(4)=KA(4)+1                                                    DSL02470
      KA(5)=240                                                        DSL02480
      GO TO 4                                                          DSL02490
    3 KA(5)=KA(5)+1                                                    DSL02500
    4 CALL PACK                                                        DSL02510
      NSYMB=NSYMB+1                                                    DSL02520
      KSYMB(NSYMB)=KA(1)                                               DSL02530
      KSYMB(NSYMB+200)=KA(2)                                           DSL02540
      KOMN(M+6)=NSYMB                                                  DSL02550
      GO TO 204                                                        DSL02560
    8 KCMN(M+6)=KOUT(J)                                                DSL02570
      KOUT(J)=-KOUT(J)                                                 DSL02580
  204 IF(M-NINT)206,15,15                                              DSL02581
  206 M=M+1                                                            DSL02582
      I=I+1                                                            DSL02583
      IF(I-NINT)224,224,205                                           DSL02584
  224 J=LINT(I)                                                        DSL02587
      IF(J)205,7,205                                                   DSL02588
   10 CONTINUE                                                         DSL02590
  205 NKS=NKS+1                                                        DSL02592
      IF(NKS-NTSEQ)222,222,15                                         DSL02594
   15 NCOM=NINT+6                                                      DSL02600
      IF(NING)115,115,110                                             DSL02610
C...  ADD DUMMY WORDS IN COMMON FOR INTEGERS TO MATCH SYMBOL TABLE     DSL02620
  110 KOUNT=IRES(151)                                                  DSL02630
```

254

```
          DO 112   I=1,NING                                              DSL02640
          DO 112   J=7,NPAR                                              DSL02650
          IF(INGER(I)-KPAR(J))112,111,112                               DSL02660
   111    KPAR(J)=KPAR(NPAR)                                            DSL02670
          NPAR=NPAR-1                                                    DSL02680
   112 CONTINUE                                                         DSL02690
          DO 113   I=1,NING                                              DSL02700
          NPAR=NPAR+2                                                    DSL02710
          KPAR(NPAR-1)=-1                                                DSL02720
   113 KPAR(NPAR)=INGER(I)                                              DSL02730
   115 DO  17   I=1,NOUT                                                 DSL02740
          KVAR=KOUT(I)                                                   DSL02750
          IF(KVAR)17,17,16                                              DSL02760
C...    SKIP IF ALSO A PARAMETER                                        DSL02762
    16 DO  13   K=7,NPAR                                                DSL02764
          IF(KVAR-KPAR(K))13,17,13                                     DSL02766
    13 CONTINUE                                                         DSL02768
          IF(KVAR-500)116,116,17                                       DSL02770
   116 DO  12   K=1,NCOM                                                DSL02780
          IF(KVAR-KOMN(K))12,17,12                                     DSL02790
    12 CONTINUE                                                         DSL02800
          IF(NSTOR)215,215,214                                         DSL02833
   214 DO 211   K=2,NSTOR                                               DSL02834
          IF(KVAR-KSTOR(K))211,17,211                                  DSL02835
   211 CONTINUE                                                         DSL02836
   215 NCOM=NCOM+1                                                      DSL02840
          KOMN(NCOM)=KVAR                                               DSL02850
          IF(NCOM-110)17,17,300                                        DSL02860
    17 CONTINUE                                                         DSL02870
   264 IF(KPAR(NPAR))263,263,262                                       DSL02880
   263 NPAR=NPAR-1                                                      DSL02882
          IF(NPAR)262,262,264                                          DSL02884
C...    OUTPUT CONTROL CARDS FOR UPDATE SUBROUTINE                      DSL02888
   262 WRITE(L,250)                                                     DSL02890
   250 FORMAT(/6H// JOB/6H// FOR/9H*LIST ALL/18H*ONE  WORD INTEGERS/40H**DSL02900
      1SL/1800 (DIGITAL SIMULATION LANGUAGE)/22H        SUBROUTINE UPDAT )DSL02910
C...    OUTPUT COMMON VARIABLE LIST                                     DSL02920
          KGO=0                                                         DSL02930
          NB1=6                                                         DSL02940
          KGO1=0                                                        DSL02950
          NWDS=10                                                       DSL02960
          KOUN=0                                                        DSL02965
          KBUFR(1)=IBLNK                                                DSL02970
          KBUFR(2)=IBLNK                                                DSL02980
          IF(NREAL)51,51,9                                             DSL02990
C...    REMOVE INTEGERS FROM REAL VARIABLE TABLE                        DSL03000
     9 IF(NING)11,11,14                                                DSL03010
    14 DO 29    I=1,NING                                                DSL03020
          DO 27    J=1,NREAL                                            DSL03030
          IF(INGER(I)-KREAL(J))27,20,27                                DSL03040
    20 KREAL(J)=KREAL(NREAL)                                           DSL03050
          NREAL=NREAL-1                                                 DSL03060
          IF(NREAL)51,51,29                                            DSL03070
    27 CONTINUE                                                         DSL03080
    29 CONTINUE                                                         DSL03090
    11 KBUFR(4)=IRES(152)                                              DSL03100
          KBUFR(5)=IRES(153)                                           DSL03110
          KBUFR(6)=IBLNK                                                DSL03120
          KSW=1                                                         DSL03130
          N3=NREAL                                                      DSL03140
    18 N1=1                                                             DSL03150
          KBUFR(3)=IBLNK                                                DSL03160
    19 NBUF=NB1                                                         DSL03170
          N2=N1+NWDS-1                                                  DSL03180
    26 IF(N2-N3)22,22,21                                               DSL03190
    21 N4=N3-N1+1                                                       DSL03200
          N2=N3                                                         DSL03210
    22 DO 30    I=N1,N2                                                 DSL03220
          IF(KSW-3)23,24,80                                            DSL03230
    80 IF(KSW-7)81,25,25                                               DSL03240
    23 J=KREAL(I)                                                       DSL03250
          GO TO 28                                                      DSL03260
    24 J=INGER(I)                                                       DSL03270
          GO TO 28                                                      DSL03280
    81 J=KPAR(I)                                                        DSL03290
          IF(J)128,128,28                                              DSL03300
   128 NBUF=NBUF+3                                                      DSL03310
          KBUFR(NBUF-2)=IRES(3)                                         DSL03320
          KBUFR(NBUF-1)=IRES(149)                                       DSL03330
          KBUFR(NBUF)=KOUNT                                             DSL03340
          KOUNT=KOUNT+256                                               DSL03350
          GO TO 30                                                      DSL03360
```

```
    25 J=KCMN(I)                                                    DSL03370
    28 KA(1)=KSYMB(J)                                               DSL03380
       KA(2)=KSYMB(J+200)                                           DSL03390
       CALL UNPAK                                                   DSL03400
       NBUF=NBUF+3                                                  DSL03410
       KBUFR(NBUF-2)=KA(1)                                          DSL03420
       KBUFR(NBUF-1)=KA(2)                                          DSL03430
       KBUFR(NBUF)=KA(3)                                            DSL03440
    30 CONTINUE                                                     DSL03450
       IF(N2-N3)31,35,35                                            DSL03460
    31 KEND=36                                                      DSL03470
    32 IF(KGO)33,33,83                                              DSL03480
    33 K1=NB1+3                                                     DSL03490
       IF(K1-KEND)75,75,40                                          DSL03500
    75 DO 34   K=K1,KEND,3                                          DSL03510
    34 KBUFR(K)=KBUFR(K)+43                                         DSL03520
       GO TO 40                                                     DSL03530
    35 KEND=NBUF-3                                                  DSL03540
       KSW=KSW+1                                                    DSL03550
       IF(KSW-7)32,36,36                                            DSL03560
    36 IF(NPAR-6)38,38,37                                           DSL03570
    37 IF(NBUF-36)39,31,31                                          DSL03580
    38 KSW=6                                                        DSL03590
       GO TO 32                                                     DSL03600
    39 N2=NWDS-N4+6                                                 DSL03610
       KSW=5                                                        DSL03620
       N3=NPAR                                                      DSL03630
       N1=7                                                         DSL03640
       GO TO 26                                                     DSL03650
    40 IF(KGO1)59,52,54                                             DSL03660
    54 WRITE(L,254) (KBUFR(I),I=16,NBUF)                            DSL03670
   254 FORMAT(30H          COMMON NALRM,IZZZ( 833)  , 21A2)         DSL03680
       KGO1=0                                                       DSL03690
       KOUN=7                                                       DSL03700
       GO TO 53                                                     DSL03710
    59 WRITE(L,255)                                                 DSL03720
   255 FORMAT(12H          COMMON)                                  DSL03730
       KGO1=0                                                       DSL03740
       KOUN=0                                                       DSL03750
       GO TO 57                                                     DSL03760
    52 IF(KOUN-33)57,56,56                                          DSL03770
    56 IF(KEND-36)83,84,84                                          DSL03775
    84 KBUFR(KEND)=KBUFR(KEND)-43                                   DSL03780
       KGO1=-1                                                      DSL03790
    57 KOUN=KOUN+11                                                 DSL03800
    83 WRITE(L,251) (KBUFR(I),I=1,NBUF)                             DSL03810
   251 FORMAT(36A2)                                                 DSL03820
    53 GO TO(41,51,41,42,41,44,41,43),KSW                          DSL03830
    41 N1=N2+1                                                      DSL03840
       IF(KGO)49,49,19                                              DSL03850
    49 KBUFR(3)=IONE                                                DSL03860
       NB1=3                                                        DSL03870
       NWDS=11                                                      DSL03880
       GO TO 19                                                     DSL03890
C...   WRITE DIMENSION STMNT FOR ALL STORAG + INTGRL VARS           DSL03892
    42 IF(NSTOR)134,134,130                                         DSL03893
   130 IF(NINGL)134,134,244                                         DSL03895
   244 J=2                                                          DSL03896
   249 DO 245   I=1,NINGL                                           DSL03897
       IF(KSTOR(J)-INGRL(I))245,246,245                            DSL03898
   246 NSTOR=NSTOR-1                                                DSL03900
       IJ=INGRL(I)                                                  DSL03902
       KA(1)=KSYMB(IJ)                                              DSL03903
       KA(2)=KSYMB(IJ+200)                                          DSL03904
       CALL UNPAK                                                   DSL03905
       WRITE(L,270) KA(1),KA(2),KA(3)                              DSL03906
   270 FORMAT(16H          DIMENSION 2A2,A1,3H(1) )                 DSL03907
       DO 248   J2=J,NSTOR                                          DSL03910
       KSTOR(J2)=KSTOR(J2+1)                                        DSL03911
   248 LSTOR(J2)=LSTOR(J2+1)                                        DSL03912
       GO TO 247                                                    DSL03913
   245 CONTINUE                                                     DSL03914
       J=J+1                                                        DSL03915
   247 IF(J-NSTOR)249,249,134                                       DSL03916
   134 IF(NSPEC)141,141,140                                         DSL03918
   140 WRITE(L,251)(KSPEC(I),I=1,36)                               DSL03919
   141 KGO1=1                                                       DSL03920
       NB1=15                                                       DSL03930
       NWDS=7                                                       DSL03940
    46 KSW=7                                                        DSL03950
```

256

```
      N3=NCOM                                                   DSL03960
      GO TO 18                                                  DSL03970
   43 KSW=5                                                     DSL03980
      N3=NPAR                                                   DSL03990
      N1=7                                                      DSL04000
      GO TO 19                                                  DSL04010
   51 IF(NING)42,42,55                                          DSL04020
   55 NB1=9                                                     DSL04030
      NWDS=9                                                    DSL04040
      KSW=3                                                     DSL04050
      N3=NING                                                   DSL04060
      KBUFR(4)=IRES(154)                                        DSL04070
      KBUFR(5)=IRES(155)                                        DSL04080
      KBUFR(6)=IRES(156)                                        DSL04090
      KBUFR(7)=IRES(157)                                        DSL04100
      KBUFR(8)=IBLNK                                            DSL04110
      KBUFR(9)=IBLNK                                            DSL04120
      GO TO 18                                                  DSL04130
   44 IF(KGO)45,45,67                                           DSL04140
   45 KPOIN=NCOM+NPAR-NING-5                                    DSL04150
      KPOIT=KPOIN                                               DSL04160
C...  MOVE STORAG VARIABLES TO PARAMETER TABLE                 DSL04170
      IF(NSTOR)66,66,61                                         DSL04180
C...  WRITE STORAG VARIABLES  (ADD TO COMMON)                  DSL04190
   61 DO 62  J=2,NSTOR                                          DSL04200
      NPAR=NPAR+1                                               DSL04210
   62 KPAR(NPAR)=KSTOR(J)                                       DSL04220
      LSTOR(1)=NSTOR-1                                          DSL04230
      KPOIT=KPOIT+LSTOR(1)                                      DSL04240
      KBUFR(1)=IRES(149)                                        DSL04250
      KBUFR(13)=IRES(150)                                       DSL04260
      KBUFR(25)=IRES(151)                                       DSL04270
      DO 98  I=2,NSTOR                                          DSL04280
      KPOIT=KPOIT+LSTOR(I)                                      DSL04290
      J=KSTOR(I)                                                DSL04300
      KA(1)=KSYMB(J)                                            DSL04310
      KA(2)=KSYMB(J+200)                                        DSL04320
      CALL UNPAK                                                DSL04330
      KBUFR(I)=KA(1)                                            DSL04340
      KBUFR(I+12)=KA(2)                                         DSL04350
   98 KBUFR(I+24)=KA(3)                                         DSL04360
      N1=1                                                      DSL04370
      ICOMA=IBLNK+43                                            DSL04380
      IF(NSTOR-5)91,91,90                                       DSL04390
   90 N2=5                                                      DSL04400
      GO TO 92                                                  DSL04410
   91 N2=NSTOR                                                  DSL04420
   92 WRITE(L,260) (ICOMA,KBUFR(I),KBUFR(I+12),KBUFR(I+24),     DSL04430
     1 LSTOR(I),I=N1,N2)                                        DSL04440
  260 FORMAT(6H     1 5(    3A2,A1,1H( I3,1H)))                 DSL04450
      IF(N2-NSTOR)93,66,66                                      DSL04460
   93 N1=6                                                      DSL04470
      GO TO 91                                                  DSL04480
C...  WRITE FORTRAN STATEMENTS                                 DSL04490
   66 DO 60  I=1,NTSEQ                                          DSL04500
      J=KSEQ(I)                                                 DSL04510
      N1=LFORT(J)                                               DSL04520
      N2=NFORT(J)                                               DSL04530
      N3=IABS(N2)                                               DSL04540
      IF(N2)74,60,74                                            DSL04550
   74 READ(1'N1)(KFORT(M),M=1,N3)                              DSL04560
      IF(N2)47,60,58                                            DSL04570
   47 N4=1                                                      DSL04580
      DO 50  M=1,N3                                             DSL04590
      IF(KFORT(M)-IDOLR)50,48,50                                DSL04600
   48 N5=M-1                                                    DSL04610
      WRITE(L,251) (KFORT(K),K=N4,N5)                           DSL04620
      N4=M+1                                                    DSL04630
   50 CONTINUE                                                  DSL04640
      GO TO 60                                                  DSL04650
   58 WRITE(L,251) (KFORT(K),K=1,N3)                            DSL04660
   60 CONTINUE                                                  DSL04670
C...  END OF UPDAT SUBROUTINE                                  DSL04680
      WRITE(L,252)                                              DSL04690
C...  WRITE SYMBOL TABLE                                       DSL04770
  252 FORMAT('     RETURN'/'      END'/'// DUP'/ '*DELETE            U
     1PDAT'/'*STORE     WS  UA  UPDAT'//'*DELETE              SIMUL'/
     2'*DUMP     UA  WS  INTEG'/'*STORECI    WS  UA  SIMUL'/'// XEQ SI
```

```
      3MX')
      IF(NING)165,165,160                                      DSL04780
160   I=7                                                       DSL04790
163   IF(KPAR(I))161,161,164                                    DSL04800
161   NPAR=NPAR-1                                               DSL04810
      DO 162   J=I,NPAR                                         DSL04820
162   KPAR(J)=KPAR(J+1)                                         DSL04830
164   I=I+1                                                     DSL04840
      IF(I-NPAR)163,165,165                                     DSL04850
165   NSYM=NCOM+NPAR-6                                          DSL04860
      WRITE(L,253) NINT,NSYM,KPOIT                              DSL04870
253   FORMAT(3I4)                                               DSL04880
      KGO=1                                                     DSL04890
      NB1=0                                                     DSL04900
      NWDS=12                                                   DSL04910
      IF(NSYM-140)46,46,155                                     DSL04920
155   WRITE(KPT,156)NSYM                                        DSL04930
156   FORMAT(I4,23H SYMBOLS, REDUCE TO 140)                     DSL04940
      GO TO 46                                                  DSL04950
67    IF(NSTOR)65,65,63                                         DSL04960
C...  WRITE PARAM CARD TO LOCATE STORAG VARIABLE TABLES         DSL04970
63    DO 94   I=2,NSTOR                                         DSL04980
      J=KSTOR(I)                                                DSL04990
      KA(1)=KSYMB(J)                                            DSL05000
      KA(2)=KSYMB(J+200)                                        DSL05010
      CALL UNPAK                                                DSL05020
      KBUFR(I-1)=ICOMA                                          DSL05030
      KBUFR(I+8)=KA(1)                                          DSL05040
      KBUFR(I+17)=KA(2)                                         DSL05050
      KBUFR(I+26)=KA(3)                                         DSL05060
      KPOIN=KPOIN+LSTOR(I-1)                                    DSL05070
94    KSTOR(I)=KPOIN                                            DSL05080
      KBUFR(1)=IBLNK                                            DSL05090
      N1=1                                                      DSL05100
      IF(NSTOR-7)69,69,64                                       DSL05110
64    N2=6                                                      DSL05120
      GO TO 71                                                  DSL05130
69    N2=NSTOR-1                                                DSL05140
71    WRITE(L,261) (KBUFR(I),KBUFR(I+9),KBUFR(I+18),KBUFR(I+27),DSL05150
     1 KSTOR(I+1),I=N1,N2)                                      DSL05160
261   FORMAT(6HPARAM 6(3A2,A1,1H= I3))                          DSL05170
      IF(N2-NSTOR+1)72,65,65                                    DSL05180
72    N1=7                                                      DSL05190
      GO TO 69                                                  DSL05200
65    N3=1                                                      DSL05210
      NDAT=NDATA                                                DSL05220
      DO 70   I=1,NDAT                                          DSL05230
      READ(2'I)LETRS(N3)                                        DSL05240
      IF(LETRS(N3)-IDOLR)73,68,73                               DSL05250
68    N3=N3-1                                                   DSL05260
      WRITE(L,251)(LETRS(K),K=1,N3)                             DSL05270
      N3=1                                                      DSL05280
      GO TO 70                                                  DSL05290
73    N3=N3+1                                                   DSL05300
70    CONTINUE                                                  DSL05310
150   STOP                                                      DSL05360
      END                                                       DSL05370
// DUP
*DELETE             OUTP
// FOR REMSW                                                    DSL05610
*ONE WORD INTEGERS                                             DSL05620
      SUBROUTINE REMSW(J,K)                                     DSL05630
      CALL DATSW(J,K)                                           DSL05640
      RETURN                                                    DSL05650
      END                                                       DSL05660
*STORE       UA       REMSW                                     DSL05680
// FOR READR                                                    DSL05400
*LIST SYMBOL TABLE                                              DSL05410
*LIST SUBPROGRAM NAMES                                          DSL05420
*LIST ALL
*ONE WORD INTEGERS                                             DSL05430
**READ ROUTINE FOR DSL/1800 TRANSLATOR                         DSL05440
      SUBROUTINE READR                                          DSL05450
      COMMON NALRM,IERR,ICONT,JPT,KEY,ISTRT,KA(6),LETRS(72),ID(1810),DSL05460
     1 KEYB,KARD                                                DSL05465
      CALL REMSW(2,IRD)                                         DSL05480
      GO TO(4,5),IRD                                            DSL05490
4     L=KEYB                                                    DSL05500
```

258

```
          GO TO 6                                            DSL05504
    5 L=KARD                                                 DSL05506
    6 READ(  L,10)(LETRS(K),K=1,36)                          DSL05510
   10 FCRMAT(36A2)                                           DSL05520
      CALL REMSW(6,K)                                        DSL05530
      GO TO(12,13),K                                         DSL05540
   13 KPT = JPT                                              DSL05545
      WRITE (KPT,11) (LETRS(K),K=1,36)                       DSL05550
   11 FORMAT(1H   36A2)                                      DSL05560
   12 RETURN                                                 DSL05570
      END                                                    DSL05580
// DUP
*DELETE                READR
*STORE          WS  UA  READR
// FOR SCAN                                                  DSL05690
*LIST SYMBOL TABLE                                           DSL05700
*LIST SUBPROGRAM NAMES                                       DSL05710
*LIST ALL
*ONE WORD INTEGERS                                           DSL05720
**DSL/1130 AND 1800 SCAN ROUTINE                            DSL05730
      SUBROUTINE SCAN                                        DSL05740
      COMMON NALRM,IERR,ICONT,JPT,KEY,ISTRT,KA(6),LETRS(73),ICHAR(12), DSL05750
     1 IRES(158),KSYMB(400),NSYMB,ITYPE,NOUT,NINP,NPAR,NINT,NDATA,NFOR, DSL05760
     2 NSORT,ISORT,NMEM,NING,NINGL,NSTOR,NTSEQ,NREAL,MFOR,  DSL05770
     3 KOUT(100),LFORT(100),NFORT(100),LIN(100),NIN(100),INPUT(300), DSL05780
     4 KPAR( 60),KSTOR(10),LINT(50),INGER(10),MEMRY(15),KMEM(15), DSL05790
     5 LSTOR(10),INGRL(10),KREAL(20),KSORT(20)              DSL05800
      COMMON ID(204),KEY1,KPREV,KEXP,KPT,J                  DSL05805
C... SUBROUTINE TO SELECT ALPHA-NUMERIC NAME FROM LETRS(ISTRT) WITH  DSL05810
C... EQUALS, COMMA, LEFT OR RIGHT PARENTHESIS, END OF CARD, OR  DSL05820
C... AN ARITHMETIC OPERATOR AS SEPARATOR                     DSL05830
C... (... WILL FORCE READ + CONTINUATION)                    DSL05840
      IF(ICONT)50,80,50                                      DSL05850
   50 ICONT=0                                                DSL05860
      KEY=KEY1                                               DSL05870
      GO TO 51                                               DSL05880
   80 J=0                                                    DSL05890
      KEY=0                                                  DSL05900
      KPT=0                                                  DSL05910
      KEXP=0                                                 DSL05920
   51 I=ISTRT                                                DSL05930
  100 LETER=LETRS(I)                                         DSL05940
C... TEST FOR BLANK (64)                                     DSL05950
      IF(LETER-64)85,2C0,101                                 DSL05960
C... TEST IF EOF FROM KEYBOARD                               DSL05964
   85 IF(LETER-21)99,86,99                                   DSL05965
   86 I=73                                                   DSL05966
      GO TO 145                                              DSL05967
C... TEST IF SPECIAL CHAR OR ALPHA-NUMERIC-EQUALS(126), $(91) - 029 KP DSL05970
  101 IF(LETER-126)108,102,120                               DSL05980
  102 KPREV=K                                                DSL05985
      K=6                                                    DSL05990
      GO TO 6                                                DSL06000
  108 IF(LETER-91)109,145,109                                DSL06010
  109 KPREV=K                                                DSL06020
C... TEST FOR SPECIAL CHARACTER                              DSL06030
      DO 110  K=1,12                                         DSL06040
      IF(LETER-ICHAR(K))110,115,110                          DSL06050
  110 CONTINUE                                               DSL06060
   99 NALRM=1                                                DSL06070
      ISTRT=I                                                DSL06080
      CALL ERR1                                              DSL06090
      IERR=1                                                 DSL06100
      GO TO 200                                              DSL06110
C... COMMA (107), LEFT PAREN (77/108), RIGHT PAREN (93/76),  DSL06120
C... EQUALS (123), PLUS (78/80), MINUS (96), ASTERISK (92),  DSL06130
C... DECIMAL (75), SLASH (97)                                DSL06140
  115 GO TO(1,2,2,4,4,6,7,7,7,10,11,12),K                    DSL06150
    1 IF(KEY-1)200,21,150                                    DSL06160
   21 ISTRT=I+1                                              DSL06170
      GO TO 80                                               DSL06180
    2 IF(KEY-1)200,99,24                                     DSL06190
C... ALPHA NAME FOLLOWED BY (                                DSL06200
   24 KEY=4                                                  DSL06210
C... TEST FOR IF STATEMENT                                   DSL06220
      IF(J-2)150,22,150                                      DSL06230
   22 IF(KA(1)-201)150,23,150                                DSL06240
   23 IF(KA(2)-198)150,121,150                               DSL06250
```

259

```
      4 K=4
        IF(KEY-1)200,21,150                                        DSL06255
      6 IF(KEY-2)104,25,99                                         DSL06260
    104 IF(KPREV-4)99,21,99                                        DSL06270
C...    ALPHA NAME FOLLOWED BY =                                   DSL06275
     25 KEY=3                                                      DSL06280
        GO TO 160                                                  DSL06290
      7 IF(KEY-1)60,26,150                                         DSL06300
     60 IF(KPREV-6)200,200,99                                      DSL06310
     26 IF(KEXP)21,21,27                                           DSL06320
     27 KEXP=-1                                                    DSL06330
        GO TO 200                                                  DSL06340
     10 IF(KEY-1)28,21,150                                         DSL06350
     28 IF(KPREV-10)40,29,99                                       DSL06360
     29 K=0                                                        DSL06370
        GO TO 21                                                   DSL06380
     11 IF(KEY-1)30,37,31                                          DSL06390
     30 KEY=1                                                      DSL06400
        J=-1                                                       DSL06410
        KSAV=KPREV                                                 DSL06420
        GO TO 32                                                   DSL06425
     31 IF(KPT)99,32,34                                            DSL06430
     32 KPT=1                                                      DSL06440
        KPT1=I                                                     DSL06450
        GO TO 200                                                  DSL06460
     37 IF(KPT-1)32,119,34                                         DSL06470
    119 IF(KPREV-11)32,62,32                                       DSL06480
     34 IF(KPREV-11)99,62,99                                       DSL06490
     62 KPT=KPT+1                                                  DSL06500
        IF(KPT-3)200,200,36                                        DSL06510
     36 KPT=1                                                      DSL06520
        KPT1=KPT1+1                                                DSL06530
        GO TO 105                                                  DSL06540
     12 IF(KEY-1)40,21,150                                         DSL06550
     40 IF(KPREV-5)61,200,99                                       DSL06560
     61 IF(KPREV-4)99,200,99                                       DSL06570
    120 K=LETER                                                    DSL06580
C...    TEST IF NUMERIC OR ALPHA CHARACTER                         DSL06590
        IF(LETER-240)123,130,130                                   DSL06600
    123 IF(KEY-1)127,124,128                                       DSL06610
    124 IF(LETER-197)99,125,99                                     DSL06620
    125 IF(KEXP)99,126,99                                          DSL06630
    126 KEXP=1                                                     DSL06640
        GO TO 200                                                  DSL06650
C...    FIRST CHARACTER ALPHA NAME                                 DSL06660
    127 KEY=2                                                      DSL06670
    128 J=J+1                                                      DSL06680
        IF(J-5)122,122,121                                         DSL06690
C...    GREATER THAN 5 CHARACTERS                                  DSL06700
    121 KEY=5                                                      DSL06710
        GO TO 300                                                  DSL06720
    122 KA(J)=LETER                                                DSL06730
    200 I=I+1                                                      DSL06740
        IF(I-72)100,100,145                                        DSL06750
C...    TEST FOR CONTINUATION                                      DSL06760
    145 IF(KPT-3)148,146,99                                        DSL06770
    146 KPT=0                                                      DSL06780
        K=KSAV                                                     DSL06790
        IF(J)147,147,105                                           DSL06795
    147 KEY=0                                                      DSL06800
        J=0                                                        DSL06810
    105 KEY1=KEY                                                   DSL06820
        KEY=7                                                      DSL06830
        KA(6)=KPT1                                                 DSL06840
        GO TO 300                                                  DSL06850
    148 IF(KEY-1)149,149,150                                       DSL06860
C...    END OF CARD, NO ALPHA NAME                                 DSL06870
    149 KEY=6                                                      DSL06880
        GO TO 300                                                  DSL06890
    130 IF(KEY-1)131,200,128                                       DSL06900
C...    FIRST NUMERIC DIGIT                                        DSL06910
    131 J=1                                                        DSL06920
        KEY=1                                                      DSL06930
        GO TO 200                                                  DSL06940
C...    TEST FOR KEY WORDS (DO)                                    DSL06950
    160 IF(J-3)150,150,153                                         DSL06960
    153 IF(KA(1)-196)150,154,150                                   DSL06970
    154 IF(KA(2)-214)150,155,150                                   DSL06990
```

```
      155 IF(KA(3)-240)150,156,156                                   DSL07000
      156 IF(KA(J)-240)121,150,150                                   DSL07010
      150 IF(KPT)63,63,64                                            DSL07020
       64 KPT=0                                                      DSL07030
          GO TO 99                                                   DSL07040
       63 IF(J-6)70,152,99                                           DSL07050
       70 J=J+1                                                      DSL07055
          KA(J)=64                                                   DSL07060
          GO TO 63                                                   DSL07070
      152 ISTRT=I+1                                                  DSL07080
C...     TEST IF 1ST CHAR I TO N                                     DSL07090
          NR1=0                                                      DSL07100
          IF(KA(1)-201)158,159,157                                   DSL07110
      157 IF(KA(1)-213)159,159,158                                   DSL07120
      159 NR1=1                                                      DSL07130
C...     STORE NAME IN SYMBOL TABLE                                  DSL07140
      158 CALL PACK                                                  DSL07150
          DO 58  L=1,NSYMB                                           DSL07160
          IF(KSYMB(L)-KA(1))58,54,58                                 DSL07170
       54 IF(KSYMB(L+200)-KA(2))58,55,58                             DSL07180
       55 KA(4)=L                                                    DSL07190
          GO TO 307                                                  DSL07200
       58 CONTINUE                                                   DSL07210
          IF(KEY-4)59,301,99                                         DSL07220
      301 IF(ITYPE-9)302,59,56                                       DSL07230
      302 IF(ITYPE-8)56,59,56                                        DSL07240
       56 IF(NR1)305,305,59                                          DSL07250
       59 IF(NSYMB-200)303,304,304                                   DSL07260
      304 NALRM=10                                                   DSL07270
          IERR=1                                                     DSL07280
          CALL ERR1                                                  DSL07290
      305 KA(4)=-1                                                   DSL07300
          GO TO 300                                                  DSL07310
      303 NSYMB=NSYMB+1                                              DSL07320
          KA(4)=NSYMB                                                DSL07330
          KSYMB(NSYMB)=KA(1)                                         DSL07340
          KSYMB(NSYMB+200)=KA(2)                                     DSL07350
      307 IF(NR1)300,300,306                                         DSL07360
      306 DO 309  L=1,NREAL                                          DSL07370
          IF(KREAL(L)-KA(4))309,300,309                              DSL07380
      309 CONTINUE                                                   DSL07390
          IF(NREAL-20)308,304,304                                    DSL07400
      308 NREAL=NREAL+1                                              DSL07410
          KREAL(NREAL)=KA(4)                                         DSL07420
      300 RETURN                                                     DSL07430
          END                                                        DSL07440
// DUP
*DELETE           SCAN
*STORE        WS  UA  SCAN
// FOR SORT                                                          DSL07470
*LIST SYMBOL TABLE                                                   DSL07480
*LIST SUBPROGRAM NAMES                                               DSL07490
*LIST ALL
*ONE WORD INTEGERS                                                   DSL07500
*NONPROCESS PROGRAM                                                  DSL07510
*NAME SCRTX                                                          DSL07520
*IOCS(TYPEWRITER,1132 PRINTER)                                       DSL07530
**DSL/1800 STATEMENT SEQUENCING ROUTINE                              DSL07540
      COMMON NALRM,IERR,ICONT,JPT,KEY,ISTRT,KA(6),LETRS(73),ICHAR(12),  DSL07550
     1 IRES(158),KSYMB(400),NSYMB,ITYPE,NOUT,NINP,NPAR,NINT,NDATA,NFOR, DSL07560
     2 NSORT,ISORT,NMEM,NING,NINGL,NSTOR,NTSEQ,NREAL,MFOR,             DSL07570
     3 KOUT(100),LFORT(100),NFORT(100),LIN(100),NIN(100),INPUT(300),   DSL07580
     4 KPAR( 60),KSTOR(10),LINT(50),INGER(10),MEMRY(15),KMEM(15),      DSL07590
     5 LSTCR(10),INGRL(10),KREAL(20),KSORT(20),KFORT(202)              DSL07600
      COMMON KSEQ(120),KSEQ2(30),LIN2(30),NIN2(30),KSEQ3(10)          DSL07602
      KPT=JPT                                                         DSL07604
      JINT=1                                                          DSL07606
      J2=0                                                            DSL07608
      KGO=1                                                          DSL07610
      NSOR1=1                                                        DSL07612
      NOUT1=1                                                        DSL07614
      MFOR=MFOR-1                                                    DSL07616
      NOUT=NOUT-1                                                    DSL07618
      NDATA=NDATA-1                                                  DSL07620
      IF(NSORT)101,101,18                                            DSL07622
  101 NTSEQ=NOUT                                                      DSL07624
      DO 102  I=1,NOUT                                               DSL07626
  102 KSEQ(I)=I                                                      DSL07628
```

```
C...   TEST IF ALL PARAMETERS INPUT                                      DSL07630
       DO 104  I=1,NINP                                                   DSL07632
       KVAR=INPUT(I)                                                      DSL07634
       DO 105  J=1,NOUT                                                   DSL07636
       IF(KOUT(J)-KVAR)105,104,105                                       DSL07638
  105 CONTINUE                                                            DSL07640
C...   TEST IF PARAMETER OR STORAG VARIABLE                              DSL07642
       DO 106  J=1,NPAR                                                   DSL07644
       IF(KPAR(J)-KVAR)106,104,106                                       DSL07646
  106 CONTINUE                                                            DSL07648
       IF(NSTOR)117,117,108                                               DSL07650
  108 DO 109  J=1,NSTOR                                                   DSL07652
       IF(KSTOR(J)-KVAR)109,104,109                                      DSL07654
  109 CONTINUE                                                            DSL07656
  117 IF(KVAR)104,104,118                                                 DSL07658
  118 KA(1)=KSYMB(KVAR)                                                   DSL07660
       KA(2)=KSYMB(KVAR+200)                                             DSL07662
       CALL UNPAK                                                         DSL07664
       WRITE(KPT,200) KA(1),KA(2),KA(3)                                  DSL07666
       IF(NPAR-60)107,210,210                                            DSL07668
  107 NPAR=NPAR+1                                                         DSL07672
       KPAR(NPAR)=KVAR                                                    DSL07674
  104 CONTINUE                                                            DSL07676
       GO TO 400                                                          DSL07678
C...   ESTABLISH CORRECT SEQUENCE OF OUTPUTS                             DSL07680
   18 IF(ISORT)28,28,19                                                  DSL07687
   19 KSORT(NSOR1)=NOUT                                                  DSL07688
C...   SET LOCATIONS OF 1ST AND LAST OUTPUT WITHIN NEXT SORT SECTION     DSL07690
   28 NSOR2=KSORT(NSOR1)                                                 DSL07700
       NSOR3=KSORT(NSOR1+1)                                              DSL07710
       NTIN1=LIN(NSOR2)                                                  DSL07720
       IF(NSOR3-NOUT)30,29,30                                            DSL07730
   29 NTIN2=NINP                                                         DSL07740
       GO TO 31                                                          DSL07750
   30 NSOR4=NSOR3                                                        DSL07753
  170 IF(LIN(NSOR4))171,171,172                                          DSL07754
  171 NSOR4=NSOR4-1                                                      DSL07755
       GO TO 170                                                         DSL07756
  172 NTIN2=LIN(NSOR4)+NIN(NSOR4)-1                                      DSL07760
   31 IF(NSOR2-NOUT1)32,35,32                                            DSL07770
   32 NOUT2=NSOR2-1                                                      DSL07780
   36 DO 33  I=NOUT1,NOUT2                                               DSL07790
       IF(LFORT(I))34,33,34                                              DSL07800
   34 J2=J2+1                                                            DSL07810
       KSEQ(J2)=I                                                        DSL07820
   33 CONTINUE                                                           DSL07830
       GO TO(35,130),KGO                                                 DSL07840
C...   REMOVE VARIABLES FROM INPUT TABLE WHICH ARE NOT COMPUTED         DSL07850
   35 DO 4  I=NTIN1,NTIN2                                                DSL07860
       KVAR=INPUT(I)                                                     DSL07870
       IF(KVAR)65,4,16                                                   DSL07875
   16 DO 5  J=NSOR2,NSOR3                                                DSL07880
       IF(KOUT(J)-KVAR)5,3,5                                             DSL07890
C...   REMOVE FROM INPUT TABLE IF INTEGRAL                              DSL07900
    3 IF(NINT)2,2,9                                                      DSL07910
    9 DO 10  K=1,NINT                                                    DSL07920
       IF(LINT(K)-J)10,8,10                                             DSL07930
   10 CONTINUE                                                           DSL07940
    2 INPUT(I)=J                                                         DSL07950
       GO TO 4                                                           DSL07960
    5 CONTINUE                                                           DSL07970
C...   TEST IF PARAMETER OR STORAG VARIABLE OR OUTPUT IN PREV. SECTION  DSL07980
       DO 6  J=1,NPAR                                                    DSL07990
       IF(KPAR(J)-KVAR)6,8,6                                            DSL08000
    6 CONTINUE                                                           DSL08010
       IF(NSOR2-1)14,14,13                                              DSL08020
   13 DO 17  J=1,NSOR2                                                   DSL08030
       IF(KOUT(J)-KVAR)17,8,17                                          DSL08040
   17 CONTINUE                                                           DSL08050
   14 IF(NSTOR)15,15,11                                                  DSL08060
   11 DO 12  J=1,NSTOR                                                   DSL08070
       IF(KSTOR(J)-KVAR)12,8,12                                         DSL08080
   12 CONTINUE                                                           DSL08090
   15 KA(1)=KSYMB(KVAR)                                                  DSL08110
       KA(2)=KSYMB(KVAR+200)                                            DSL08120
       CALL UNPAK                                                        DSL08130
       WRITE(KPT,200) KA(1),KA(2),KA(3)                                 DSL08140
  200 FORMAT(25H PARAM NOT INPUT, SET=0   3A2)                          DSL08150
```

```
      IF(NPAR-60)7,300,300                                          DSL08160
300   NALRM=10                                                       DSL08170
301   CALL ERR1                                                      DSL08180
      IERR=1                                                         DSL08190
      GO TO 8                                                        DSL08195
7     NPAR=NPAR+1                                                    DSL08200
      KPAR(NPAR)=KVAR                                                DSL08210
8     INPUT(I)=0                                                     DSL08220
4     CONTINUE                                                       DSL08230
      IF(NINT)100,100,40                                             DSL08240
40    KSW=1                                                          DSL08250
      NSEQ=J2+1                                                      DSL08260
C...  BEGIN EACH SEQUENCE WITH INTEGRATOR OUTPUT AND WORK BACK       DSL08270
41    JLOC=LINT(JINT)                                                DSL08280
      IF(JLOC)95,95,46                                               DSL08290
C...  TEST IF INTEGRAL IS WITHIN SORT SECTION                        DSL08300
46    IF(JLOC-NSOR3)42,42,100                                        DSL08310
42    IF(JLOC-NSOR2)95,43,43                                         DSL08320
43    KSEQ2(1)=JLOC                                                  DSL08330
44    J=J2                                                           DSL08340
      J1=J+1                                                         DSL08350
      J3=J1                                                          DSL08360
      LEV=1                                                          DSL08370
      NSEQ3=0                                                        DSL08380
C...  TEST FOR MULTIPLE OUTPUTS, PLACE ALL IN SAME LOOP AT FIRST USE DSL08390
C...  TEST IF DUMMY ENTRY (INTEGRATOR WITHIN PROCEDURE)              DSL08400
50    IF(LFORT(JLOC))52,52,51                                        DSL08410
51    IF(LFORT(JLOC+1))58,53,58                                      DSL08420
52    JLOC=JLOC-1                                                    DSL08430
      IF(LFORT(JLOC))52,52,53                                        DSL08440
53    IF(JLOC-NOUT)153,58,58                                         DSL08450
153   IF(KOUT(JLOC+1))154,154,45                                     DSL08460
154   NSEQ3=NSEQ3+1                                                  DSL08470
      KSEQ3(NSEQ3)=JLOC+1                                            DSL08480
      IF(NSEQ3-10)58,58,210                                          DSL08490
45    JL1=JLOC                                                       DSL08500
54    J=J+1                                                          DSL08510
      IF(J-120)56,56,55                                              DSL08520
55    WRITE(KPT,201)                                                 DSL08530
201   FORMAT(30HOMORE THAN 120 OUTPUTS IN SORT)                     DSL08540
      GO TO 65                                                       DSL08550
56    KSEQ(J)=JLOC                                                   DSL08560
      JLOC=JLOC+1                                                    DSL08570
      IF(LFORT(JLOC))57,155,57                                       DSL08580
155   IF(JLOC-NOUT)54,54,57                                          DSL08585
57    JLOC=JL1                                                       DSL08590
      GO TO 60                                                       DSL08600
58    J=J+1                                                          DSL08610
      IF(J-120)59,59,55                                              DSL08620
59    KSEQ(J)=KSEQ2(LEV)                                             DSL08630
60    LIN2(LEV)=LIN(JLOC)                                            DSL08640
      NIN2(LEV)=NIN(JLOC)                                            DSL08650
C...  TEST IF MORE INPUTS TO BLOCK ON PRESENT LEVEL                  DSL08660
61    IF(NIN2(LEV))65,78,66                                          DSL08670
C...  TEST IF INTEGRATOR, PARAMETER, OR OUTPUT NEXT                  DSL08730
66    K=LIN2(LEV)                                                    DSL08740
      IF(INPUT(K))73,80,72                                           DSL08750
72    JLOC=INPUT(K)                                                  DSL08760
C...  TEST IF WITHIN SORT SECTION                                    DSL08770
73    IF(JLOC-NSOR2)80,75,74                                         DSL08780
74    IF(JLOC-NSOR3)75,75,80                                         DSL08790
C...  INCREASE LEVEL                                                 DSL08800
75    LEV=LEV+1                                                      DSL08810
      IF(LEV-30)77,77,76                                             DSL08820
76    WRITE(KPT,203)                                                 DSL08830
203   FORMAT(36HOMORE THAN 30 OUTPUTS IN SINGLE LOOP)               DSL08840
      GO TO 65                                                       DSL08850
77    KSEQ2(LEV)=INPUT(K)                                            DSL08860
      GO TO 50                                                       DSL08870
C...  DECREASE LEVEL                                                 DSL08880
78    LEV=LEV-1                                                      DSL08890
      IF(LEV)180,180,80                                              DSL08900
80    NIN2(LEV)=NIN2(LEV)-1                                          DSL08910
      LIN2(LEV)=LIN2(LEV)+1                                          DSL08920
      GO TO 61                                                       DSL08930
180   IF(NSEQ3)81,81,181                                             DSL08940
181   JLOC=KSEQ3(NSEQ3)                                              DSL08950
      NSEQ3=NSEQ3-1                                                  DSL08960
```

263

```
            KSEQ2(1)=JLOC                                           DSL08970
            LEV=1                                                   DSL08980
C...     REMOVE COMPLETED ELEMENTS FROM INPUT FOR PARTIAL LOOP      DSL08990
            DO 182  I=J3,J                                          DSL09000
            KVAR=KSEQ(I)                                            DSL09010
            DO 182  K=NTIN1,NTIN2                                   DSL09020
            IF(KVAR-INPUT(K))182,183,182                            DSL09030
      183 INPUT(K)=0                                                DSL09040
      182 CONTINUE                                                  DSL09050
            J3=J+1                                                  DSL09060
            GO TO 58                                                DSL09070
C...     END OF SEQUENCE FOR ONE LOOP, ELIMINATE DUPLICATES         DSL09080
       81 IF(J-J1)82,82,83                                          DSL09090
       82 J2=J2+1                                                   DSL09100
            GO TO 91                                                DSL09110
       83 J5=J-1                                                    DSL09120
            DO 86  I=J1,J5                                          DSL09130
            KVAR=KSEQ(I)                                            DSL09140
            I2=I+1                                                  DSL09150
            DO 84  K=I2,J                                           DSL09160
            IF(KVAR-KSEQ(K))84,85,84                                DSL09170
       84 CONTINUE                                                  DSL09180
            GO TO 86                                                DSL09190
       85 KSEQ(I)=0                                                 DSL09200
       86 CONTINUE                                                  DSL09210
C...     CONDENSE, THEN REVERSE SEQUENCE                            DSL09220
            DO 88  I=J1,J                                           DSL09230
            IF(KSEQ(I))87,88,87                                     DSL09240
       87 J2=J2+1                                                   DSL09250
            KSEQ(J2)=KSEQ(I)                                        DSL09260
       88 CONTINUE                                                  DSL09270
            J3=(J2-J1+1)/2+J1-1                                     DSL09280
            J4=J2                                                   DSL09290
            DO 90  I=J1,J3                                          DSL09300
            KSAVE=KSEQ(J4)                                          DSL09310
            KSEQ(J4)=KSEQ(I)                                        DSL09320
            KSEQ(I)=KSAVE                                           DSL09330
       90 J4=J4-1                                                   DSL09340
C...     REMOVE COMPLETED ELEMENTS FROM INPUT                       DSL09350
       91 DO 92  I=J1,J2                                            DSL09360
            KVAR=KSEQ(I)                                            DSL09370
            DO 92  K=NTIN1,NTIN2                                    DSL09380
            IF(KVAR-INPUT(K))92,93,92                               DSL09390
       93 INPUT(K)=0                                                DSL09400
       92 CONTINUE                                                  DSL09410
            GO TO(95,121),KSW                                       DSL09420
C...     TEST IF LAST INTEGRATOR                                    DSL09430
       95 JINT=JINT+1                                               DSL09440
            IF(JINT-NINT)160,160,100                                DSL09450
C...     TEST IF ALREADY IN SEQUENCE                                DSL09453
      160 KVAR=LINT(JINT)                                           DSL09454
            DO 162  K=1,J2                                          DSL09455
            IF(KVAR-KSEQ(K))162,95,162                              DSL09456
      162 CONTINUE                                                  DSL09457
            GO TO 41                                                DSL09458
      100 KSW=2                                                     DSL09460
C...     TRACE BRANCHES NOT WITHIN INTEGRATION OR MEMORY LOOPS      DSL09470
      121 LTEST=0                                                   DSL09480
            DO 126  I=NSOR2,NSOR3                                   DSL09490
            IF(LFORT(I))126,126,122                                 DSL09500
      122 DO 123  K=NSEQ,J2                                         DSL09510
            IF(KSEQ(K)-I)123,126,123                                DSL09520
      123 CONTINUE                                                  DSL09530
C...     TEST IF INPUT TO ANOTHER BLOCK                             DSL09540
            DO 124  K=NTIN1,NTIN2                                   DSL09550
            IF(INPUT(K)-I)124,125,124                               DSL09560
      124 CONTINUE                                                  DSL09570
            KSEQ2(1)=I                                              DSL09580
            JLOC=I                                                  DSL09590
            GO TO 44                                                DSL09600
      125 LTEST=1                                                   DSL09610
      126 CONTINUE                                                  DSL09620
            IF(LTEST)127,127,128                                    DSL09630
      127 NSOR1=NSOR1+2                                             DSL09640
            NOUT1=NSOR3+1                                           DSL09650
            IF(NSOR1-NSORT)28,28,129                                DSL09660
      129 IF(NOUT1-NOUT)131,131,130                                 DSL09670
      131 NOUT2=NOUT                                                DSL09680
```

264

```
      KGO=2                                                            DSL09690
      GO TO 36                                                         DSL09700
  210 NALRM=10                                                         DSL09702
      CALL ERR1                                                        DSL09704
      IERR=1                                                           DSL09706
      GO TO 400                                                        DSL09708
  128 WRITE(KPT,204)                                                   DSL09710
  204 FORMAT(25HOUNDEFINED IMPLICIT LOOP )                             DSL09720
   65 IERR=1                                                           DSL09730
      J2=J                                                             DSL09740
  130 NTSEQ=J2                                                         DSL09750
      WRITE(KPT,205) (KSEQ(I),I=1,J2)                                  DSL09760
  205 FORMAT(25HOOUTPUT VARIABLE SEQUENCE /(2014))                     DSL09770
  400 WRITE(KPT,900) NINT,NINP,NOUT,NPAR,NSYMB,MFOR,NDATA              DSL09772
  900 FORMAT(21HOSTORAGE USED/MAXIMUM /6H INTGR I4,12H/50, IN VARS I4, DSL09773
     1 14H/300, OUT VARS I4,12H/100, PARAMS I3,12H/60, SYMBOLS I4,     DSL09774
     2 14H/200, FORT WDS I5,15H/5000, DATA WDS I4,5H/1000)             DSL09775
      IF(IERR)150,151,150                                              DSL09777
  150 STOP                                                             DSL09778
  151 ISTRT=INPT(2)                                                    DSL09779
      CALL LINK(OUTPX)                                                 DSL09780
      END                                                              DSL09790
// DUP
*STORECI    WS   UA   SORTX
// ASM                                                                 DSL09820
*LIST                                                                  DSL09830
                   ENT      START                                      DSL09840
           START   DC       0                                          DSL09850
                   STX    1 SAVEX+1                                    DSL09860
                   STX    2 SAVEX+3                                    DSL09870
                   LDX    1 -105                                       DSL09880
                   LDX    L2 IRES                                      DSL09890
           LOOP1   LD     L1 DATA+105                                  DSL09900
                   STO    2 0                                          DSL09910
                   LD     L1 DATA+106                                  DSL09920
                   STO    2 -35                                        DSL09930
                   LD     L1 DATA+107                                  DSL09940
                   STO    2 -70                                        DSL09950
                   MDX    2 -1                                         DSL09960
                   MDX    1 3                                          DSL09970
                   MDX      LOOP1                                      DSL09980
                   LDX    1 -53                                        DSL09990
                   LDX    L2 IRES-105                                  DSL10000
           LOOP4   LD     L1 DATA2+53                                  DSL10010
                   STO    2 0                                          DSL10020
                   MDX    2 -1                                         DSL10030
                   MDX    1 1                                          DSL10040
                   MDX      LOOP4                                      DSL10050
                   LDX    1 -66                                        DSL10060
                   LDX    L2 KSYMB                                     DSL10070
           LOOP3   LD     L1 DATA1+66                                  DSL10080
                   STO    L  KA-1                                      DSL10090
                   SRA       8                                         DSL10100
                   STO    L  KA                                        DSL10110
                   LD     L1 DATA1+67                                  DSL10120
                   STO    L  KA-3                                      DSL10130
                   SRA       8                                         DSL10140
                   STO    L  KA-2                                      DSL10150
                   LD     L1 DATA1+68                                  DSL10160
                   STO    L  KA+5                                      DSL10170
                   SRA       8                                         DSL10180
                   STO    L  KA-4                                      DSL10190
                   LD     L  KA-2                                      DSL10200
                   SRT       6                                         DSL10210
                   LD     L  KA-1                                      DSL10220
                   SRT       6                                         DSL10230
                   LD     L  KA                                        DSL10240
                   SLT       9                                         DSL10250
                   AND       MASK                                      DSL10260
                   STO    L  KA                                        DSL10270
                   LD     L  KA-4                                      DSL10280
                   SRT       6                                         DSL10290
                   LD     L  KA-3                                      DSL10300
                   SRT       6                                         DSL10310
                   LD     L  KA-2                                      DSL10320
                   SLT       12                                        DSL10330
                   AND       MASK                                      DSL10340
                   STO    L  KA-1                                      DSL10350
```

```
                LD    L   KA                                   DSL10360
                STO   2   0                                    DSL10370
                LD    L   KA-1                                 DSL10380
                STO   L2  -200                                DSL10390
                MDX   2   -1                                   DSL10400
                MDX   1   3                                    DSL10410
                MDX       LOOP3                                DSL10420
                LDX   1   -10                                  DSL10430
LOOP5           LD    L1  CON1+10                              DSL10440
                STO   L1  KMEM+1                               DSL10450
                MDX   1   1                                    DSL10460
                MDX       LOOP5                                DSL10470
                LDX   1   -12                                  DSL10480
                LDX   L2  ICHAR                                DSL10490
LOOP2           LD    L1  CONST+12                             DSL10500
                STO   2   0                                    DSL10510
                MDX   2   -1                                   DSL10520
                MDX   1   1                                    DSL10530
                MDX       LOOP2                                DSL10540
                LDX   1   -4                                   DSL10542
                LDX   L2  KA                                   DSL10543
LOOP6           LD    L1  DATA3+4                              DSL10544
                STO   2   0                                    DSL10545
                MDX   2   -1                                   DSL10546
                MDX   1   1                                    DSL10548
                MDX       LOOP6                                DSL10549
SAVEX           LDX   L1  *-*                                  DSL10550
                LDX   L2  *-*                                  DSL10560
                BSC   I   START                                DSL10570
MASK            DC    /7FFF                                    DSL10580
DATA            EBC   .PARAM CONST INCON AFGEN NLFGEN.         DSL10590
                EBC   .TABLE PROCEDMEMORYSTORAGINTGER.         DSL10600
                EBC   .RENAMEINTGRLPRINT CONTRLRELERR.         DSL10610
                EBC   .ABSERRRANGE FINISHEND    STOP .         DSL10620
                EBC   .SORT  NOSORTENDPROTITLE SCOPE .         DSL10630
                EBC   .GRAPH LABEL SCALE CONTININTEG .         DSL10640
                EBC   .RESET CHART TYPE  DUMP  *LOCAL.         DSL10650
DATA2           EBC   .$$       IF(NALRM) 9200, 9200, .        DSL10660
                EBC   .9300$$9300  CONTINUE$$       GO.        DSL10670
                ERC   . TO 9100$$9200  CONTINUE$$.             DSL10680
                EBC   .ZZ990 REAL INTEGER   1.                 DSL10690
DATA1           EBC   .TIME  DELT  DELMI FINTI DELTP .         DSL10700
                EBC   .DELTC INTGR REALP MODIN TRNFR .         DSL10710
                EBC   .CMPXP LEDLG .                           DSL10720
                EBC   .DEBUG DERIV HSTRS IMPUL PULSE .         DSL10730
                EBC   .RST   ZHOLD IMPL  IMPLC DELAY .         DSL10740
DATA3           EBC   .CALLD .                                 DSL10744
                DC    /FF00                                    DSL10746
CON1            DC    0        DELAY                           DSL10750
                DC    3        IMPLC                           DSL10760
                DC    2        IMPL                            DSL10770
                DC    1        ZHOLD                           DSL10780
                DC    1        RST                             DSL10790
                DC    2        PULSE                           DSL10800
                DC    3        IMPUL                           DSL10810
                DC    2        HSTRS                           DSL10820
                DC    6        DERIV                           DSL10830
                DC    1        DEBUG                           DSL10840
CONST           DC    107      COMMA                           DSL10850
                DC    77       LEFT PAREN   (029 KP)           DSL10860
                DC    108      LEFT PAREN   (026 KP)           DSL10870
                DC    93       RIGHT PAREN  (029)              DSL10880
                DC    76       RIGHT PAREN  (026)              DSL10890
                DC    123      EQUALS  (026)                   DSL10900
                DC    78       PLUS   (029)                    DSL10910
                DC    80       PLUS   (026)                    DSL10920
                DC    96       MINUS                           DSL10930
                DC    92       ASTERISK                        DSL10940
                DC    75       DECIMAL                         DSL10950
                DC    97       SLASH                           DSL10960
KA              EQU   /7FF9                                    DSL10970
ICHAR           EQU   /7FAA                                    DSL10980
IRES            EQU   /7F9E                                    DSL10990
KSYMB           EQU   /7F00                                    DSL11000
KMEM            EQU   /79AE                                    DSL11010
                END                                            DSL11020
*STORE     UA   START                                         DSL11040
// FOR TRAN1                                                   DSL11050
```

266

```
*LIST SYMBOL TABLE                                              DSL11060
*LIST SUBPROGRAM NAMES                                          DSL11070
*LIST ALL
*ONE WORD INTEGERS                                              DSL11080
      SUBROUTINE TRAN1                                          DSL11130
      DIMENSION KDUM(1020)                                      DSL11135
      COMMON NALRM,IERR,ICONT,JPT,KEY,ISTRT,KA(6),LETRS(73),ICHAR(12),  DSL11140
     1 IRES(158),KSYMB(400),NSYMB,ITYPE,NOUT,NINP,NPAR,NINT,NDATA,NFOR, DSL11150
     2 NSORT,ISORT,NMEM,NING,NINGL,NSTOR,NTSEQ,NREAL,MFOR,      DSL11160
     3 KOUT(100),LFORT(100),NFORT(100),LIN(100),NIN(100),INPUT(300),    DSL11170
     4 KPAR( 60),KSTOR(10),LINT(50),INGER(10),MEMRY(15),KMEM(15),       DSL11180
     5 LSTOR(10),INGRL(10),KREAL( 20),KSORT(20)                 DSL11190
      COMMON KFORT(128),KSPEC(36),LOCAL(36),NLOC,NSPEC,KEYB,KARD DSL11200
      EQUIVALENCE (KDUM(1),KOUT(1))                             DSL11210
      JPT=INPT(3)                                               DSL11215
      WRITE(JPT,800)                                            DSL11220
  800 FORMAT(1H1 20X,33H*** DSL/1800 TRANSLATOR INPUT ***)      DSL11222
      IERR=0                                                    DSL11225
      NINT=0                                                    DSL11230
      NDATA=1                                                   DSL11240
      NINP=0                                                    DSL11250
      NING=0                                                    DSL11260
      NINGL=0                                                   DSL11270
      NSTOR=0                                                   DSL11280
      NREAL=0                                                   DSL11290
      NLOC=0                                                    DSL11300
      NSPEC=0                                                   DSL11310
      NOUT=1                                                    DSL11320
      NFOR=1                                                    DSL11330
      MFOR=1                                                    DSL11340
      NSORT=2                                                   DSL11350
      NPAR=6                                                    DSL11360
      NSYMB=22                                                  DSL11370
      NMEM=10                                                   DSL11380
      ISORT=1                                                   DSL11440
      KEYB=INPT(6)                                              DSL11450
      KARD=INPT(8)                                              DSL11452
C...  CLEAR TO ZERO KOUT(1) THRU KSORT(20)                      DSL11460
      DO 3  I=1,1020                                            DSL11470
    3 KDUM(I)=0                                                 DSL11480
      KSORT(1)=1                                                DSL11540
      DO 1   I=1,6                                              DSL11550
    1 KPAR(I)=I                                                 DSL11560
      DO 13  I=13,22                                            DSL11570
   13 MEMRY(I-12)=I                                             DSL11572
      RETURN                                                    DSL11577
      END                                                       DSL11578
// DUP
*DELETE               TRAN1
*STORE        WS  UA  TRAN1
// FOR TRANS                                                    DSL11580
*LIST SYMBOL TABLE                                              DSL48180
*LIST SUBPROGRAM NAMES                                          DSL48190
*LIST ALL
*ONE WORD INTEGERS                                              DSL11583
*NAME DSL                                                       DSL11585
*IOCS(KEYBOARD,TYPEWRITER,CARD,1132 PRINTER,DISK)              DSL11586
**DSL/1800 LANGUAGE TRANSLATOR                                  DSL11587
      DEFINE FILE 1(5000,1,U,MFOR),2(1000,1,U,NDATA)           DSL11588
      COMMON NALRM,IERR,ICONT,JPT,KEY,ISTRT,KA(6),LETRS(73),ICHAR(12),  DSL11589
     1 IRES(158),KSYMB(400),NSYMB,ITYPE,NOUT,NINP,NPAR,NINT,NDATA,NFOR, DSL11590
     2 NSORT,ISORT,NMEM,NING,NINGL,NSTOR,NTSEQ,NREAL,MFOR,      DSL11591
     3 KOUT(100),LFORT(100),NFORT(100),LIN(100),NIN(100),INPUT(300),    DSL11592
     4 KPAR( 60),KSTOR(10),LINT(50),INGER(10),MEMRY(15),KMEM(15),       DSL11593
     5 LSTOR(10),INGRL(10),KREAL(20),KSORT(20)                 DSL11594
      COMMON KFORT(128),KSPEC(36),LOCAL(36),NLOC,NSPEC,KEYB,KARD,ISC(5) DSL11595
      EQUIVALENCE (IRES(106),IDOLR),(IRES(107),IBLNK),(KA(4),LOC),      DSL11596
     1 (KA(6),NMBR),(IRES(158),IONE)                           DSL11597
      KEND=0                                                    DSL11598
      IPRO=0                                                    DSL11600
      NIMP=0                                                    DSL11601
      IMP1=600                                                  DSL11602
      KDUMP=0                                                   DSL11604
      MPL=0                                                     DSL11606
C...  INITIALIZE DATA AREAS, STORE SYMBOL TABLES IN COMMON      DSL11609
      CALL TRAN1                                                DSL11610
      CALL START                                                DSL11611
      MAXN=INPT(4)                                              DSL11612
```

```
          ICA=KA(1)                                                      DSL11613
          LL=KA(2)                                                       DSL11614
          ID=KA(3)                                                       DSL11615
          MASK=KA(4)                                                     DSL11616
C...   START OF LOOP FOR EACH CARD,   READ AND PRINT INPUT               DSL11620
    4 ICONT=0                                                            DSL11622
          ITYPE=0                                                        DSL11624
          MEMR=0                                                         DSL11626
          JINTG=0                                                        DSL11627
    5 CALL READR                                                         DSL11628
          ISTRT=7                                                        DSL11630
          IF(ICONT)63,2,172                                             DSL11640
    2 IF(NFOR-1)15,15,12                                                DSL11642
   12 N2=0                                                               DSL11643
          IF(MPL)320,320,310                                             DSL11644
  310 MPL=0                                                              DSL11645
C...   MULTIPLE OUTPUT BLOCK, INSERT OUTPUTS IN CALL STATEMENT           DSL11646
C...   TEST FOR LEADING BLANK OR ALPHA IN PAIR, OR RIGHT PAREN           DSL11647
          IF(KFORT(NFOR-1))323,323,322                                  DSL11649
C...   INSERT BLANK-COMMA(16491), RIGHT PAREN-BLANK(23872). + COMMA(107) DSL11651
  322 KFORT(NFOR-1)=16491                                                DSL11652
          GO TO 324                                                      DSL11653
  323 KFORT(NFOR-1)=IOR(IAND(KFORT(NFOR-1),MASK),107)                    DSL11656
  324 DO 325  NF= 94,NF8                                                 DSL11657
          KFORT(NFOR)=KFORT(NF)                                          DSL11658
  325 NFOR=NFOR+1                                                        DSL11659
          KFORT(NFOR)=23872                                              DSL11660
          GO TO 450                                                      DSL11661
  320 NFOR=NFOR-1                                                        DSL11662
  450 WRITE(1'MFOR)(KFORT(K),K=1,NFOR)                                  DSL11670
          NFOR=1                                                         DSL11680
          IF(MFOR-5000)15,15,100                                        DSL11685
   15 IF(LETRS(1)-IBLNK)6,70,6                                          DSL11690
C...   DETERMINE DATA STATEMENT TYPE   (1-35)                            DSL11700
C...   PARAM,CONST,INCON,AFGEN,NLFGEN,TABLE,PROCED,MEMORY,STORAG,INTGER, DSL11710
C...   RENAME,INTGRL,PRINT,CONTRL,RELERR,ABSERR,RANGE,FINISH,END,STOP,   DSL11720
C...   SORT,NOSORT,ENDPRO,TITLE,SCOPE,GRAPH,LABLE,SCALE,CONTIN,INTEG,    DSL11730
C...   RESET,CHART,TYPE,DUMP,*LOCAL                                      DSL11740
    6 DO 7  I=1,35                                                       DSL11750
          IF(LETRS(1)-IRES(I))7,8,7                                      DSL11760
    8 IF(LETRS(2)-IRES(I+35))7,9,7                                       DSL11770
    9 IF(LETRS(3)-IRES(I+70))7,10,7                                      DSL11780
    7 CONTINUE                                                           DSL11790
C...   TEST FOR D IN COL 1                                               DSL11800
          IF(LETRS(1)-ID)70,28,70                                        DSL11805
   10 ITYPE=I                                                            DSL11810
   63 IF(ITYPE-7)14,21,64                                                DSL11820
   64 IF(ITYPE-12)21,21,66                                               DSL11830
   66 IF(ITYPE-21)14,51,67                                               DSL11840
   67 IF(ITYPE-23)52,53,25                                               DSL11850
   25 IF(ITYPE-34)14,36,26                                               DSL11860
   36 KDUMP=1                                                            DSL11864
          GO TO 4                                                        DSL11866
   26 NLOC=1                                                             DSL11870
          DO 27  I=1,36                                                  DSL11880
   27 LOCAL(I)=LETRS(I)                                                  DSL11890
          GO TO 4                                                        DSL11900
   28 DO 29  I=2,36                                                      DSL11910
   29 KSPEC(I)=LETRS(I)                                                  DSL11920
          KSPEC(1)=IBLNK                                                 DSL11930
          NSPEC=1                                                        DSL11940
          GO TO 4                                                        DSL11950
C...   STORE DATA CARD AS VARIABLE WORD RECORD WITH $$ LAST WORD         DSL11960
   14 IEND=36                                                            DSL11970
   16 IF(LETRS(IEND)-IBLNK)18,17,18                                      DSL11980
   17 IEND=IEND-1                                                        DSL11990
          GO TO 16                                                       DSL12000
   18 IEND=IEND+1                                                        DSL12010
          LETRS(IEND)=IDOLR                                              DSL12020
          WRITE(2'NDATA)(LETRS(K),K=1,IEND)                             DSL12030
          IF(NDATA-1000)20,20,100                                        DSL12040
   20 IF(KEND)61,11,24                                                   DSL12050
C...   NOTE - ADD 30 TO KEY NAME SEQUENCE NO. TO GET FORTRAN STATEMENT NODSL12060
   11 IF(ITYPE-19)21,49,24                                               DSL12070
   24 IF(ITYPE-20)4,50,4                                                 DSL12080
   21 CALL CPEN                                                          DSL12090
          IF(ITYPE-12)22,42,43                                           DSL12100
   22 GO TO(31,31,31,34,34,43,37,38,39,40,41),ITYPE                      DSL12110
```

268

```
   31 CALL SCAN                                                    DSL12120
      L=1                                                          DSL12130
      GO TO(61,61,62,61,61,4,48),KEY                               DSL12140
   61 NALRM=6                                                      DSL12150
      GO TO 99                                                     DSL12160
   62 DO 122  M=1,NPAR                                             DSL12170
      IF(LCC-KPAR(M))122,123,122                                   DSL12180
  122 CONTINUE                                                     DSL12190
      IF(NPAR-60)108,100,100                                       DSL12200
  108 NPAR=NPAR+1                                                  DSL12210
      KPAR(NPAR)=LOC                                               DSL12220
  123 IF(L-1)61,31,43                                              DSL12230
   34 IF(ICONT)43,65,61                                            DSL12240
   65 CALL SCAN                                                    DSL12250
      L=2                                                          DSL12260
      IF(KEY-3)65,62,61                                            DSL12270
C... PROCEDURAL BLOCK                                              DSL12280
   37 IPRO=1                                                       DSL12290
      NFOR1=MFOR                                                   DSL12300
      GO TO 68                                                     DSL12310
C... MEMORY BLOCK SPECIFICATION                                    DSL12320
   38 CALL SCAN                                                    DSL12330
      IF(KEY-4)61,150,149                                          DSL12340
  149 IF(KEY-5)61,61,4                                             DSL12350
  150 IF(NMEM-15)151,100,100                                       DSL12360
  151 NMEM=NMEM+1                                                  DSL12370
      MEMRY(NMEM)=LOC                                              DSL12380
      CALL NUMRC                                                   DSL12390
      KMEM(NMEM)=NMBR                                              DSL12400
      GO TO 38                                                     DSL12410
C... STORAG BLOCK SPECIFICATION                                    DSL12420
   39 CALL SCAN                                                    DSL12430
      IF(KEY-4)61,160,159                                          DSL12440
  159 IF(KEY-5)61,61,4                                             DSL12450
  160 IF(NSTOR)162,162,163                                         DSL12460
  162 NSTOR=2                                                      DSL12470
      GO TO 164                                                    DSL12480
  163 IF(NSTOR-10)161,100,100                                      DSL12490
  161 NSTOR=NSTOR+1                                                DSL12500
  164 KSTOR(NSTOR)=LOC                                             DSL12510
      CALL NUMRC                                                   DSL12520
      LSTOR(NSTOR)=NMBR                                            DSL12530
      GO TO 39                                                     DSL12540
   40 CALL SCAN                                                    DSL12550
      IF(KEY-2)4,110,4                                             DSL12560
  110 IF(NING-10)111,100,100                                       DSL12570
  111 NING=NING+1                                                  DSL12580
      INGER(NING)=LOC                                              DSL12590
      GO TO 40                                                     DSL12600
C... RENAME                                                        DSL12610
   41 CALL SCAN                                                    DSL12620
      IF(KEY-3)61,112,61                                           DSL12630
  112 IF(LCC-6)113,113,61                                          DSL12640
  113 I=LOC                                                        DSL12650
      CALL SCAN                                                    DSL12660
      KSYMB(I)=KSYMB(NSYMB)                                        DSL12670
      KSYMB(I+200)=KSYMB(NSYMB+200)                                DSL12680
      NSYMB=NSYMB-1                                                DSL12690
      IF(ISTRT-72)41,4,4                                           DSL12700
C... INTEGRAL OUTPUT VARIABLE SPECIFICATION                        DSL12710
   42 CALL SCAN                                                    DSL12720
      IF(KEY-2)61,170,61                                           DSL12730
  170 IF(NINGL-10)171,100,100                                      DSL12740
  171 NINGL=NINGL+1                                                DSL12750
      INGRL(NINGL)=LOC                                             DSL12760
      IF(ISTRT-72)72,4,4                                           DSL12770
C... TEST FOR CONTINUATION OF DATA CARD                            DSL12780
   43 KPT=0                                                        DSL12790
      I=IEND+IEND-2                                                DSL12800
   44 IF(LETRS(I)-64)4,45,46                                       DSL12810
   45 I=I-1                                                        DSL12820
      GO TO 44                                                     DSL12830
   46 IF(LETRS(I)-75)4,47,4                                        DSL12840
   47 KPT=KPT+1                                                    DSL12850
      IF(KPT-3)45,48,48                                            DSL12860
   48 ICONT=-1                                                     DSL12870
      GO TO 5                                                      DSL12880
C... END AND STOP CARDS                                            DSL12890
```

```
     49 KEND=1                                              DSL12900
        GO TO 4                                             DSL12910
     51 IF(ISORT)264,264,4                                  DSL12950
    264 ISORT=1                                             DSL12960
        IF(NOUT-1)4,4,175                                   DSL12970
    175 NSORT=NSORT+2                                        DSL12980
        KSORT(NSORT-1)=NOUT                                 DSL12990
        GO TO 4                                             DSL13000
     52 IF(ISORT)265,4,265                                  DSL13005
    265 ISORT=0                                             DSL13010
        IF(NOUT-1)176,176,177                               DSL13020
    176 NSORT=0                                             DSL13030
        GO TO 4                                             DSL13040
    177 KSORT(NSORT)=NOUT-1                                 DSL13050
        GO TO 4                                             DSL13060
     53 IPRO=0                                             DSL13070
        NFORT(NOUT1)=NFOR1-MFOR                             DSL13080
        LFORT(NOUT1)=NFOR1                                 DSL13090
        LIN(NOUT1)=NINP1                                   DSL13100
        NIN(NOUT1)=NINP-NINP1+1                             DSL13110
        GO TO 4                                             DSL13120
C...    START SCAN OF DSL STRUCTURE STATEMENT               DSL13130
C...    CONTINUATION OF STRUCTURE STATEMENT                 DSL13140
    172 NB1=NMBR/2                                          DSL13150
        NFOR=NFORA+NB1+MEMR+1                               DSL13160
        MEMR=0                                              DSL13164
        KFORT(NFOR-1)=IDOLR                                 DSL13170
        IF(NMBR-NB1-NB1)94,174,70                           DSL13180
    174 KFORT(NFOR-2)=KFORT(NFOR-2)-11                      DSL13190
     70 IEND=36                                             DSL13200
     71 IF(LETRS(IEND)-IBLNK)73,72,73                       DSL13210
     72 IEND=IEND-1                                         DSL13220
        GO TO 71                                            DSL13230
     73 NFORA=NFOR                                          DSL13240
        DO 74  K=1,IEND                                     DSL13250
        KFORT(NFOR)=LETRS(K)                               DSL13260
     74 NFOR=NFOR+1                                         DSL13270
        IF(NFOR- 200)116,116,100                            DSL13280
    116 CALL OPEN                                           DSL13290
C...    TEST FOR COMMENT (ASTERISK COL. 1), REFERENCE NUMBER, CONTINUATION DSL13300
        IF(LETRS(1)-92)92,91,92                             DSL13310
     91 NFOR=NFOR-IEND                                      DSL13320
        GO TO 5                                             DSL13330
     92 DO 95  K=1,5                                        DSL13340
        IF(LETRS(K)-64)94,95,93                             DSL13350
     93 IF(LETRS(K)-240)94,95,95                            DSL13360
     95 CONTINUE                                            DSL13370
        IF(ICONT)94,210,211                                 DSL13380
    210 IF(LETRS(6)-64)94,96,94                             DSL13390
     94 NALRM=9                                             DSL13400
        GO TO 99                                            DSL13410
    211 KFORT(NFORA+2)=IONE                                 DSL13420
     96 IF(IPRO)120,121,121                                 DSL13430
    120 KFORT(NFOR)=IDOLR                                   DSL13440
        NFOR=NFOR+1                                         DSL13450
        GO TO 4                                             DSL13460
    121 IF(ICONT)94,101,102                                 DSL13470
    102 KFORT(NFOR)=IDOLR                                   DSL13480
        NFOR=NFOR+1                                         DSL13490
        NFORT(NOUT1)=NFOR1-NFOR                             DSL13500
        IF(L-2)77,94,82                                     DSL13510
C...    FIND ALL OUTPUT VARIABLES                           DSL13520
    101 LFORT(NOUT)=MFOR                                   DSL13530
        NFORT(NOUT)=IEND                                   DSL13540
        NFOR1=NFORA                                        DSL13550
     68 NOUT1=NOUT                                          DSL13560
        L=1                                                 DSL13570
     77 CALL SCAN                                           DSL13580
C...    BRANCH ON NUMERIC,OPERATOR,=,(,GT 5 CHAR.,EOR,CONTINUE DSL13585
        GO TO(94,79,78,263,130,94,103),KEY                 DSL13590
C...    TEST IF END OF CARD BEFORE COMMA OR EQUALS          DSL13600
     79 IF(ISTRT-72)350,190,190                             DSL13610
    350 MPL=1                                               DSL13614
        GO TO 355                                           DSL13616
    263 L=3                                                 DSL13618
        GO TO 355                                           DSL13619
    190 IF(NSYMB-LOC)94,191,94                              DSL13620
    191 NSYMB=NSYMB-1                                       DSL13630
```

270

```
      130 IF(ISORT)94,131,94                                      DSL13640
      131 KOUT(NOUT)=0                                            DSL13642
          NOUT=NOUT+1                                            DSL13643
          GO TO 4                                                DSL13644
       78 L=2                                                    DSL13645
      355 KSTR=1                                                 DSL13647
C...     TEST IF OUTPUT IS STORAG VAR                            DSL13649
          IF(NSTOR)353,353,351                                  DSL13650
      351 DO 352 I=2,NSTOR                                       DSL13651
          IF(LOC-KSTOR(I))352,354,352                           DSL13652
      354 MPL=1                                                  DSL13653
          KSTR=LSTOR(I)                                         DSL13654
          GO TO 353                                             DSL13655
      352 CONTINUE                                               DSL13656
C...     TEST IF OUTPUT VARIABLE IS AN INTEGRAL                  DSL13660
      353 IF(NINGL)196,196,192                                  DSL13661
      192 DO 193   I=1,NINGL                                     DSL13662
          IF(INGRL(I)-LOC)193,194,193                           DSL13663
      194 IF(NINT- 50)195,100,100                               DSL13664
      195 LINT(NINT+1)=NOUT                                      DSL13665
          NINT=NINT+KSTR                                        DSL13666
          JINTG=1                                               DSL13667
          GC TO 196                                             DSL13668
      193 CONTINUE                                               DSL13669
      196 GO TO(76,356,262),L                                   DSL13672
      356 IF(MPL)76,76,272                                      DSL13674
      262 MPL=0                                                  DSL13678
          GO TO 225                                             DSL13680
      272 NF1=(ISTRT-1)/2                                       DSL13685
          NF8=NF1+90                                            DSL13686
          IF(ISTRT-NF1-NF1-2)276,275,100                        DSL13687
C...     ADD BLANK-ZERO (16384)                                  DSL13689
      275 KFORT(NF1+1)=LETRS(ISTRT)+16384                        DSL13690
          GO TO 277                                             DSL13691
      276 KFORT(NF1)=IOR(256*LETRS(ISTRT-2),64)                 DSL13697
      277 DO 273   NF=4,NF1                                      DSL13698
      273 KFORT(NF+ 90)=KFORT(NF)                                DSL13699
          IF(NF1-5)271,279,278                                  DSL13700
      271 NALRM=8                                                DSL13701
          GO TO 99                                              DSL13702
      278 NF3=5                                                  DSL13703
          DO 274   NF=NF1,NFOR                                   DSL13704
          KFORT(NF3)=KFORT(NF)                                  DSL13705
      274 NF3=NF3+1                                              DSL13706
          NF3=NF1-5                                             DSL13707
          MEMR=MEMR-NF3                                         DSL13709
          NFOR=NFOR-NF3                                         DSL13710
C...     INSERT CA-LL IN 4 AND 5                                 DSL13711
      279 KFORT(4)=ICA                                           DSL13712
          KFORT(5)=LL                                           DSL13713
          NFORT(NOUT1)=NFORT(NOUT1)+3                            DSL13714
       76 IF(NOUT-100)109,109,100                               DSL13715
      109 IF(NIMP)94,225,220                                    DSL13718
      220 IF(LOC-IMP3)225,221,225                               DSL13720
      221 LCC=IMP2                                               DSL13730
          NIMP=-2                                               DSL13740
      225 KOUT(NOUT)=LOC                                         DSL13750
          NOUT=NOUT+1                                           DSL13850
          IF(L-1)94,77,81                                       DSL13860
C...     FIND ALL INPUT VARIABLES                                DSL13870
       81 NINP1=NINP+1                                           DSL13880
       82 CALL SCAN                                              DSL13890
          L=3                                                   DSL13900
C...     BRANCH ON NUMERIC,OPERATOR,=,(,GT 5 CHAR.,EOR,CONTINUE  DSL13905
          GO TO(94,84,94,85,94,89,103),KEY                      DSL13910
      103 ICONT=1                                                DSL13920
          GO TO 5                                               DSL13930
       84 IF(NINP1-NINP)138,138,106                             DSL13960
      138 DO 104   M=NINP1,NINP                                  DSL13970
          IF(INPUT(M)-LOC)104,82,104                            DSL13980
      104 CONTINUE                                               DSL13990
          IF(NINP-300)106,100,100                               DSL14000
      106 IF(NIMP)228,230,228                                   DSL14010
      228 IF(LCC-IMP2)230,229,230                               DSL14020
      229 LCC=IMP1                                               DSL14030
      230 NINP=NINP+1                                            DSL14040
          INPUT(NINP)=LOC                                       DSL14050
          GO TO 82                                              DSL14060
```

271

```
C...      TEST IF INTEGRAL BLOCK (INTGR,REALP,MODIN,TRNFR,CMPXP,LEDLG)     DSL14070
  85  IF(LOC-7)82,141,148                                                  DSL14090
 148  IF(LCC-12)141,145,180                                                DSL14100
 141  IF(JINTG)266,266,82                                                  DSL14103
 266  IF(NINT)143,143,147                                                  DSL14105
 147  IF(LINT(NINT)-NOUT1)240,142,143                                      DSL14110
 240  IF(LINT(NINT)+NOUT1)143,244,143                                      DSL14120
 143  ICH=ISTRT-6                                                          DSL14130
 246  ICH=ICH-1                                                            DSL14140
      IF(LETRS(ICH)-64)103,246,248                                         DSL14150
 248  IF(LETRS(ICH)-126)249,247,246                                        DSL14155
 249  IF(LETRS(ICH)-123)241,247,241                                        DSL14160
 241  NIN(NOUT1)=NINP-NINP1+1                                              DSL14170
      LIN(NOUT1)=NINP1                                                     DSL14180
      NOUT1=NOUT                                                           DSL14190
      NOUT=NOUT+1                                                          DSL14200
      NINP1=NINP+1                                                         DSL14210
      GO TO 245                                                            DSL14220
 244  LINT(NINT)=0                                                         DSL14230
 245  LINT(NINT+1)=-NOUT1                                                  DSL14240
      GO TO 107                                                            DSL14250
 142  LINT(NINT)=0                                                         DSL14260
 247  LINT(NINT+1)=NOUT1                                                   DSL14270
 107  NINT=NINT+1                                                          DSL14280
      IF(LOC-10)146,144,145                                                DSL14290
 144  CALL NUMRC                                                           DSL14300
      CALL NUMRC                                                           DSL14310
      NINT=NINT+NMBR-1                                                     DSL14320
 146  IF(NINT-50)82,82,100                                                 DSL14330
 145  NINT=NINT+1                                                          DSL14340
      LINT(NINT)=NOUT1+1                                                   DSL14350
      GO TO 146                                                            DSL14360
C...      TEST IF SYSTEM MEMORY BLOCK, DELAY BLOCK,                        DSL14370
C...      OR IMPLICIT FUNCTION (IMPL, IMPLC)                               DSL14380
 180  DO 181  I=1,NMEM                                                     DSL14390
      IF(MEMRY(I)-LOC)181,183,181                                         DSL14400
 181  CONTINUE                                                             DSL14410
      GO TO 82                                                             DSL14420
 183  IF(NFORT(NOUT1))255,256,256                                         DSL14422
 255  NFORT(NOUT1)=NFORT(NOUT1)-2                                         DSL14424
      GO TO 257                                                            DSL14426
 256  NFORT(NOUT1)=NFORT(NOUT1)+2                                         DSL14430
 257  NF1=(ISTRT-1)/2                                                      DSL14440
      NF2=NFORA+NF1+MEMR-1                                                 DSL14450
      MEMR=MEMR+2                                                          DSL14455
      NF3=NFOR-1                                                           DSL14460
      NFOR=NFOR+2                                                          DSL14470
 184  KFORT(NF3+2)=KFORT(NF3)                                             DSL14480
      NF3=NF3-1                                                            DSL14490
      IF(NF3-NF2)185,185,184                                              DSL14500
 185  IF(ISTRT-NF1-NF1-2)179,178,100                                      DSL14510
 178  KA(5)=-NF2                                                           DSL14520
C...      TEST IF PAREN FROM 026 OR 029                                    DSL14530
      IF(KFORT(NF2+3)-20000)281,280,280                                   DSL14540
 280  KFORT(NF2+3)=KFORT(NF2+3)-256                                       DSL14550
      GO TO 188                                                            DSL14560
 281  KFORT(NF2+3)=KFORT(NF2+3)+7680                                      DSL14570
      GO TO 188                                                            DSL14580
 179  KA(5)=NF2                                                            DSL14590
C...      TEST IF DELAY BLOCK                                              DSL14598
 188  IF(I-10)187,186,187                                                 DSL14600
 186  CALL NUMRC                                                           DSL14610
      KMEM(I)=NMBR+NMBR+5                                                 DSL14620
 187  MAXN=MAXN-KMEM(I)                                                   DSL14630
      NMBR=MAXN+1                                                          DSL14640
      CALL BNDEC                                                           DSL14650
C...      TEST IF IMPLICIT FUNCTION BLOCK                                  DSL14658
      IF(I-8)82,88,189                                                    DSL14660
 189  IF(I-9)82,88,82                                                     DSL14670
  88  KFORT(NFORA)=IRES(138)                                              DSL14680
      KFORT(NFORA+1)=IRES(139)                                           DSL14690
      DO 250  KR=106,131                                                  DSL14700
      KFORT(NFOR)=IRES(KR)                                                DSL14710
 250  NFOR=NFOR+1                                                         DSL14720
      NFORT(NOUT1)=-NFORT(NOUT1)-26                                      DSL14730
      IMP1=IMP1+1                                                         DSL14740
      IMP2=KOUT(NOUT1)                                                    DSL14750
      KOUT(NOUT1)=IMP1                                                   DSL14760
```

```
      NIMP=-1                                                          DSL14770
      GO TO 82                                                         DSL14780
  89  IF(NIMP)267,270,270                                              DSL14790
 267  IF(NIMP+2)94,268,269                                             DSL14800
 268  NIMP=0                                                           DSL14810
      IF(NFORT(NOUT1))223,224,224                                      DSL14820
 223  NFOR=NFOR-1                                                      DSL14830
      NFORT(NOUT1)=NFORT(NOUT1)-17                                     DSL14840
      GO TO 226                                                        DSL14850
 224  NFORT(NOUT1)=-NFORT(NOUT1)-18                                    DSL14860
 226  DO 222  KR=131,148                                              DSL14870
      KFORT(NFOR)=IRES(KR)                                             DSL14880
 222  NFOR=NFOR+1                                                      DSL14890
      IRES(125)=IRES(125)+1                                            DSL14900
      IRES(139)=IRES(139)+1                                            DSL14910
      IRES(142)=IRES(142)+1                                            DSL14920
      IRES(116)=IRES(142)                                              DSL14930
      IRES(119)=IRES(142)                                              DSL14940
      IRES(122)=IRES(125)                                              DSL14950
      GO TO 270                                                        DSL14960
 269  NIMP=1                                                           DSL14970
      IMP3=INPUT(NINP)                                                 DSL14980
      INPUT(NINP)=0                                                    DSL14990
 270  IF(IPRO)4,97,98                                                  DSL15000
  97  NIN(NOUT1)=NINP-NINP1+1                                          DSL15010
      LIN(NOUT1)=NINP1                                                 DSL15020
      GO TO 4                                                          DSL15030
  98  IPRO=-1                                                          DSL15040
      GO TO 4                                                          DSL15050
 100  NALRM=10                                                         DSL15060
      NSORT=0                                                          DSL15070
      IERR=1                                                           DSL15080
      CALL ERR1                                                        DSL15090
      GO TO 50                                                         DSL15100
  99  CALL ERR1                                                        DSL15110
      GO TO 4                                                          DSL15120
  50  IF(KDUMP)55,56,55                                                DSL15130
  55  CALL LINK(TRDMP)                                                 DSL15140
  56  CALL LINK(SORTX)                                                 DSL15170
      END
// DUP
*DELETE            TRANS
*STORE        WS  UA  TRANS
// FOR TRDMP                                                           DSL15200
*LIST ALL                                                             DSL15210
*ONE WORD INTEGERS                                                    DSL15220
*NONPROCESS PROGRAM                                                   DSL15230
*NAME TRDMP                                                           DSL15240
*IOCS(TYPEWRITER,1132 PRINTER)                                        DSL15250
**DSL/1800 AND 1130 TABLE DUMP ROUTINE                                DSL15260
      COMMON NALRM,IERR,ICONT,JPT,KEY,ISTRT,KA(6),LETRS(73),ICHAR(12), DSL15270
    1 IRES(158),KSYMB(400),NSYMB,ITYPE,NOUT,NINP,NPAR,NINT,NDATA,NFOR, DSL15280
    2 NSORT,ISORT,NMEM,NING,NINGL,NSTOR,NTSEQ,NREAL,MFCR,              DSL15290
    3 KOUT(100),LFORT(100),NFORT(100),LIN(100),NIN(100),INPUT(300),    DSL15300
    4 KPAR( 60),KSTOR(10),LINT(50),INGER(10),MEMRY(15),KMEM(15),       DSL15310
    5 LSTOR(10),INGRL(10),KREAL(20),KSORT(20),KFORT(202)               DSL15320
      M=JPT                                                            DSL15340
      WRITE(M,880) (INPUT(I),  I=1,NINP)                               DSL15350
 880  FORMAT(20I6)                                                     DSL15360
      WRITE(M,860) (KOUT(I),LFORT(I),NFORT(I),LIN(I),NIN(I),I=1,NOUT)  DSL15370
 860  FORMAT(5I6)                                                      DSL15380
      WRITE(M,880) (LINT(I),I=1,NINT),(KPAR(I),I=1,NPAR)               DSL15390
      WRITE(M,880) (INGER(I),I=1,NING),(INGRL(I),I=1,NINGL),           DSL15400
    1 (KSORT(I),I=1,NSORT)                                             DSL15410
      WRITE(M,880) (MEMRY(I),KMEM(I),I=1,NMEM)                         DSL15420
      IF(NSTOR)890,890,885                                             DSL15430
 885  WRITE(M,880) (KSTOR(I),LSTOR(I),I=1,NSTOR)                       DSL15440
 890  J=0                                                             DSL15445
      DO 840  I=1,NSYMB                                               DSL15450
      KA(1)=KSYMB(I)                                                  DSL15460
      KA(2)=KSYMB(I+200)                                              DSL15470
      CALL UNPAK                                                       DSL15480
      J=J+1                                                           DSL15485
      KFORT(J)=KA(1)                                                  DSL15490
      KFORT(J+15)=KA(2)                                               DSL15500
      KFORT(J+30)=KA(3)                                               DSL15510
      IF(J-15)840,830,830                                            DSL15520
 830  WRITE(M,850)(KFORT(K),KFORT(K+15),KFORT(K+30),K=1 ,15)          DSL15525
```

```
      850 FORMAT(15(2X,3A2))                                        DSL15530
          J=0                                                       DSL15532
      840 CONTINUE                                                  DSL15534
          IF(J)870,870,855                                          DSL15536
      855 WRITE(M,850)(KFORT(K),KFORT(K+15),KFORT(K+30),K=1,J)      DSL15538
      870 CALL LINK(SORTX)                                          DSL15540
          END                                                       DSL15550
// DUP
*DELETE              TRDMP
*STORECI     WS  UA  TRDMP
// FOR ALPHA                                                        DSL30000
**DSL/1800 - SUBROUTINE TO EXTRACT NEXT ALPHA-NUMERIC WORD         DSL30010
*LIST SYMBCL TABLE                                                  DSL30020
*LIST SUBPROGRAM NAMES                                              DSL30030
*LIST ALL
*ONE WORD INTEGERS                                                  DSL30040
          SUBROUTINE ALPHA                                          DSL30050
          INTEGER WORD(3)                                           DSL30060
          COMMON NALRM,IDUMY(14),IERR,KDUMY(205),ISTRT,WORD,KA(6),LETRS(72) DSL30070
C...  SUBROUTINE TO SELECT ALPHA-NUMERIC NAME FROM LETRS(ISTRT) WITH DSL30080
C...  EQUALS, COMMA, OR LEFT PARENTHESIS AS SEPARATOR              DSL30090
C...  (... WILL FORCE READ + CONTINUATION)                        DSL30100
          J=0                                                       DSL30110
          I=ISTRT                                                   DSL30120
C...  TEST FOR BLANK (64) TO SKIP LETTER                          DSL30130
      100 IF(LETRS(I)-64)108,200,101                                DSL30140
C...  TEST FOR EOF FROM KEYBOARD (21)                             DSL30150
      108 IF(LETRS(I)-21)99,109,99                                  DSL30160
      109 I=73                                                      DSL30170
          GO TO 150                                                 DSL30180
C...  TEST SEPARATORS EQUALS (126/123), COMMA (107), LEFT PAREN (77/108)DSL30190
C...  RIGHT PAREN (93/76)                                          DSL30191
      101 IF(LETRS(I)-126)140,150,116                               DSL30200
      140 IF(LETRS(I)-123)102,150,119                               DSL30210
      102 IF(LETRS(I)-107)106,150,141                               DSL30220
      141 IF(LETRS(I)-108)99,150,119                                DSL30230
      106 IF(LETRS(I)-77)103,150,111                                DSL30240
      111 IF(LETRS(I)-93)119,150,119                                DSL30245
C...  TEST FOR THREE CONSECUTIVE DEC POINTS (75) = CONTINUATION    DSL30250
      103 IF(LETRS(I)-75)119,104,150                                DSL30260
      104 KCUNT=-2                                                  DSL30270
      107 I=I+1                                                     DSL30280
          IF(LETRS(I)-75)115,110,99                                 DSL30290
      110 KCUNT=KOUNT+1                                             DSL30300
          IF(KOUNT)107,105,105                                      DSL30310
      115 IF(LETRS(I)-64)99,107,99                                  DSL30320
      116 IF(LETRS(I)-240)120,120,119                               DSL30330
       99 NALRM=15                                                  DSL30340
          ISTRT=I                                                   DSL30345
          CALL ERR                                                  DSL30350
          I=72                                                      DSL30360
          GO TO 300                                                 DSL30370
C...  READ AND PRINT CONTINUATION CARD                            DSL30380
      105 CALL READ                                                 DSL30390
C...  ALTER FORMAT TO 72 RIGHT-JUSTIFIED EBCDIC CHARS             DSL30400
          CALL SPLIT                                                DSL30410
          I=1                                                       DSL30420
          GO TO 100                                                 DSL30430
      119 IF(J)99,99,120                                            DSL30440
      120 J=J+1                                                     DSL30450
          IF(J-6)121,121,99                                         DSL30460
      121 KA(J)=LETRS(I)                                            DSL30470
      200 I=I+1                                                     DSL30480
          IF(I-72)100,100,150                                       DSL30490
C...  END OF WORD (EQUALS, COMMA, OR COL 72 OF CARD)              DSL30500
      150 IF(J-6)151,152,99                                         DSL30510
      151 J=J+1                                                     DSL30520
          KA(J)=64                                                  DSL30530
          GO TO 150                                                 DSL30540
      152 CALL BUILD                                                DSL30550
      300 ISTRT=I+1                                                 DSL30560
          RETURN                                                    DSL30570
          END                                                       DSL30580
*STORE       UA      ALPHA                                          DSL30600
// FOR ERASE                                                        DSL30650
*ONE WORD INTEGERS                                                  DSL30660
          SUBRCUTINE ERASE                                          DSL30670
          RETURN                                                    DSL30680
```

274

```
        END                                                          DSL30690
*STORE        UA        ERASE                                        DSL30710
// FOR ERR                                                           DSL30910
**DSL/1800 - ERROR PRINT ROUTINE                                     DSL30920
*PUNCH                                                               DSL30930
*LIST ALL                                                            DSL30940
*ONE WORD INTEGERS                                                   DSL30950
        SUBROUTINE ERR                                               DSL30960
        DIMENSION IERRS(10)                                          DSL30965
        COMMON NALRM,NINTG,NSYMB,JPT,ID(217),ISTRT,JD(78),NCT,K(533),C(2) DSL30970
        L=JPT                                                        DSL30974
        IF(NALRM)2,5,3                                               DSL30980
    3   IF(NALRM-30)2,4,4                                            DSL30982
    4   NO=NALRM-29                                                  DSL30984
        GO TO(12,12,2,12,12,2),NO                                    DSL30986
   12   IF(NCT-10)13,14,14                                           DSL30988
   13   NCT=NCT+1                                                    DSL30990
   14   IERRS(NCT)=NALRM                                             DSL30992
        GO TO 15                                                     DSL30994
    5   IF(NCT)9,9,6                                                 DSL30996
    6   DO 10  J=1,NCT                                               DSL30998
   10   WRITE(L,7) IERRS(J)                                          DSL31000
    7   FORMAT(8H DSL ERR I4)                                        DSL31002
        NCT=0                                                        DSL31004
        GO TO 9                                                      DSL31006
    2   CALL REMSW(2,N)                                              DSL31008
        IF(N-1)21,20,21                                              DSL31010
   20   L=INPT(1)                                                    DSL31012
   21   WRITE(L,8) NALRM,ISTRT                                       DSL31015
    8   FORMAT(8H DSL ERR I4,6H   COL. I4)                           DSL31018
   15   NALRM=0                                                      DSL31030
    9   RETURN                                                       DSL31040
        END                                                          DSL31050
*STORE        UA        ERR                                          DSL31070
// FOR FXINT                                                         DSL31080
**    DSL/1800 - FIXED INTEGRATION SIMULATION PROGRAM                DSL31085
*LIST SYMBOL TABLE                                                   DSL31090
*LIST SUBPROGRAM NAMES                                               DSL31100
*LIST ALL
*ONE WORD INTEGERS                                                   DSL31110
        SUBROUTINE FXINT(PO,XABS,R,YS,Z)                             DSL31120
        INTEGER TITLE(60),HDNG(75),SYMB(420)                         DSL31130
        DIMENSION Y(1),XABS(1),R(1),PO(1),KC(1),PD(50),DELY(50),YS(1),Z(1) DSL31140
        COMMON NALRM,NINTG,NSYMB,JPT,KPRNT,KRANG,KFINI,KTITL,KREL,KABS, DSL31150
    1   KMEM,KPOIN,KLOCK,INTYP,INDX1,KEEP,DELTA,STEPC,NX,NY,RLAST,ALAST, DSL31160
    2   FINVA(5),TITLE,HDNG,INDXP(25),IRANG(15),IFINL(5),JFINL(5),   DSL31170
    3   ISTRT,RMIN(15),RMAX(15),KWORD(11),KAFGN,LINE,KLINE,LINE1,NLINS, DSL31180
    4   KODE,I2,LOCY,LOCX,KOUNT,IRES(101),DELTG,SCAL1,SCAL2,YPOS,INDX2, DSL31190
    5   ISCOP,SYMB,C(7)                                              DSL31200
        EQUIVALENCE (C(1),TIME),(C(2),DELT),(C(3),XINT),(C(7),Y(1)), DSL31210
    1   (DELTP,C(5)),(FINTI,C(4)),(C(1),KC(1)),(DELTA,H),(DELTC,C(6)), DSL31220
    4   (IRES(57),INDY)                                             DSL31230
  200   NALRM=0                                                      DSL31240
        LOCY=0                                                       DSL31245
        LINK=1                                                       DSL31250
        LINE=60                                                      DSL31260
        TPRNT=0.0                                                    DSL31270
        DPT=0.                                                       DSL31280
        TPLOT=0.0                                                    DSL31290
        DPL=0.                                                       DSL31300
        KPT = JPT                                                    DSL31310
        KAFGN=-10                                                    DSL31320
        PLAST=C(KMEM)                                                DSL31330
        W=PLAST+DELTP-TIME                                           DSL31340
        IF(DELTP)209,209,207                                         DSL31345
  207   IF(W+DELTP*DPT)208,208,209                                   DSL31350
  208   DPT=DPT+1.0                                                  DSL31360
        GO TO 207                                                    DSL31370
  209   TLAST=C(KMEM+1)                                              DSL31380
C...    INITIALIZE DERIVATIVE SUBROUTINE                             DSL31390
        KEEP = 0                                                     DSL31400
        KOUNT=0                                                      DSL31410
        CALL UPDAT                                                   DSL31420
        CALL PLOTS(ISCOP)                                            DSL31422
        IF(ISCOP)206,206,205                                         DSL31424
  205   CALL ORGSC(2,2.)                                             DSL31426
C...    TEST IF CONTINUATION OF PARAMETER STUDY                      DSL31430
  206   IF(KLOCK)220,221,220                                         DSL31440
```

```
      220 KEEP=-1                                                       DSL31450
          KLOCK=0                                                       DSL31455
          IF(DELTG)223,223,211                                          DSL31460
      211 CALL FPLOT(-2,SCAL1*C(INDX1),SCAL2*C(INDY))                   DSL31470
          GO TO 223                                                     DSL31480
      221 KEEP=1                                                        DSL31490
          TLAST=0.0                                                     DSL31500
          PLAST=0.0                                                     DSL31510
          IF(DELTG)223,223,219                                          DSL31515
      219 CALL FPLOT(2,0.,0.)                                           DSL31516
      223 CALL UPDAT                                                    DSL31520
          IF(KFINI)277,277,225                                         DSL31530
      225 DO 226  I=1,KFINI                                             DSL31540
          I2=IFINL(I)                                                   DSL31550
          IF(C(I2)- FINVA(I))227,227,228                               DSL31560
      227 JFINL(I)=1                                                    DSL31570
          GO TO 226                                                     DSL31580
      228 JFINL(I)=-1                                                   DSL31590
      226 CONTINUE                                                      DSL31600
      277 KTITL=IABS(KTITL)                                             DSL31610
          LINE1=KTITL/20+2                                             DSL31620
          KLINE=52                                                      DSL31630
          IF( KPRNT-10)46,46,66                                         DSL31640
       66 NLINS=(KPRNT+2)/4+1                                           DSL31650
          KLINE=54-NLINS                                                DSL31660
          GO TO 46                                                      DSL31670
C... (INTEGRATION TYPE - RKSFX,RKSFX,RKSFX,ADAMS,TRAPZ,NON-CENTRAL)     DSL31675
       31 GO TO(92,92,92,94,95,98),INTYP                                DSL31680
C... RUNGE-KUTTA INTEGRATION    (FIXED INTERVAL)                        DSL31690
       92 KEEP=0                                                        DSL31700
          N=NINTG                                                       DSL31710
          DO 27 I=1,N                                                   DSL31720
          DELY(I)=DELT*PO(I)                                            DSL31730
       27 PD(I)=DELY(I)                                                 DSL31740
          TIME=TIME+0.5*DELT                                           DSL31750
          DO 68 I=1,N                                                   DSL31760
          YS(I)=Y(I)                                                    DSL31770
       68 Y(I)=YS(I)+0.5*DELY(I)                                        DSL31780
          CALL UPDAT                                                    DSL31790
          DO 63 I=1,N                                                   DSL31800
          DELY(I)=DELT*PO(I)                                            DSL31810
          PD(I)=PD(I)+2.*DELY(I)                                        DSL31820
       63 Y(I)=YS(I)+0.5*DELY(I)                                        DSL31830
          CALL UPDAT                                                    DSL31840
          DO 62 I=1,N                                                   DSL31850
          DELY(I)=DELT*PO(I)                                            DSL31860
          PD(I)=PD(I)+2.*DELY(I)                                        DSL31870
       62 Y(I)=YS(I)+DELY(I)                                            DSL31880
          TIME=TIME+0.5*DELT                                           DSL31890
          CALL UPDAT                                                    DSL31900
          DO 280 I=1,N                                                  DSL31910
          DELY(I)=DELT*PO(I)                                            DSL31920
          PD(I)=PD(I)+DELY(I)                                           DSL31930
      280 Y(I)=YS(I)+PD(I)/6.                                           DSL31940
          KEEP=1                                                        DSL31950
          CALL UPDAT                                                    DSL31960
          GO TO 46                                                      DSL31970
C... ADAMS-MOULTON INTEGRATION                                          DSL31980
       94 IF(LINK)377,375,377                                           DSL31990
      375 DO 376  I=1,NINTG                                             DSL32000
          Y(I) =Y(I)+0.5*DELT*(3.0*PO(I)-PD(I))                        DSL32010
      376 PD(I)=PO(I)                                                   DSL32020
          TIME=TIME+DELT                                               DSL32030
          GO TO 390                                                     DSL32040
      377 LINK=0                                                        DSL32050
C... TRAPEZOIDAL INTEGRATION                                            DSL32060
C... STORE HISTORY                                                      DSL32070
       95 DO 360  I=1,NINTG                                             DSL32080
          PD(I)=PO(I)                                                   DSL32090
      360 DELY(I)=Y(I)                                                  DSL32100
C... PREDICT Y AT T + DELT                                              DSL32110
          TIME=TIME+DELT                                               DSL32120
          DO 380  I=1,NINTG                                             DSL32130
      380 Y(I)=Y(I)+DELT*PO(I)                                          DSL32140
C... UPDATE INPUTS                                                      DSL32150
          CALL UPDAT                                                    DSL32160
C... COMPUTE OUTPUTS OF INTEGRATORS, CORRECT PREVIOUS COMPUTATION       DSL32170
          DO 385  I=1,NINTG                                             DSL32180
```

276

```
  385 Y(I)=DELY(I)+0.5*DELT*(PO(I)+PD(I))                                    DSL32190
      GO TO 390                                                            DSL32200
C...  NON-CENTRALIZED INTEGRATION                                          DSL32210
   98 TIME=TIME+DELT                                                       DSL32220
C...  KEEP IS SET = 1 FOR DERIV AND DELAY TO IDENTIFY POINT TO STORE       DSL32230
  390 KEEP=1                                                               DSL32240
      CALL UPDAT                                                           DSL32250
      KEEP=0                                                               DSL32260
   46 TIMEX=TIME+0.5*DELT                                                  DSL32270
      IF(KRANG)42,42,167                                                   DSL32280
  167 DO 168  I=1,KRANG                                                    DSL32290
      I2=IRANG(I)                                                          DSL32300
      IF(RMIN(I)-C(I2))170,170,169                                        DSL32310
  169 RMIN(I)=C(I2)                                                        DSL32320
      IF(TIME)170,170,168                                                  DSL32330
  170 IF(RMAX(I)-C(I2))171,168,168                                        DSL32340
  171 RMAX(I)=C(I2)                                                        DSL32350
  168 CONTINUE                                                             DSL32360
   42 IF( KPRNT)44,44,45                                                   DSL32370
   45 IF(TIMEX-TPRNT)44,34,34                                              DSL32380
   34 KCDE=1                                                               DSL32390
      CALL PRINT                                                           DSL32400
      DPT=DPT+1.                                                           DSL32410
      TPRNT =DPT*DELTP+PLAST                                               DSL32420
   44 IF(DELTG)48,48,33                                                    DSL32430
   33 IF(TIMEX-TPLOT)48,36,36                                              DSL32440
   36 CALL FPLOT(0,SCAL1*C(INDX1),SCAL2*C(INDY))                           DSL32450
      DPL=DPL+1.                                                           DSL32460
      TPLOT=DPL*DELTG+TLAST                                                DSL32470
   48 IF(TIMEX-FINTI)229,3,3                                               DSL32480
  229 IF(KFINI)224,224,222                                                 DSL32490
  222 DO 235  I=1,KFINI                                                    DSL32500
      I2=IFINL(I)                                                          DSL32510
      IF( JFINL(I))231,232,232                                            DSL32520
  231 IF(C(I2)- FINVA(I))240,240,235                                      DSL32530
  232 IF(C(I2)- FINVA(I))235,240,240                                      DSL32540
  235 CONTINUE                                                             DSL32550
C...  TEST IF INTERRUPT FOR TYPEWRITER INPUT                               DSL32560
  224 CALL REMSW(1,IRD)                                                    DSL32570
      GO TO(3,31),IRD                                                      DSL32580
  240 KCDE=2                                                               DSL32590
      GO TO 215                                                           DSL32600
    3 KODE=3                                                               DSL32605
  215 IF(TPRNT)213,213,212                                                 DSL32610
  213 C(KMEM)=TIME                                                         DSL32620
      GO TO 214                                                           DSL32630
  212 C(KMEM)=TPRNT-DELTP                                                  DSL32640
  214 C(KMEM+1)=TIME                                                       DSL32650
      CALL PRINT                                                          DSL32655
      IF(DELTG)37,37,35                                                    DSL32660
   35 CALL FPLOT(0,SCAL1*C(INDX1),SCAL2*C(INDY))                           DSL32670
      CALL FPLOT(1,0.,0.)                                                  DSL32680
   37 CALL REMSW(3,IRD)                                                    DSL32705
      GO TO(236,237),IRD                                                   DSL32706
  236 CALL TYPIN                                                           DSL32707
      GO TO 200                                                           DSL32708
  237 RETURN                                                               DSL32710
      END                                                                 DSL32720
*STORE       UA       FXINT                                               DSL32730
// FOR INITL                                                              DSL32740
**DSL/1800 - INITIALIZES DATA AND COMMON AREAS                            DSL32750
*LIST SYMBOL TABLE                                                        DSL32760
*LIST SUBPROGRAM NAMES                                                    DSL32770
*LIST ALL                                                                 DSL32780
*ONE WORD INTEGERS                                                        DSL32790
      SUBROUTINE INITL                                                    DSL32800
      INTEGER TITLE(60),HDNG(75),SYMB(420),WORD(3)                        DSL32810
      COMMON NALRM,NINTG,NSYMB,KPT,KPRNT,KRANG,KFINI,KTITL,KREL,KABS,     DSL32820
     1 KMEM,KPOIN,KLOCK,INTYP,INDX1,KEEP,DELTA,STEPC,NX,NY,RLAST,ALAST,   DSL32830
     2 FINVA(5),TITLE,HDNG,INDXP(25),IRANG(15),IFINL(5),JFINL(5),         DSL32840
     3 ISTRT,WORD,KA(6),LETRS(72),IRES(101),DELTG,SCAL1,SCAL2,YPOS,       DSL32850
     4 INDX2,ISCOP,SYMB,C(1)                                              DSL32860
      NALRM=0                                                             DSL32870
      YPOS=0.0                                                            DSL32880
      KLOCK=0                                                             DSL32890
      KPRNT=0                                                             DSL32900
      KRANG=0                                                             DSL32910
      KFINI=0                                                             DSL32920
```

277

```
            KTITL=0                                                    DSL32930
            KREL=0                                                     DSL32940
            KABS=0                                                     DSL32950
            DELTG=0.                                                   DSL32960
            ISCOP=0                                                    DSL32970
            SCAL1=1.0                                                  DSL32980
            SCAL2=1.0                                                  DSL32990
C...    STORE SYSTEM SYMBOL TABLE IN COMMON                            DSL33000
            CALL SETUP                                                 DSL33010
C...    READ IN SYMBOL TABLE                                           DSL33020
            L=INPT(8)                                                  DSL33025
            M=INPT(4)                                                  DSL33026
            READ (L,11) NINTG,NSYMB,KMEM                               DSL33030
       11 FORMAT(3I4)                                                  DSL33040
            READ (L,10) (SYMB(K),SYMB(K+140),SYMB(K+280),K=1,NSYMB)    DSL33050
       10 FORMAT(36A2)                                                 DSL33060
            KPOIN=KMEM+NINTG                                           DSL33070
            DO 4  I=1,M                                                DSL33080
        4 C(I)=0.0                                                     DSL33090
            IF(NINTG-1)8,6,9                                           DSL33100
        8 INTYP=6                                                      DSL33110
            KPOIN=KPOIN+2                                              DSL33120
            GO TO 7                                                    DSL33130
C...    SET INTEGRATION TYPE TO 2 FOR RUNGE-KUTTA VAR STEP IF NOT SET  DSL33140
        6 KPOIN=KPOIN+1                                                DSL33150
        9 INTYP=2                                                      DSL33160
        7 RETURN                                                       DSL33170
            END                                                        DSL33180
*STORE           UA        INITL                                       DSL33200
// FOR INTEG                                                           DSL37180
*PUNCH
**MAIN ROUTINE FOR INTEGRATION CONTROL ***USES RKS***                 DSL37240
*ONE WORD INTEGERS                                                     DSL37210
*LIST ALL
**ALTERNATE ROUTINE FOR INTEGRATION CONTROL ***USES MILNE***
*NONPROCESS PROGRAM                                                    DSL37220
*IOCS(KEYBOARD,TYPEWRITER,1132 PRINTER,PLOTTER)                        DSL33250
            COMMON NALRM,NINTG,NSYMB,KPT,KPRNT,KRANG,KFINI,KTITL,KREL,KABS, DSL37260
          1 KMEM,KPOIN,IDUM(401),ISCOP,IDUM2(420),C(500)              DSL37270
            K1=KPOIN                                                   DSL37280
            K2=KPOIN+NINTG                                             DSL37290
            IF(ISCOP)6,6,5                                             DSL37292
CC THE FOLLOWING STATEMENT USED IF 611 SCOPE USED                     DSL37293
C     5 CALL DOUTN(1,14,1)                                            DSL37294
        5     CONTINUE                                                 DSL37295
        6 CALL MILNE(C(KMEM),C(KABS),C(KREL),C(K1),C(K2))             DSL37300
C THE FOLLOWING STATEMENT USED FOR 611 SCOPE                          DSL37303
C         CALL DOUTN(1,14,0)                                          DSL37304
            CALL LINK(SIMX)                                            DSL37310
            END                                                       DSL37320
// DUP
*DELETE           INTEG
*STORE       WS UA INTEG
// *DATE 5-18-67
// FOR INTEG                                                          DSL33210
*LIST ALL                                                             DSL33220
*ONE WORD INTEGERS                                                    DSL33230
*NONPROCESS PROGRAM                                                   DSL33240
*IOCS(KEYBOARD,TYPEWRITER,1403 PRINTER,PLOTTER)                       DSL33250
**MAIN ROUTINE FOR INTEGRATION CONTROL                                DSL33260
*NAME SIMUL                                                           DSL33270
            COMMON NALRM,NINTG,NSYMB,KPT,KPRNT,KRANG,KFINI,KTITL,KREL,KABS, DSL33280
          1 KMEM,KPOIN,IDUMY(822),C(500)                              DSL33290
            K1=KPOIN                                                   DSL33300
            K2=KPOIN+NINTG                                             DSL33310
            CALL   RKS(C(KMEM),C(KABS),C(KREL),C(K1),C(K2))           DSL33320
            CALL LINK(SIMX)                                            DSL33330
            END                                                       DSL33340
// DUP
*DELETE           INTEG
*STORE       WS UA INTEG
// FOR INTRA                                                          DSL33540
**DSL/1800 SUBROUTINE TO READ AND TRANSLATE SIMULATION DATA           DSL33550
*LIST SYMBOL TABLE                                                    DSL33560
*LIST SUBPROGRAM NAMES                                                DSL33570
*LIST ALL
*ONE WORD INTEGERS                                                    DSL33580
*NONPROCESS PROGRAM                                                   DSL33590
```

```
      SUBROUTINE INTRA                                                  DSL33600
      INTEGER TITLE(60),HDNG(75),SYMB(420),WORD(3),AFLAS(3)             DSL33610
      DIMENSION KC(2),RMIN(2),RMAX(2)                                   DSL33620
      COMMON NALRM,NINTG,NSYMB,JPT,KPRNT,KRANG,KFINI,KTITL,KREL,KABS,   DSL33630
     1 KMEM,KPCIN,KLOCK,INTYP,INDX1,IERR,DELTA,STEPC,NX,NY,RLAST,ALAST, DSL33640
     2 FINVA(5),TITLE,HDNG,INDXP(25),IRANG(15),IFINL(5),JFINL(5),       DSL33650
     3 ISTRT,WORD,KA(6),LETRS(72),IRES(101),DELTG,SCAL1,SCAL2,YPOS,     DSL33660
     4 INDX2,ISCOP,SYMB,C(6)                                            DSL33670
      EQUIVALENCE (TIME,C(1)),(DELT,C(2)),(DELMI,C(3)),(FINTI,C(4)),    DSL33680
     1 (DELTP,C(5)),(C(1),KC(1)),(WORD(2),DWORD,KWORD),(C(6),DELTC),    DSL33690
     2 (RMIN(1),WORD(2)),(RMAX(1),LETRS(23))                            DSL33700
      KPT = JPT                                                         DSL33710
      LPT=INPT(1)                                                       DSL33715
      WRITE(KPT,813)                                                    DSL33720
  813 FORMAT(1H1 20X,32H*** DSL/1800 SIMULATION DATA ***)               DSL33730
      IERR=0                                                            DSL33740
      AFLAS(1)=0                                                        DSL33750
      CSAVE=DELT                                                        DSL33760
      DELT=RLAST                                                        DSL33770
C...  START OF LOOP FOR EACH CARD                                       DSL33780
C...  READ AND PRINT DATA CARD                                          DSL33790
    4 CALL READ                                                         DSL33800
      DO 5  I=1,25                                                      DSL33810
      IF(LETRS(1)-IRES(I))5,2,5                                         DSL33820
    2 IF(LETRS(2)-IRES(I+31))5,3,5                                      DSL33830
    3 IF(LETRS(3)-IRES(I+62))5,6,5                                      DSL33840
    5 CONTINUE                                                          DSL33850
      WRITE(LPT,802) LETRS(1),LETRS(2),LETRS(3)                         DSL33860
  802 FORMAT(31H0DATA TYPE NOT A RESERVE WORD   3A2)                    DSL33870
      GO TO 4                                                           DSL33880
    6 ITYPE=I                                                           DSL33890
C...  TEST FOR LABEL OR TITLE CARD                                      DSL33900
      IF(ITYPE-13)58,113,59                                             DSL33910
   58 IF(ITYPE-11)59,111,59                                             DSL33920
   59 ISTRT=7                                                           DSL33930
C...  ALTER FORMAT TO 72 RIGHT-JUSTIFIED EBCDIC CHARS                   DSL33940
      CALL SPLIT                                                        DSL33950
C...  PARAM,INCON,CONST,PRINT,SCOPE ,GRAPH,RANGE,FINISH,RELERR,ABSERR,  DSL33960
C...  LABEL,CHART,TITLE,CONTRL,CONTIN,INTEG,AFGEN,NLFGEN,STOP,END,      DSL33970
C...  SCALE,TABLE,RESET,TYPE,END*              (*=EOF=/15)              DSL33980
      GO TO (101,101,101,104,105,106,107,108,109,110,111,112,113,101,  DSL33990
     1 115,116,117,119,120,121,122,123,124,120),ITYPE                  DSL34000
C...  PARAM,INCON,CONTRL,CONST                                          DSL34010
  101 IF(ISTRT-72)30,30,4                                               DSL34020
   30 KEY=1                                                             DSL34030
   60 CALL ALPHA                                                        DSL34040
      DO 9  I=1,NSYMB                                                   DSL34050
      IF(WORD(1)-SYMB(I))9,8,9                                         DSL34060
    8 IF(WORD(2)-SYMB(I+140))9,10,9                                     DSL34070
   10 IF(WORD(3)-SYMB(I+280))9,12,9                                     DSL34080
   12 GO TO(7,11,146,183,274,21,131,78,79,240),KEY                      DSL34090
    9 CONTINUE                                                          DSL34100
      WRITE(LPT,804)WORD(1),WORD(2),WORD(3)                            DSL34110
  804 FORMAT(21H UNDEFINED VARIABLE   3A2)                              DSL34120
      GO TO(15,52,290,15,15,16,16,265,265,242),KEY                      DSL34130
   15 CALL NUMER                                                        DSL34140
   16 IERR=1                                                            DSL34150
      GO TO(101,4,4,291,271,4,4),KEY                                    DSL34160
    7 CALL NUMER                                                        DSL34170
      C(I)=DWORD                                                        DSL34180
      GO TO 101                                                         DSL34190
C...  PRINT                                                             DSL34200
  104 CALL NUMER                                                        DSL34210
      DELTP=DWORD                                                       DSL34220
      KPRNT=1                                                           DSL34230
      HDNG(1)=SYMB(1)                                                   DSL34240
      HDNG(26)=SYMB(141)                                                DSL34250
      HDNG(51)=SYMB(281)                                                DSL34260
      INDXP(1)=1                                                        DSL34270
      KEY=2                                                             DSL34280
   52 IF(ISTRT-72)60,60,4                                               DSL34290
   11 IF(KPRNT-25)80,81,81                                              DSL34300
   81 NALRM=16                                                          DSL34310
      CALL ERR                                                          DSL34320
      GO TO 4                                                           DSL34330
   80 KPRNT=KPRNT+1                                                     DSL34340
      INDXP(KPRNT)=I                                                    DSL34350
      HDNG(KPRNT)=WORD(1)                                               DSL34360
```

```
      HDNG(KPRNT+25)=WORD(2)                                          DSL34370
      HDNG(KPRNT+50)=WORD(3)                                          DSL34380
      GO TO 52                                                        DSL34390
C...  SCOPE     (X-AXIS = 8.5 UNITS,   Y-AXIS = 7.0 UNITS)            DSL34400
  105 ISCOP=1                                                         DSL34410
      GO TO 48                                                        DSL34420
C...  GRAPH     (X-AXIS = 12.0, Y-AXIS = 8.0 INCHES)                  DSL34430
  106 ISCOP=0                                                         DSL34440
   48 CALL NUMER                                                      DSL34450
      DELTG=DWORD                                                     DSL34460
      CALL NUMER                                                      DSL34470
      NX=KWORD                                                        DSL34480
      CALL NUMER                                                      DSL34490
      NY=KWORD                                                        DSL34500
      KEY=8                                                           DSL34510
      GC TO 60                                                        DSL34520
   78 INDX1=I                                                         DSL34530
      KEY=9                                                           DSL34540
      GO TO 60                                                        DSL34550
   79 INDX2=I                                                         DSL34560
      GO TO 4                                                         DSL34570
C...  RANGE     (RECORD MAX AND MIN FOR VARIABLE)                     DSL34580
  107 KRANG=0                                                         DSL34590
      KEY=3                                                           DSL34600
  290 IF(ISTRT-72)60,60,4                                             DSL34610
  146 IF(KRANG -15)142,143,143                                        DSL34620
  143 NALRM=17                                                        DSL34630
      CALL ERR                                                        DSL34640
      GO TO 4                                                         DSL34650
  142 KRANG=KRANG+1                                                   DSL34660
      IRANG(KRANG)=I                                                  DSL34670
      GC TO 290                                                       DSL34680
C...  FINISH    (SET FINAL VALUES OF VARIABLES TO END JOB)            DSL34690
  108 KFINI=0                                                         DSL34700
      KEY=4                                                           DSL34710
  291 IF(ISTRT-72)60,60,4                                             DSL34720
  183 IF(KFINI - 5)185,184,184                                        DSL34730
  184 NALRM=18                                                        DSL34740
      CALL ERR                                                        DSL34750
      GO TO 4                                                         DSL34760
  185 KFINI=KFINI+1                                                   DSL34770
      IFINL(KFINI)=I                                                  DSL34780
      CALL NUMER                                                      DSL34790
      FINVA(KFINI)=DWORD                                              DSL34800
      GO TO 291                                                       DSL34810
C...  RELERR   (RELATIVE ERROR TO CONTROL INTEG INTERVAL FOR CENTRALIZED DSL34820
C...  INTEGRATION - SPECIFIED PER INTEGRATOR OUTPUT)                  DSL34830
  109 IF(KREL)275,270,275                                             DSL34840
  270 KREL=-KPOIN                                                     DSL34850
      KPOIN=KPOIN+NINTG                                               DSL34860
      NERR=6+NINTG                                                    DSL34870
  275 I4=IABS(KREL)-7                                                 DSL34880
  271 KEY=5                                                           DSL34890
      IF(ISTRT-72)60,60,280                                           DSL34900
  274 CALL NUMER                                                      DSL34910
      I3=I4+I                                                         DSL34920
      C(I3)=DWORD                                                     DSL34930
      GO TO 271                                                       DSL34940
  280 IF(ITYPE-9)281,281,282                                          DSL34950
  281 RLAST=DWORD                                                     DSL34960
      GO TO 4                                                         DSL34970
  282 ALAST=DWORD                                                     DSL34980
      GO TO 4                                                         DSL34990
C...  ABSERR    (ABSOLUTE ERROR PER INTEGRATOR OUTPUT)                DSL35000
  110 IF(KABS)277,276,277                                             DSL35010
  276 KABS=-KPOIN                                                     DSL35020
      KPOIN=KPOIN+NINTG                                               DSL35030
      NERR=6+NINTG                                                    DSL35040
  277 I4=IABS(KABS)-7                                                 DSL35050
      GO TO 271                                                       DSL35060
C...  LABEL                                                           DSL35070
  111 DO 47   I=4,23                                                  DSL35080
   47 TITLE(I+37)=LETRS(I)                                            DSL35090
      GO TO 4                                                         DSL35100
C...  CHART     (BRUSH CHART RECORDER)                                DSL35110
  112 GO TO 4                                                         DSL35120
C...  TITLE                                                           DSL35130
  113 IF(KTITL )215,216,217                                           DSL35140
```

```
      217 IF(KTITL-60)216,4,4                                            DSL35150
      215 KTITL=0                                                        DSL35160
      216 DO 61   I=4,23                                                 DSL35170
          KTITL=KTITL+1                                                  DSL35180
       61 TITLE(KTITL)=LETRS(I)                                          DSL35190
          GO TO 4                                                        DSL35200
C...  CONTIN                                                             DSL35210
      115 KLOCK=2                                                        DSL35220
          GO TO 4                                                        DSL35230
C...      (INTEGRATION TYPE - MILNE,RKS,RKSFX,ADAMS,TRAPZ)              DSL35240
      116 CALL ALPHA                                                     DSL35250
          DO 24   I=27,31                                                DSL35260
          IF(WORD(1)-IRES(I))24,54,24                                    DSL35270
       54 IF(WORD(2)-IRES(I+31))24,55,24                                 DSL35280
       55 IF(WORD(3)-IRES(I+62))24,25,24                                 DSL35290
       24 CONTINUE                                                       DSL35300
          GO TO 15                                                       DSL35310
       25 INTYP=I-26                                                     DSL35320
          GO TO 4                                                        DSL35330
C...  AFGEN   (ARBITRARY FUNCTION GENERATOR) + NLFGEN  (NON-LINEAR)     DSL35340
      117 CALL ALPHA                                                     DSL35350
          KCVER=0                                                        DSL35360
          KEY=6                                                          DSL35370
C...  TEST FOR OVERLAY                                                   DSL35380
          IF(WORD(1)-IRES(98))125,41,125                                 DSL35390
       41 IF(WORD(2)-IRES(99))125,42,125                                 DSL35400
       42 IF(WORD(3)-IRES(100))125,43,125                                DSL35410
       43 KCVER=1                                                        DSL35420
          GO TO 60                                                       DSL35430
      125 ISTRT=7                                                        DSL35440
          GO TO 60                                                       DSL35450
       21 J=I+I-1                                                        DSL35460
          IF(WORD(1)-AFLAS(1))22,44,22                                   DSL35470
       44 IF(WORD(2)-AFLAS(2))22,45,22                                   DSL35480
       45 IF(WORD(3)-AFLAS(3))22,46,22                                   DSL35490
       46 KP1=KC(J)+KC(J)-3                                              DSL35500
          KPOIT=KC(KP1)+2                                                DSL35510
          GO TO 74                                                       DSL35520
       22 IF(KOVER)126,127,126                                           DSL35530
      126 KPOIT=KC(J)                                                    DSL35540
          GO TO 128                                                      DSL35550
      127 KPOIT=KPOIN+1                                                  DSL35560
          KC(J)=KPOIT                                                    DSL35570
      128 KP1=KPOIT+KPOIT-3                                              DSL35580
          AFLAS(1)=WORD(1)                                               DSL35590
          AFLAS(2)=WORD(2)                                               DSL35600
          AFLAS(3)=WORD(3)                                               DSL35610
       74 IF( ISTRT-72)75,75,76                                          DSL35620
       75 CALL NUMER                                                     DSL35630
          C(KPOIT)=DWORD                                                 DSL35640
          KPOIT=KPOIT+1                                                  DSL35650
          GO TO 74                                                       DSL35660
       76 KC(KP1)=KPOIT-2                                                DSL35670
          IF(KOVER)4,129,4                                               DSL35680
      129 KPOIN=KPOIT                                                    DSL35690
          GO TO 4                                                        DSL35700
C...  END           (LAST CARD OF DATA SET)                             DSL35710
      120 IF(IERR)211,210,211                                           DSL35720
      211 WRITE(LPT,810)                                                 DSL35730
      810 FORMAT(27H DATA ERR, START TO EXECUTE)                         DSL35740
          PAUSE                                                          DSL35745
          GO TO 210                                                      DSL35746
C...  STOP          (LAST CARD OF PROBLEM SET)                          DSL35750
      119 CALL EXIT                                                      DSL35760
C...  SCALE FACTORS FOR VARIABLES TO BE GRAPHED                          DSL35770
      121 CALL NUMER                                                     DSL35780
          SCAL1=DWORD                                                    DSL35790
          CALL NUMER                                                     DSL35800
          SCAL2=DWORD                                                    DSL35810
          GO TO 4                                                        DSL35820
C...  TABLE     (TABULAR DATA INPUT)                                     DSL35830
      122 KEY=7                                                          DSL35840
          GO TO 60                                                       DSL35850
      131 LOCN=KC(2*I-1)-1                                               DSL35860
          CALL NUMER                                                     DSL35870
          IF(KWORD)132,133,134                                           DSL35880
      132 IPT1=-KWORD+LOCN                                               DSL35890
          CALL NUMER                                                     DSL35900
```

```
      IPT2=KWORD+LOCN                                        DSL35910
  135 IF( ISTRT-72)136,136,133                               DSL35920
  136 CALL NUMER                                             DSL35930
      IF( ISTRT)137,133,138                                  DSL35940
  137 IPT3=KWORD                                             DSL35950
      ISTRT=-ISTRT                                           DSL35960
      CALL NUMER                                             DSL35970
      DO 139  I=1,IPT3                                       DSL35980
      C(IPT1)=DWORD                                          DSL35990
  139 IPT1=IPT1+1                                            DSL36000
      GO TO 141                                              DSL36010
  138 C(IPT1)=DWORD                                          DSL36020
      IPT1=IPT1+1                                            DSL36030
  141 IF(IPT1-IPT2)135,135,140                               DSL36040
  134 IPT1=KWORD+LOCN                                        DSL36050
      CALL NUMER                                             DSL36060
      C(IPT1)=DWORD                                          DSL36070
  140 IF(ISTRT-72)60,133,4                                   DSL36080
  133 NALRM=19                                               DSL36090
      CALL ERR                                               DSL36100
      GO TO 4                                                DSL36110
C... RESET     (SET VARIABLE VALUES TO ZERO)                 DSL36120
  123 IF( ISTRT-72)250,250,4                                 DSL36130
  250 CALL ALPHA                                             DSL36140
      IF(WORD(1)-IRES(101))49,260,49                         DSL36150
   49 DO 252  I=1,12                                         DSL36160
      IF(WORD(1)-IRES(I))252,50,252                          DSL36170
   50 IF(WORD(2)-IRES(I+31))252,51,252                       DSL36180
   51 IF(WORD(3)-IRES(I+62))252,53,252                       DSL36190
   53 GO TO(4,4,4,264,265,266,267,268,269,279,4,261),I       DSL36200
  252 CONTINUE                                               DSL36210
      GO TO 15                                               DSL36220
  260 KPRNT=0                                                DSL36230
      KRANG=0                                                DSL36240
      DELTG=0.0                                              DSL36250
  261 DELTC=0.0                                              DSL36260
      GO TO 4                                                DSL36270
  264 KPRNT=0                                                DSL36280
      GO TO 123                                              DSL36290
  265 ISCOP=0                                                DSL36300
  266 DELTG=0.0                                              DSL36310
      GO TO 123                                              DSL36320
  267 KRANG=0                                                DSL36330
      GO TO 123                                              DSL36340
  268 KFINI=0                                                DSL36350
      GO TO 123                                              DSL36360
  269 IF(KREL)123,123,153                                    DSL36370
  153 J4=KREL+NINTG-1                                        DSL36380
      J3=KREL                                                DSL36390
      KREL=-KREL                                             DSL36400
  156 DO 154  J=J3,J4                                        DSL36410
  154 C(J)=0.0                                               DSL36420
      GO TO 123                                              DSL36430
  279 IF(KABS)123,123,155                                    DSL36440
  155 J4=KABS+NINTG-1                                        DSL36450
      J3=KABS                                                DSL36460
      KABS=-KABS                                             DSL36470
      GO TO 156                                              DSL36480
C... TYPE     (KEYBOARD DATA ENTRY)                          DSL36490
  124 KEY=10                                                 DSL36500
      GO TO 60                                               DSL36510
  240 WRITE(LPT,241) C(I)                                    DSL36520
  241 FORMAT(E14.6)                                          DSL36530
  242 IF(ISTRT-72)60,60,4                                    DSL36540
  210 IF(KTITL )306,301,306                                  DSL36550
  301 TITLE(1)=IRES(94)                                      DSL36560
      TITLE(2)=IRES(95)                                      DSL36570
      TITLE(3)=IRES(96)                                      DSL36580
      TITLE(4)=IRES(97)                                      DSL36590
      DO 302  I=5,20                                         DSL36600
  302 TITLE(I)=IRES(101)                                     DSL36610
      KTITL=20                                               DSL36620
  306 IF(KRANG)64,64,63                                      DSL36630
   63 DO 310  I=1,KRANG                                      DSL36640
      RMIN(I)=1.E30                                          DSL36650
  310 RMAX(I)=-1.E30                                         DSL36660
   64 IF(DELTG)39,39,38                                      DSL36670
   39 IF(KPRNT)35,35,37                                      DSL36680
```

282

```
        35 NALRM=21                                                       DSL36690
           CALL ERR                                                       DSL36700
           DELTA=FINTI                                                    DSL36710
           GO TO 33                                                       DSL36720
        37 DELTA=DELTP                                                    DSL36730
           GO TO 33                                                       DSL36740
        38 IF(KPRNT)40,40,36                                              DSL36750
        36 IF(DELTP-DELTG)37,37,40                                        DSL36760
        40 DELTA=DELTG                                                    DSL36770
        33 IF(NINTG)90,90,370                                             DSL36780
       370 IF(INTYP-2)371,371,90                                          DSL36790
C...    SET RELATIVE AND ABSOLUTE ERRORS PER INTEGRATOR                   DSL36800
       371 IF(KREL)381,287,287                                            DSL36810
       381 KREL=-KREL                                                     DSL36820
           I4=KREL                                                        DSL36830
           I3=0                                                           DSL36840
           EMAX=RLAST                                                     DSL36850
       382 I5=I4+NINTG-1                                                  DSL36860
           DO 286  I=I4,I5                                                DSL36870
           IF(C(I))285,285,286                                           DSL36880
       285 C(I)=EMAX                                                      DSL36890
       286 CONTINUE                                                       DSL36900
           IF(I3)288,287,288                                             DSL36910
       287 IF(KABS)289,288,288                                           DSL36920
       289 I3=1                                                           DSL36930
           KABS=-KABS                                                     DSL36940
           I4=KABS                                                        DSL36950
           EMAX=ALAST                                                     DSL36960
           GO TO 382                                                      DSL36970
       288 IF(KLOCK)92,375,92                                            DSL36980
       375 IF(KABS)295,295,293                                           DSL36990
       295 KABS=KPOIN                                                     DSL37000
           KPOIN=KPOIN+NINTG                                              DSL37010
           DO 296  I=KABS,KPOIN                                          DSL37020
       296 C(I)=.0005                                                     DSL37030
       293 IF(KREL)292,292,91                                            DSL37040
       292 KREL=KPOIN                                                     DSL37050
           KPOIN=KPOIN+NINTG                                              DSL37060
           DO 294  I=KREL,KPOIN                                          DSL37070
       294 C(I)=.005                                                      DSL37080
        90 IF(KLOCK)92,91,92                                             DSL37090
        92 DELT=DSAVE                                                     DSL37100
           GO TO 93                                                       DSL37110
        91 TIME=0.0                                                       DSL37120
        93 RLAST=DELT                                                     DSL37130
           RETURN                                                         DSL37140
           END                                                           DSL37150
// DUP
*DELETE               INTRA
*STORE      WS  UA    INTRA
// *DATE 5-18-67
// FOR MILNE
**    DSL/1800 MAIN SIMULATION SUBROUTINE, INCLUDES ALL METHODS          DSL37350
*LIST SYMBOL TABLE                                                       DSL37355
*LIST SUBPROGRAM NAMES                                                   DSL37360
*LIST ALL                                                               DSL37370
*ONE WORD INTEGERS                                                       DSL37380
*PUNCH
           SUBROUTINE MILNE(P0,XABS,R,Y1,P1)                             DSL37390
           INTEGER TITLE(60),HDNG(75),SYMB(420)                         DSL37400
           DIMENSION  Y(1),XABS(1),R(1),P0(1),KC(1),                    DSL37410
         1 Y0(1),Y1(1),Y3(1),Y7(1),P1(1),P3(1),P7(1),                   DSL37420
         2 Y4(50),Y5(50),Y6(50),P4(50),P5(50),P6(50),                   DSL37430
         3 PD(50),SD(50),YS(50),DELY(50)                                DSL37440
           COMMON NALRM,NINTG,NSYMB,JPT,KPRNT,KRANG,KFINI,KTITL,KREL,KABS, DSL37450
         1 KMEM,KPOIN,KLOCK,INTYP,INDX1,KEEP,H        ,STEPC,NX,NY,RLAST,ALAST, DSL37460
         2 FINVA(5),TITLE,HDNG,INDXP(25),IRANG(15),IFINL(5),JFINL(5),   DSL37470
         3 ISTRT,RMIN(15),RMAX(15),KWORD(11),KAFGN,LINE,KLINE,LINE1,NLINS, DSL37480
         4 KODE,I2,LOCY,LOCX,KOUNT,IRES(101),DELTG,SCAL1,SCAL2,YPOS,INDX2, DSL37490
         5 ISCCP,SYMB,C(7)                                              DSL37500
           EQUIVALENCE  (C(1),TIME),(C(2),DELT),(C(3),XINT),(C(7),Y(1)), DSL37510
         1 (DELTP,C(5)),(FINTI,C(4)),(C(1),KC(1)),(DELTC,C(6)),         DSL37520
         2 (PD(1),P3(1)),(SD(1),Y3(1)),(C(7),Y0(1)),(DELY(1),P7(1)),    DSL37530
         3 (YS(1),Y7(1)),(IRES(57),INDY),(NINTG,M,N)                    DSL37540
           NALRM=0                                                      DSL37550
           LOCY=0                                                       DSL37555
           LINK=1                                                       DSL37560
           LINE=60                                                      DSL37570
```

```
       TPRNT=0.0                                                      DSL37580
       DPT=0.                                                         DSL37590
       TPLOT=0.0                                                      DSL37600
       DPL=0.                                                         DSL37610
       K = 4                                                          DSL37620
       KPT = JPT                                                      DSL37630
       KAFGN=-10                                                      DSL37640
       RL=10.**.2                                                     DSL37650
       DELTA=1.                                                       DSL37660
       IY2=KPOIN-1+NINTG+NINTG                                        DSL37670
       IP2=IY2+NINTG                                                  DSL37680
       PLAST=C(KMEM)                                                  DSL37690
       W=PLAST+DELTP-TIME                                             DSL37700
       IF(DELTP)209,209,207                                           DSL37705
   207 IF(W+DELTP*DPT)208,208,209                                     DSL37710
   208 DPT=DPT+1.0                                                    DSL37720
       GO TO 207                                                      DSL37730
   209 TLAST=C(KMEM+1)                                                DSL37740
C...   INITIALIZE DERIVATIVE SUBROUTINE                               DSL37750
       KEEP = 0                                                       DSL37760
       KOUNT=0                                                        DSL37770
       CALL UPDAT                                                     DSL37780
       CALL PLCTS(ISCOP)                                              DSL37782
       IF(ISCOP)206,206,205                                           DSL37784
   205 CALL CRGSC(2.,2.)                                              DSL37786
C...   TEST IF CONTINUATION OF PARAMETER STUDY                        DSL37790
   206 IF(KLCCK)220,221,220                                           DSL37800
   220 KEEP=-1                                                        DSL37810
       KLCCK=0                                                        DSL37815
       IF(DELTG)223,223,211                                           DSL37820
   211 CALL FPLOT(-2,SCAL1*C(INDX1),SCAL2*C(INDY))                    DSL37830
       GO TO 223                                                      DSL37840
   221 KEEP=1                                                         DSL37850
       TLAST=0.0                                                      DSL37860
       PLAST=0.0                                                      DSL37870
   223 CALL UPDAT                                                     DSL37880
       IF(KFINI)277,277,225                                           DSL37890
   225 DO 226  I=1,KFINI                                              DSL37900
       I2=IFINL(I)                                                    DSL37910
       IF(C(I2)- FINVA(I))227,227,228                                DSL37920
   227 JFINL(I)=1                                                     DSL37930
       GO TO 226                                                      DSL37940
   228 JFINL(I)=-1                                                    DSL37950
   226 CONTINUE                                                       DSL37960
   277 KTITL=IABS(KTITL)                                              DSL37970
       LINE1=KTITL/20+2                                               DSL37980
       KLINE=52                                                       DSL37990
       IF( KPRNT-10)46,46,66                                         DSL38000
    66 NLINS=(KPRNT+2)/4+1                                            DSL38010
       KLINE=54-NLINS                                                 DSL38020
       GO TO 46                                                       DSL38030
C...   LOOP FOR CENTRALIZED INTEGRATION                              DSL38040
C...   (INTEGRATION TYPE - MILNE,RKS,RKSFX,ADAMS,TRAPZ,NON-CENTRAL)  DSL38050
    31 GO TC (91,60,92,94,95,98),INTYP                               DSL38060
C...   MILNE 5TH ORDER PREDICTOR-CORRECTOR   (MIDAS)                  DSL38070
C      TEST FOR INITIAL ENTRY, CONTINUATION, OR NORMAL ENTRY          DSL38080
    91 IF(LINK)90,405,90                                              DSL38090
    90 Q=1.0                                                          DSL38100
       IND=4                                                          DSL38120
       W=TIME/H                                                       DSL38130
       D=W+1.0                                                        DSL38140
       DO 210 I=1,M                                                   DSL38150
       P4(I)=P0(I)                                                    DSL38160
   210 Y4(I)=Y0(I)                                                    DSL38170
C...   COMPUTE STARTING VALUES, DETERMINE INTERVAL OF INTEGRATION     DSL38180
       IF(DELT)304,304,8                                              DSL38190
     8 IF(DELT-.25*H)305,305,304                                      DSL38200
   304 DELT=0.25*H                                                    DSL38210
   305 K=1                                                            DSL38220
       INTYP=2                                                        DSL38230
       GO TO 92                                                       DSL38240
   300 DO 250 I=1,M                                                   DSL38250
       P5(I)=P0(I)                                                    DSL38260
   250 Y5(I)=Y0(I)                                                    DSL38270
       K=2                                                            DSL38280
       INTYP=1                                                        DSL38290
       GO TO 92                                                       DSL38300
   310 DO 330 I=1,M                                                   DSL38310
```

284

```
         P6(I)=P0(I)                                                        DSL38320
  330 Y6(I)=Y0(I)                                                           DSL38330
         K=3                                                                DSL38340
         GO TO 92                                                           DSL38350
  320 DC 40   I=1,M                                                         DSL38360
         P7(I)=P0(I)                                                        DSL38370
   40 Y7(I)=Y0(I)                                                           DSL38380
         A=DELT/H                                                           DSL38390
         S=3.*A+W                                                           DSL38400
         LINK=0                                                             DSL38405
         GO TO 700                                                          DSL38410
  405 IF(7-IND)410,41C,500                                                  DSL38420
C     COMPUTE AND TEST DOUBLE INTERVAL                                      DSL38430
  410 IF(D-S-A-.00001)500,500,420                                           DSL38440
  420 DO 430 I=1,M                                                          DSL38450
  430 Y0(I)=Y5(I)+(2.0*DELT/3.0)*(8.0*P7(I)-5.0*P5(I)+4.0*P3(I)-P1(I))      DSL38460
         KEEP=0                                                             DSL38470
         TIME=(A+A+S)*H                                                     DSL38480
         CALL UPDAT                                                         DSL38490
         DO 440 I=1,M                                                       DSL38500
         YC=0.125*Y7(I)+0.875*Y5(I)+(DELT/96.0)*(65.0*P0(I)+243.0*P7(I)+    DSL38510
    1 51.0*P5(I)+P3(I))                                                     DSL38520
         E=      ABS(YC-Y0(I))                                             DSL38530
         Z=ABS(YC)                                                          DSL38540
         IF(Z-Q)12,13,13                                                    DSL38550
   12 IF(E-XABS(I))440,440,500                                              DSL38560
   13 IF(E-R(I)*Z)440,440,500                                               DSL38570
  440 Y0(I)=0.96116*YC+0.03884*Y0(I)                                        DSL38580
C     DOUBLE THE INTERVAL                                                   DSL38590
         DO 460 I=1,M                                                       DSL38600
         Y4(I)=Y3(I)                                                        DSL38610
         P4(I)=P3(I)                                                        DSL38620
         Y3(I)=Y1(I)                                                        DSL38630
         P3(I)=P1(I)                                                        DSL38640
         Y6(I)=Y7(I)                                                        DSL38650
  460 P6(I)=P7(I)                                                           DSL38660
         A=2.0*A                                                            DSL38670
         DELT = A*H                                                         DSL38680
         KEEP=1                                                             DSL38690
         TIME=(S+A)*H                                                       DSL38700
         CALL UPDAT                                                         DSL38710
         DO 50   I=1,M                                                      DSL38720
         P7(I)=P0(I)                                                        DSL38730
   50 Y7(I)=Y0(I)                                                           DSL38740
         S=S+A                                                              DSL38750
         IND=5                                                              DSL38760
         GO TO 700                                                          DSL38770
C     COMPUTE AND TEST SINGLE INTERVAL                                      DSL38780
  500 CO 510 I=1,M                                                          DSL38790
  510 Y0(I)=Y6(I)+(DELT/3.0)*(8.0*P7(I)-5.0*P6(I)+4.0*P5(I)-P4(I))          DSL38800
         KEEP=0                                                             DSL38810
         TIME=(S+A)*H                                                       DSL38820
         CALL UPDAT                                                         DSL38830
         DO 520 I=1,M                                                       DSL38840
         YC=0.125*Y7(I)+0.875*Y6(I)+(DELT/192.0)*(65.0*P0(I)+243.0*P7(I)+   DSL38850
    1 51.0*P6(I)+P5(I))                                                     DSL38860
         E=      ABS(YC-Y0(I))                                             DSL38870
         Z=ABS(YC)                                                          DSL38880
         IF(Z-Q)14,15,15                                                    DSL38890
   14 IF(E-XABS(I))520,520,515                                              DSL38900
   15 IF(E-R(I)*Z)520,520,515                                               DSL38910
  515 IF(DELT-XINT)518,600,600                                             DSL38920
  518 NALRM=28                                                             DSL38930
         CALL ERR                                                           DSL38940
         XINT=0.5*XINT                                                      DSL38945
         GC TO 600                                                          DSL38950
  520 Y0(I)=0.96116*YC+0.03884*Y0(I)                                        DSL38960
         KEEP=1                                                             DSL38970
         CALL UPDAT                                                         DSL38980
         DO 540 I=1,M                                                       DSL38990
         J2=IP2+I                                                           DSL39000
         J3=IY2+I                                                           DSL39010
         P1(I)=C(J2)                                                        DSL39020
         Y1(I)=C(J3)                                                        DSL39030
         C(J2)=P3(I)                                                        DSL39040
         C(J3)=Y3(I)                                                        DSL39050
         P3(I)=P4(I)                                                        DSL39060
         Y3(I)=Y4(I)                                                        DSL39070
```

```
      P4(I)=P5(I)                                                    DSL39080
      Y4(I)=Y5(I)                                                    DSL39090
      P5(I)=P6(I)                                                    DSL39100
      Y5(I)=Y6(I)                                                    DSL39110
      P6(I)=P7(I)                                                    DSL39120
      Y6(I)=Y7(I)                                                    DSL39130
      P7(I)=P0(I)                                                    DSL39140
  540 Y7(I)=Y0(I)                                                    DSL39150
      S=S+A                                                          DSL39160
      IND=IND+1                                                      DSL39170
      GO TO 700                                                      DSL39180
C     HALVE THE INTERVAL                                             DSL39190
  600 DO 610 I=1,M                                                   DSL39200
      Y3(I)=Y5(I)                                                    DSL39210
      Y4(I)=Y7(I)-(3.0*DELT/128.0)*(15.0*P7(I)+51.0*P6(I)-3.0*P5(I)+ DSL39220
     1 P4(I))                                                        DSL39230
      Y5(I)=Y6(I)                                                    DSL39240
  610 Y0(I)=Y7(I)-(DELT/384.0)*(119.0*P7(I)+107.0*P6(I)-43.0*P5(I)+  DSL39250
     1 9.0*P4(I))                                                    DSL39260
      A=0.5*A                                                        DSL39270
      DELT = A*H                                                     DSL39280
      KEEP=-1                                                        DSL39290
      TIME=(S-A)*H                                                   DSL39300
      CALL UPDAT                                                     DSL39310
      DO 55  I=1,M                                                   DSL39320
      P3(I)=P5(I)                                                    DSL39330
      P5(I)=P6(I)                                                    DSL39340
      Y6(I)=Y0(I)                                                    DSL39350
      P6(I)=P0(I)                                                    DSL39360
   55 Y0(I)=Y4(I)                                                    DSL39370
      TIME=(S-3.0*A)*H                                               DSL39380
      CALL UPDAT                                                     DSL39390
      DO 52  I=1,M                                                   DSL39400
   52 P4(I)=P0(I)                                                    DSL39410
      IND=5                                                          DSL39420
      KEEP=0                                                         DSL39430
      GO TO 500                                                      DSL39440
C     TEST FOR OUTPUT POINT                                          DSL39460
  700 IF(D-S-.00001)710,711,405                                     DSL39470
  711 DO 713  I=1,M                                                  DSL39480
  713 Y0(I)=Y7(I)                                                    DSL39490
      GO TO 712                                                      DSL39500
  710 W=(D-S)/A                                                      DSL39510
      DO 720 I=1,M                                                   DSL39520
      A2=(11.0*P7(I)-18.0*P6(I)+9.0*P5(I)-2.0*P4(I))/12.0            DSL39530
      A3=(2.0*P7(I)-5.0*P6(I)+4.0*P5(I)-P4(I))/6.0                   DSL39540
      A4=(P7(I)-3.0*P6(I)+3.0*P5(I)-P4(I))/24.0                      DSL39550
  720 Y0(I)=Y7(I)+DELT*W*(P7(I)+W*(A2+W*(A3+W*A4)))                  DSL39560
  712 TIME=D*H                                                       DSL39570
      D=D+1.0                                                        DSL39580
      KEEP=0                                                         DSL39590
      CALL UPDAT                                                     DSL39600
      GO TO 46                                                       DSL39610
   60 IF(KPRNT)92,92,350                                            DSL39620
  350 IF(TIME+1.5*DELT-TPRNT)92,92,61                               DSL39630
   61 DELT=TPRNT-TIME                                                DSL39640
      DELTA=1.                                                       DSL39650
C...  RUNGE-KUTTA INTEGRATION     (RKS)                              DSL39660
   92 KEEP=0                                                         DSL39670
      TSTRT=TIME                                                     DSL39680
      GO TO(21,190,21),INTYP                                         DSL39690
  190 IF(DELTA)192,192,22                                           DSL39700
  192 IF(AMAX-.75) 260,22,265                                       DSL39710
  260 IF(AMAX-.075) 262,22,22                                       DSL39720
   21 DEL=DELT                                                       DSL39730
      GO TO 24                                                       DSL39740
  265 DELT=DELT/RL                                                   DSL39750
   25 IF(DELT-XINT)150,22,22                                        DSL39760
  150 NALRM=28                                                       DSL39764
      CALL ERR                                                       DSL39765
      XINT=0.5*XINT                                                  DSL39766
      GO TO 22                                                       DSL39767
  262 DELT=DELT*RL                                                   DSL39770
      IF(DELT-H)22,22,263                                           DSL39774
  263 DELT=H                                                         DSL39776
   22 DEL=0.5*DELT                                                   DSL39780
      DO 23 I=1,N                                                    DSL39790
   23 SD(I)=0.0                                                      DSL39800
```

```
      IFLAG=1                                                      DSL39810
      S=1.                                                         DSL39820
   24 DO 26 I=1,N                                                  DSL39830
      Y1(I)=Y(I)                                                   DSL39840
   26 P1(I)=P0(I)                                                  DSL39850
  264 DO 27 I=1,N                                                  DSL39860
      DELY(I)=DEL*P0(I)                                            DSL39870
   27 PD(I)=DELY(I)                                                DSL39880
      GO TO(80,70,80),INTYP                                        DSL39890
   70 DO 71 I=1,N                                                  DSL39900
   71 SD(I)=SD(I)+S*P0(I)                                          DSL39910
   80 TIME=TIME+0.5*DEL                                            DSL39920
      DO 68 I=1,N                                                  DSL39930
      YS(I)=Y(I)                                                   DSL39940
   68 Y(I)=YS(I)+0.5*DELY(I)                                       DSL39950
      CALL UPDAT                                                   DSL39960
      DO 63 I=1,N                                                  DSL39970
      DELY(I)=DEL*P0(I)                                            DSL39980
      PD(I)=PD(I)+2.*DELY(I)                                       DSL39990
   63 Y(I)=YS(I)+0.5*DELY(I)                                       DSL40000
      CALL UPDAT                                                   DSL40010
      DO 62 I=1,N                                                  DSL40020
      DELY(I)=DEL*P0(I)                                            DSL40030
      PD(I)=PD(I)+2.*DELY(I)                                       DSL40040
   62 Y(I)=YS(I)+DELY(I)                                           DSL40050
      TIME=TIME+0.5*DEL                                            DSL40060
      CALL UPDAT                                                   DSL40070
      DO 280 I=1,N                                                 DSL40080
      DELY(I)=DEL*P0(I)                                            DSL40090
      PD(I)=PD(I)+DELY(I)                                          DSL40100
  280 Y(I)=YS(I)+PD(I)/6.                                          DSL40110
      GO TO(270,120,270),INTYP                                     DSL40120
  120 CALL UPDAT                                                   DSL40130
      GO TO (130,285),IFLAG                                        DSL40140
  130 S=4.                                                         DSL40150
      IFLAG=2                                                      DSL40160
      GO TO 264                                                    DSL40170
  285 AMAX =0.0                                                    DSL40180
      DO 180 I=1,N                                                 DSL40190
      SD(I)=SD(I)+P0(I)                                            DSL40200
      E=ABS(Y(I) -(Y1(I)+DEL*SD(I)/3.))/(XABS(I)+R(I)*ABS(Y(I)))   DSL40210
      IF(AMAX-E)175,180,180                                        DSL40220
  175 AMAX=E                                                       DSL40230
  180 CONTINUE                                                     DSL40240
      IF(AMAX-1.) 270,270,290                                      DSL40250
  290 J=1                                                          DSL40260
  251 AM=AMAX/10.**J                                               DSL40270
      IF(1.-AM) 255,257,257                                        DSL40280
  255 J=J+1                                                        DSL40290
      GO TO 251                                                    DSL40300
  257 TIME=TSTRT                                                   DSL40310
      DELT=DELT/(RL**J)                                            DSL40320
      DO 295 I=1,N                                                 DSL40330
      P0(I)=P1(I)                                                  DSL40340
  295 Y(I)=Y1(I)                                                   DSL40350
      GO TO 25                                                     DSL40360
  270 KEEP=1                                                       DSL40370
      CALL UPDAT                                                   DSL40380
      GO TO(300,310,320,245),K                                     DSL40410
  245 DELTA=0.                                                     DSL40420
      GO TO 46                                                     DSL40430
C...  ADAMS-MOULTON INTEGRATION                                    DSL40440
   94 IF(LINK)377,375,377                                          DSL40450
  375 DO 376  I=1,NINTG                                            DSL40460
      Y(I) =Y(I)+0.5*DELT*(3.0*P0(I)-PD(I))                        DSL40470
  376 PD(I)=P0(I)                                                  DSL40480
      TIME=TIME+DELT                                               DSL40490
      GO TO 390                                                    DSL40500
  377 LINK=0                                                       DSL40510
C...  TRAPEZOIDAL INTEGRATION                                      DSL40520
C...  STORE HISTORY                                                DSL40530
   95 DO 360  I=1,NINTG                                            DSL40540
      PD(I)=P0(I)                                                  DSL40550
  360 YS(I)=Y(I)                                                   DSL40560
C...  PREDICT Y AT T + DELT                                        DSL40570
      TIME=TIME+DELT                                               DSL40580
      DO 380  I=1,NINTG                                            DSL40590
  380 Y(I)=Y(I)+DELT*P0(I)                                         DSL40600
```

```
C...   UPDATE INPUTS                                                     DSL40610
       CALL UPDAT                                                        DSL40620
C...   COMPUTE OUTPUTS OF INTEGRATORS, CORRECT PREVIOUS COMPUTATION      DSL40630
       DO 385  I=1,NINTG                                                 DSL40640
  385  Y(I)=YS(I)+0.5*DELT*(P0(I)+PD(I))                                 DSL40650
       GO TO 390                                                         DSL40660
C...   NON-CENTRALIZED INTEGRATION                                       DSL40670
   98  TIME=TIME+DELT                                                    DSL40680
C...   KEEP IS SET = 1 FOR DERIV AND DELAY TO IDENTIFY POINT TO STORE    DSL40690
  390  KEEP=1                                                            DSL40700
       CALL UPDAT                                                        DSL40710
       KEEP=0                                                            DSL40720
   46  TIMEX=TIME+0.5*DELT                                               DSL40730
       IF(KRANG)42,42,167                                                DSL40740
  167  DO 168  I=1,KRANG                                                 DSL40750
       I2=IRANG(I)                                                       DSL40760
       IF(RMIN(I)-C(I2))170,170,169                                     DSL40770
  169  RMIN(I)=C(I2)                                                     DSL40780
       IF(TIME)170,170,168                                               DSL40790
  170  IF(RMAX(I)-C(I2))171,168,168                                     DSL40800
  171  RMAX(I)=C(I2)                                                     DSL40810
  168  CONTINUE                                                          DSL40820
   42  IF( KPRNT)44,44,45                                                DSL40930
   45  IF(TIMEX-TPRNT)44,34,34                                           DSL40840
   34  KCDE=1                                                            DSL40850
       CALL PRINT                                                        DSL40860
       DPT=DPT+1.                                                        DSL40870
       TPRNT =DPT*DELTP+PLAST                                            DSL40880
   44  IF(DELTG)48,48,33                                                 DSL40890
   33  IF(TIMEX-TPLOT)48,36,36                                           DSL40900
   36  CALL FPLOT(0,SCAL1*C(INDX1),SCAL2*C(INDY))                       DSL40910
       DPL=DPL+1.                                                        DSL40920
       TPLOT=DPL*DELTG+TLAST                                             DSL40930
   48  IF(TIMEX-FINTI)229,3,3                                            DSL40940
  229  IF(KFINI)224,224,222                                              DSL40950
  222  DO 235  I=1,KFINI                                                 DSL40960
       I2=IFINL(I)                                                       DSL40970
       IF( JFINL(I))231,232,232                                         DSL40980
  231  IF(C(I2)- FINVA(I))240,240,235                                   DSL40990
  232  IF(C(I2)- FINVA(I))235,240,240                                   DSL41000
  235  CONTINUE                                                          DSL41010
C...   TEST IF INTERRUPT FOR TYPEWRITER INPUT                            DSL41020
  224  CALL REMSW(1,IRD)                                                 DSL41030
       GO TO(3,31),IRD                                                   DSL41040
  240  KCDE=2                                                            DSL41050
       GO TO 215                                                         DSL41060
    3  KODE=3                                                            DSL41065
  215  IF(TPRNT)213,213,212                                             DSL41070
  213  C(KMEM)=TIME                                                      DSL41080
       GO TO 214                                                         DSL41090
  212  C(KMEM)=TPRNT-DELTP                                               DSL41100
  214  C(KMEM+1)=TIME                                                    DSL41110
       CALL PRINT                                                        DSL41115
       IF(DELTG)37,37,35                                                 DSL41120
   35  CALL FPLOT(0,SCAL1*C(INDX1),SCAL2*C(INDY))                       DSL41130
       CALL FPLOT(1,0.,0.)                                               DSL41140
   37  RETURN                                                            DSL41170
       END                                                              DSL41180
// DUP
*DELETE            MILNE
*STORE          WS  UA  MILNE
// FOR NUMER                                                            DSL41210
**DSL/1800 SUBROUTINE TO CONVERT NUMERIC DATA FOR SIMULATION            DSL41220
*LIST SYMBOL TABLE                                                      DSL41230
*LIST SUBPROGRAM NAMES                                                  DSL41240
*LIST ALL
*ONE WORD INTEGERS                                                      DSL41250
       SUBROUTINE NUMER                                                 DSL41260
       COMMON NALRM,IDUMY(14),IERR,KDUMY(205),ISTRT,DWORD,KA(7),NUMBR(72)DSL41270
       EQUIVALENCE (KWORD,DWORD)                                        DSL41280
C...   SUBROUTINE TO CONVERT BCD INTEGER OR FLOATING POINT NUMBER IN     DSL41290
C...   F OR E FORMAT TO BINARY, BEGINNING AT NUMBR(ISTRT)               DSL41300
C...   RECOGNIZES COMMA (107), RIGHT PAREN (93/76) WITH EQUALS (126/123),DSL41310
C...   MINUS (96), ASTERISK (92), OR COL. 72 AS END OF NUMBER           DSL41320
C...   IF SEPARATOR IS MINUS OR ASTERISK DWORD OR ISTRT RESPECTIVELY AREDSL41330
C...   SET NEGATIVE                                                     DSL41340
C...      (... WILL FORCE READ + CONTINUATION)                         DSL41350
       DWORD=0.0                                                        DSL41360
```

288

```
              LEAST=0                                                          DSL41370
              NEXP=0                                                           DSL41380
              FRACT=0.0                                                        DSL41390
              IPT=0                                                            DSL41400
              ISIGN=-1                                                         DSL41410
              IPARN=0                                                          DSL41420
        309   I=ISTRT                                                          DSL41430
      C...     TEST FOR EOF FROM KEYBOARD (21)                                 DSL41440
        310   IF(NUMBR(I)-21)307,306,307                                       DSL41450
        306   I=73                                                             DSL41460
              GO TO 332                                                        DSL41470
        307   IF (LEAST)313,313,361                                            DSL41480
        313   IF (NEXP) 315,315,370                                           DSL41490
      C...     COMPUTATIONS PRIOR TO DECIMAL POINT                             DSL41500
        315   IF(NUMBR(I)-240)330,320,320                                      DSL41510
        320   C=NUMBR(I)-240                                                   DSL41520
              CWORD=DWORD*10.0+C                                               DSL41530
              GOTO 400                                                         DSL41540
      C...     TEST FOR BLANK (64)                                             DSL41550
        330   IF(NUMBR(I)-64)99,400,347                                        DSL41560
      C...     TEST IF DECIMAL POINT (75), PLUS (78/80), OR EQUALS (126/123)   DSL41570
        347   IF(NUMBR(I)-75)99,335,344                                        DSL41580
        344   IF(NUMBR(I)-78)250,400,381                                       DSL41590
        381   IF(NUMBR(I)-80)99,400,383                                        DSL41600
        383   IF(NUMBR(I)-126)384,382,99                                       DSL41610
        384   IF(NUMBR(I)-123)341,382,99                                       DSL41620
        382   IF(IPARN)99,99,343                                              DSL41630
      C...     TEST FOR THREE CONSECUTIVE DEC POINTS = CONTINUATION            DSL41640
        335   KCUNT=-3                                                         DSL41650
        334   I=I+1                                                            DSL41660
              IF(I-72)331,331,332                                              DSL41670
        331   IF(NUMBR(I)-75)323,322,321                                       DSL41680
        323   IF(NUMBR(I)-64)321,334,321                                       DSL41690
        321   I=I-1                                                            DSL41700
        332   IF(KCUNT+2)336,99,337                                            DSL41710
        322   KCUNT=KOUNT+1                                                    DSL41720
              IF(KCUNT)334,338,338                                             DSL41730
        338   LEAST=1                                                          DSL41740
              DIV=1.0                                                          DSL41750
              IPT=1                                                            DSL41760
      C...     READ AND PRINT CONTINUATION CARD                                DSL41770
        337   CALL READ                                                        DSL41780
              ISTRT=1                                                          DSL41790
      C...     ALTER FORMAT TO 72 RIGHT-JUSTIFIED EBCDIC CHARS                 DSL41800
              CALL SPLIT                                                       DSL41810
              GO TO 309                                                        DSL41820
        336   LEAST=1                                                          DSL41830
              DIV=1.0                                                          DSL41840
              IPT=1                                                            DSL41850
              GO TO 400                                                        DSL41860
      C...     TEST FOR INTEGER SEPARATOR )=, -, OR *                          DSL41870
        341   IF(NUMBR(I)-93)325,380,326                                       DSL41880
        250   IF(NUMBR(I)-76)99,380,99                                         DSL41890
        380   IPARN=1                                                          DSL41900
              GO TO 400                                                        DSL41910
      C...     TEST FOR ASTERISK (92)                                          DSL41920
        325   IF(NUMBR(I)-92)99,385,99                                         DSL41930
        385   ISTRT=-I-1                                                       DSL41940
              KWORD=DWORD                                                      DSL41950
              GO TO 203                                                        DSL41960
      C...     TEST IF NEGATIVE (96)                                           DSL41970
        326   IF(NUMBR(I)-96)99,348,342                                        DSL41980
        348   IF(DWORD)99,327,390                                             DSL41990
        390   KWORD=-DWORD                                                     DSL42000
              GO TO 202                                                        DSL42010
        327   ISIGN=ISIGN+1                                                    DSL42020
              IF(ISIGN)400,400,99                                              DSL42030
      C...     TEST FOR COMMA (107)- INTEGER IF YES                            DSL42040
        342   IF(NUMBR(I)-107)99,343,99                                        DSL42050
        343   IF(ISIGN)210,390,99                                              DSL42055
        210   KWORD=DWORD                                                      DSL42060
              GO TO 202                                                        DSL42070
      C...     COMPUTATIONS FOR FRACTION                                       DSL42080
        361   IF(NUMBR(I)-240)363,360,360                                      DSL42090
        360   C=NUMBR(I)-240                                                   DSL42100
              DIV=DIV/10.0                                                     DSL42110
              FRACT=FRACT+C*DIV                                                DSL42120
              GOTO 400                                                         DSL42130
```

289

```
C...    TEST FOR BLANK (64) OR COMMA (107)- END OF NUMBER (F FORMAT)        DSL42140
   363  IF(NUMBR(I)-64)99,400,349                                           DSL42150
   349  IF(NUMBR(I)-107)362,350,340                                         DSL42160
C...    TEST FOR RIGHT PAREN  (93/76)                                       DSL42170
   362  IF(NUMBR(I)-93)364,400,99                                           DSL42180
   364  IF(NUMBR(I)-76)351,400,99                                           DSL42190
C...    TEST FOR E  (197)                                                   DSL42200
   340  IF(NUMBR(I)-197)99,345,99                                           DSL42210
   345  NEXP=1                                                              DSL42220
        EXP=0.0                                                             DSL42230
        LEAST=0                                                             DSL42240
        NEGEX=-1                                                            DSL42250
        GOTO 400                                                            DSL42260
C...    TEST FOR CONTINUATION - 3 CONSECUTIVE DEC POINTS (75)               DSL42270
   351  KOUNT=-3                                                            DSL42280
   352  IF(NUMBR(I)-75)355,353,99                                           DSL42290
   353  KOUNT=KOUNT+1                                                       DSL42300
        IF(KOUNT)354,337,337                                               DSL42310
   354  I=I+1                                                               DSL42320
        IF(I-72)352,352,99                                                  DSL42330
   355  IF(NUMBR(I)-64)99,354,99                                           DSL42340
C...    COMPUTATIONS FOR EXPONENT                                           DSL42350
C...    TEST FOR BLANK (64)                                                 DSL42360
   370  IF(NUMBR(I)-64)376,400,376                                          DSL42370
   376  IF(NUMBR(I)-240)374,373,373                                         DSL42380
C...    TEST IF NEGATIVE (96) OR POSITIVE (78/80) SIGN ON EXPONENT          DSL42390
   374  IF(NUMBR(I)-78)351,400,280                                          DSL42400
   280  IF(NUMBR(I)-80)99,400,371                                           DSL42410
   371  IF(NUMBR(I)-96)99,372,375                                           DSL42420
C...    TEST FOR COMMA (107)      = END OF NUMBER (E FORMAT)                DSL42430
   375  IF(NUMBR(I)-107)99,350,99                                           DSL42440
   372  NEGEX=NEGEX+1                                                       DSL42450
        IF(NEGEX)400,400,99                                                 DSL42460
   373  C=NUMBR(I)-240                                                      DSL42470
        EXP=10.0*EXP+C                                                      DSL42480
   400  I=I+1                                                               DSL42490
        IF(I-72)310,310,350                                                 DSL42500
C...    END OF COMPUTATION LOOP                                             DSL42510
   350  IF (NEXP) 455,455,405                                               DSL42520
   405  IF(NEGEX)450,410,99                                                 DSL42530
   410  EXP=-EXP                                                            DSL42540
   450  DWORD=(DWORD+FRACT)*10.0**EXP                                       DSL42550
        GO TO 200                                                           DSL42560
    99  NALRM=20                                                            DSL42570
        ISTRT=I                                                             DSL42575
        CALL ERR                                                            DSL42580
        ISTRT=73                                                            DSL42590
        GO TO 203                                                           DSL42600
   455  IF(LEAST)343,343,456                                                DSL42610
   456  DWORD=DWORD+FRACT                                                   DSL42620
   200  IF(ISIGN)202,201,99                                                 DSL42630
   201  DWORD=-DWORD                                                        DSL42640
   202  ISTRT=I+1                                                           DSL42650
   203  RETURN                                                              DSL42660
        END                                                                 DSL42670
*STORE        UA        NUMER                                               DSL42690
// FOR PRINT                                                                DSL44790
*LIST SYMBOL TABLE                                                          DSL44800
*LIST SUBPROGRAM NAMES                                                      DSL44810
*LIST ALL
*ONE WORD INTEGERS                                                          DSL44820
**SIMULATION PRINT OUTPUT ROUTINE                                          DSL44830
        SUBROUTINE PRINT                                                    DSL44840
        INTEGER TITLE(60),HDNG(75),SYMB(420)                               DSL44850
        DIMENSION XOUT(25)                                                  DSL44860
        COMMON NALRM,NINTG,NSYMB,JPT,KPRNT,KRANG,KFINI,KTITL,KREL,KABS,    DSL44870
      1 KMEM,KPCIN,KLOCK,INTYP,INDX1,KEEP,DELTA,STEPC,NX,NY,RLAST,ALAST,   DSL44880
      2 FINVA(5),TITLE,HDNG,INDXP(25),IRANG(15),IFINL(5),JFINL(5),         DSL44890
      3 ISTRT,RMIN(15),RMAX(15),KWORD(11),KAFGN,LINE,KLINE,LINE1,NLINS,    DSL44900
      4 KODE,I2,LOCY,LOCX,KOUNT,IRES(101),DELTG,SCAL1,SCAL2,YPOS,INDX2,    DSL44910
      5 ISCOP,SYMB,C(1)                                                     DSL44920
        KPT=JPT                                                            DSL44930
        NALRM=0                                                            DSL44934
        CALL ERR                                                           DSL44936
        IF(LINE-KLINE)28,28,29                                             DSL44940
    29  WRITE(KPT,805)(TITLE(I),I=1,KTITL)                                  DSL44950
   805  FORMAT(1H1 20A2/(1H 20A2))                                         DSL44960
        LINE=LINE1                                                          DSL44970
```

290

```
          IF( KPRNT-10)84,84,28                                        DSL44980
   84 WRITE(KPT,811)(HDNG(I),HDNG(I+25),HDNG(I+50),I=1,KPRNT)           DSL44990
  811 FORMAT(1H0 2X,3A2,  9(6X,3A2))                                    DSL45000
   28 DO 30   I=1,KPRNT                                                 DSL45010
      J=INDXP(I)                                                        DSL45020
   30 XOUT(I)=C(J)                                                      DSL45030
      IF( KPRNT-10)85,85,86                                            DSL45040
   85 WRITE(KPT,806)(XOUT(I),I=1,KPRNT)                                 DSL45050
  806 FORMAT( E11.3,10E12.4)                                            DSL45060
      LINE=LINE+1                                                       DSL45070
      GO TO 87                                                          DSL45080
   86 WRITE(KPT,20)  (HDNG(I),HDNG(I+25),HDNG(I+50),XOUT(I),I=1,KPRNT)  DSL45090
   20 FORMAT(1H0 3A2,1H=,  E12.4,4(3X,3A2,1H=,E12.4)/(23X,3A2,1H=,E12.4,DSL45100
    1 3X,3A2,1H=,E12.4,3X,3A2,1H=,E12.4,3X,3A2,1H=,E12.4))             DSL45110
      LINE=LINE+NLINS                                                   DSL45120
   87 GO TO(1,2,3),KODE                                                 DSL45130
    2 WRITE(KPT,843)SYMB(I2),SYMB(I2+140),SYMB(I2+280),C(I2)           DSL45140
  843 FORMAT(20H0SIMULATION HALTED, 3A2,2H =,  E12.4)                   DSL45150
    3. KLOCK=0                                                          DSL45160
      KTITL=-KTITL                                                      DSL45170
      IF(KRANG)1,1,49                                                   DSL45180
   49 WRITE(KPT,831)                                                    DSL45190
  831 FORMAT(1H0/35H VARIABLE      MINIMUM      MAXIMUM)                DSL45200
      DO 165  I=1,KRANG                                                 DSL45210
      I2=IRANG(I)                                                       DSL45220
  165 WRITE(KPT,832)SYMB(I2),SYMB(I2+140),SYMB(I2+280),RMIN(I),RMAX(I) DSL45230
  832 FORMAT(2X,3A2,5X,  E11.4,2X,E11.4)                                DSL45240
    1 RETURN                                                           DSL45250
      END                                                              DSL45260
*STORE         UA       PRINT                                          DSL45280
// FOR READ                                                            DSL45290
**DSL/1800 SUBROUTINE TO READ DATA FOR SIMULATION                      DSL45300
*LIST SYMBOL TABLE                                                     DSL45310
*LIST SUBPROGRAM NAMES                                                 DSL45320
*LIST ALL
*ONE WORD INTEGERS                                                     DSL45330
      SUBROUTINE READ                                                  DSL45340
      COMMON NALRM,NINTG,NSYMB,JPT,IDUMY(227),LETRS(72)                DSL45350
      CALL REMSW(2,IRD)                                                DSL45370
      GO TO(4,6),IRD                                                   DSL45400
    4 JRD=INPT(6)                                                      DSL45410
      GO TO 7                                                          DSL45412
    6 JRD=INPT(8)                                                      DSL45415
    7 READ(JRD,10)(LETRS(K),K=1,36)                                    DSL45420
   10 FORMAT(36A2)                                                     DSL45430
      CALL REMSW(6,J)                                                  DSL45434
      GO TO(12,13),J                                                   DSL45435
   13 KPT = JPT                                                        DSL45438
      WRITE(KPT,11) (LETRS(K),K=1,36)                                  DSL45440
   11 FORMAT(1H  36A2)                                                 DSL45450
   12 RETURN                                                           DSL45460
      END                                                              DSL45470
*STORE         UA       READ                                           DSL45490
// FOR PLOTG
*PUNCH
*LIST ALL
*LIST SYMBOL TABLE                                                     DSL43860
*LIST SUBPROGRAM NAMES                                                 DSL43870
*ONE WORD INTEGERS                                                     DSL43880
*NONPROCESS PROGRAM                                                    DSL43890
**ROUTINE TO DRAW AND LABEL GRAPH OR SCOPE AXES                        DSL43900
*IOCS(KEYBOARD,TYPEWRITER,PLOTTER)                                     DSL43910
*NAME PLOTG
      INTEGER TITLE(60),HDNG(75),SYMB(420)                             DSL43930
      COMMON NALRM,NINTG,NSYMB,JPT,KPRNT,KRANG,KFINI,KTITL,KREL,KABS,  DSL43940
    1 KMEM,KPOIN,KLOCK,INTYP,INDX1,KEEP,DELTA,STEPC,NX,NY,RLAST,ALAST, DSL43950
    2 FINVA(5),TITLE,HDNG,INDXP(25),IRANG(15),IFINL(5),JFINL(5),       DSL43960
    3 ISTRT,RMIN(15),RMAX(15),XOUT(9),LOCY,LOCX,KOUNT,IRES(101),       DSL43970
    4 DELTG,SCAL1,SCAL2,YPOS,INDX2,ISCOP,SYMB,C(500)                   DSL43980
      EQUIVALENCE (IRES(26),INDX),(IRES(57),INDY)                      DSL43990
C...  SET LOGICAL UNIT OF PLOTTER AND TYPEWRITER                       DSL43994
      M=INPT(7)                                                        DSL43995
      M2=INPT(1)                                                       DSL43996
      IF(DELTG)353,353,348                                             DSL44000
  348 CALL FPLOT(1,0.,0.)                                              DSL44010
  351 SCALR=0.5                                                        DSL44070
  330 TW=.2*SCALR                                                      DSL44100
      TH=.3*SCALR                                                      DSL44110
```

291

```
          FO=.4*SCALR
          IF(YPOS)356,354,354                                          DSL44120
    356   CALL REMSW(5,L)                                              DSL44130
          IF(L-1)371,372,371                                          DSL44140
    372   CONTINUE                                                     DSL44150
    371   IF(INDX-INDX1)359,357,359                                    DSL44170
    357   IF(INDX2)352,352,358                                         DSL44180
    358   IF(INDY-INDX2)360,380,360                                    DSL44190
    380   IF(KLOCK)352,370,352                                         DSL44200
    359   W=NX+3                                                       DSL44210
          CALL FPLOT(1,W,0.)                                           DSL44220
          CALL SCALF(1.,1.,0.,0.)                                      DSL44230
    370   YPOS=0.                                                      DSL44240
    354   RX=NX                                                        DSL44250
          CALL FGRID(0,0.,0.,1.,NX)                                    DSL44280
          CALL FCHAR(1.,-1.5*SCALR,FO,FO,0.)                           DSL44290
          WRITE(M,100)(TITLE(I),I=41,60)                              DSL44300
    100   FORMAT(20A2)                                                 DSL44310
          AA=.5*RX-.375                                                DSL44320
          CALL FCHAR(AA,-.8*SCALR,TH,TH,0.)                            DSL44330
          WRITE(M,100) SYMB(INDX1),SYMB(INDX1+140),SYMB(INDX1+280)     DSL44340
          K=NX/2                                                       DSL44350
          DO 55   I=1,K                                                DSL44360
          W=I+I                                                        DSL44370
     55   XOUT(I)=W/SCAL1                                              DSL44380
          CALL FCHAR(0.,-TH,TW,TW,0.)                                  DSL44390
    340   WRITE(M,102)(XOUT(I),I=1,K)                                  DSL44410
    102   FORMAT(E24.2,7E20.2)                                         DSL44420
    360   INDY=IABS(INDX2)                                             DSL44460
          RY=NY                                                        DSL44470
          CALL FGRID(1,YPOS,0.,1.,NY)                                  DSL44480
          AA=.5*RY-.375                                                DSL44490
          BB = YPOS-.6*SCALR                                           DSL44500
          CALL FCHAR(BB,AA,TH,TH,1.5708)                               DSL44510
          WRITE(M,100) SYMB(INDY),SYMB(INDY+140),SYMB(INDY+280)        DSL44520
          INDX2=-INDY                                                  DSL44530
          K=NY/2                                                       DSL44540
          DO 56   I=1,K                                                DSL44550
          W=I+I                                                        DSL44560
     56   XOUT(I)=W/SCAL2                                              DSL44570
          CALL FCHAR(YPOS-.1,0.,TW,TW,1.5708)                          DSL44580
    342   WRITE(M,102)(XOUT(I),I=1,K)                                  DSL44600
    344   YPOS=YPOS-1.0                                                DSL44630
C...      TEST FOR CONTINUATION                                        DSL44640
    352   IF(KLOCK)345,333,345                                         DSL44650
    345   CALL FPLOT(1,0.,0.)                                          DSL44660
          GO TO 334                                                    DSL44670
    333   CALL FPLOT(-2,0.,0.)                                         DSL44680
    334   CALL REMSW(4,L)                                              DSL44690
          IF(L-1)353,370,353                                          DSL44700
    353   CALL LINK(SIMUL)                                             DSL44750
          END                                                          DSL44760
// DUP
*DELETE              PLOTG
*STORECI      WS  UA  PLOTG


// *DATE 7-29-71


// JOB
// DUP
*DELETE              FSHOT
// JOB
// FOR
*LIST ALL
*ONE WORD INTEGERS
**DSL/1800 (DIGITAL SIMULATION LANGUAGE)
          SUBROUTINE UPDAT
          REAL KP    ,INTGR,LIMIT
          INTEGER    SWT
          COMMON NALRM,IZZZ(833),TIME ,DELT ,DELMI,FINTI,DELTP,DELTC,ZZ001,
         1ZZ002,THDOT,TH    ,TO    ,TM   ,WN   ,D2   ,WN2  ,WN2C ,BETA ,WE   ,
         1WT    ,ERROR,VP    ,VR   ,VC   ,VCLOW,VCHI ,VRO  ,VCO  ,THDT0,TH0  ,
         1T00   ,TM0   ,TS   ,DAMP ,TAUE ,TAUT ,C    ,ALPHA,KP   ,TR   ,TD   ,
         1INZZ0,SWT
          GO TO(1,2),SWT
      1   WN=4.242
          D2=2.*WN*DAMP
          WN2 = WN*WN
```

292

```
          WN2C= WN2/C
          BETA =1./ALPHA
          WE= 1./TAUE
          WT=1./TAUT
          SWT=2
        2 CONTINUE
          ERROR =(TS - TM)*C
          VP = KP*ERROR
          VR= VP + INTGR (VR0,TR*VP)
          VC = BETA*(VR+INTGR (VC0,TD*(VR-VC)))
          THDOT=INTGR (THDT0,WN2C*VC-D2*THDOT-WN2*TH)
          TH = INTGR (THO,THDOT)
          TO= INTGR (TOO,WE*(TH-TO))
          TM = INTGR (TMO,WT*(TO-TM))
          VC = LIMIT(VCLOW,VCHI,VC)
          RETURN
          END
*DELETE              SIMUL
*STORECIX    UA    1 SIMUL INTEG
*CCEND
// XEQ SIMX
    6   41  42
TIME    DELT   DELMI FINTI DELTP DELTC ZZ001 ZZ002 THDOT TH      TO      TM
WN      D2     WN2   WN2C  BETA  WE    WT    ERROR VP    VR      VC      VCLOW
VCHI    VRO    VCO   THDTO THO   TOO   TMO   TS    DAMP  TAUE    TAUT    C
ALPHA   KP     TR    TD    SWT
TITLE   THREE MODE TEMPERATURE CONTROLLER
PARAM VCLOW =0.,VCHI = 10.
INCON VRO=0.,VCO=0.,THDTO=0.,THO=0.,TOO=0.,TMO=0.,TS=20.
CONTRL DELT =.001,FINTI=8.,SWT=1
PARAM DAMP=.7,TAUE=2.,TAUT=1.,C=.1
PARAM ALPHA = .1
SCOPE  .05,8,6,TIME,ERROR
SCALE 1.,2.
LABEL 3 MODE CONTROL
INTEG RKS
RELERR TH=.001
ABSERR TH=.01
PRINT 0.2,VC,TM
PARAM KP=1.2,TR=.625,TD=.7
END
SCOPE  .05,8,6,TIME,VC
RESET PRINT
SCALE 1.,.5
END

// FOR DOUTN
*LIST ALL
*ONE WORD INTEGERS
      SUBROUTINE DOUTN(IA,IB,IC)
      RETURN
      END
// DUP
*STORE       WS   UA   DOUTN
// FOR RKS                                                          DSL45500
*LIST ALL                                                          DSL45510
*LIST SYMBOL TABLE                                                 DSL45520
*LIST SUBPROGRAM NAMES                                             DSL45530
*ONE WORD INTEGERS                                                 DSL45540
**RUNGE-KUTTA 4TH ORDER VARIABLE STEP INTEGRATION ROUTINE          DSL45550
      SUBROUTINE   RKS(PO,XABS,R,YST,DYST)                          DSL45560
      INTEGER TITLE(60),HDNG(75),SYMB(420)                          DSL45570
      DIMENSION   Y(1),XABS(1),R(1),PO(1),KC(1),YST(1),DYST(1),     DSL45580
    3 PD(50),SD(50),YS(50),DELY(50)                                 DSL45590
      COMMON NALRM,NINTG,NSYMB,JPT,KPRNT,KRANG,KFINI,KTITL,KREL,KABS,DSL45600
    1 KMEM,KPCIN,KLOCK,INTYP,INDX1,KEEP,     H,STEPC,NX,NY,RLAST,ALAST,DSL45610
    2 FINVA(5),TITLE,HDNG,INDXP(25),IRANG(15),IFINL(5),JFINL(5),    DSL45620
    3 ISTRT,RMIN(15),RMAX(15),KWORD(11),KAFGN,LINE,KLINE,LINE1,NLINS,DSL45630
    4 KODE,I2,LOCY,LOCX,KOUNT,IRES(101),DELTG,SCAL1,SCAL2,YPCS,INDX2,DSL45640
    5 ISCCP,SYMB,C(7)                                               DSL45650
      EQUIVALENCE (C(1),TIME),(C(2),DELT),(C(3),XINT),(C(7),Y(1)),  DSL45660
    1 (DELTP,C(5)),(FINTI,C(4)),(C(1),KC(1)),       (DELTC,C(6)),   DSL45670
    4 (IRES(57),INDY),(N,NINTG)                                     DSL45680
  200 NALRM=0                                                       DSL45685
      LCCY=0                                                        DSL45690
      LINK=1                                                        DSL45700
      LINE=60                                                       DSL45710
      TPRNT=0.0
```

293

```
          DPT=0.                                                         DSL45720
          TPLOT=0.0                                                      DSL45730
          DPL=0.                                                         DSL45740
          KPT = JPT                                                      DSL45750
          KAFGN=-10                                                      DSL45760
          RL=10.**.2                                                     DSL45770
          DELTA=1.                                                       DSL45780
          PLAST=C(KMEM)                                                  DSL45790
          W=PLAST+DELTP-TIME                                             DSL45800
          IF(DELTP)209,209,207                                           DSL45805
    207   IF(W+DELTP*DPT)208,208,209                                     DSL45810
    208   DPT=DPT+1.0                                                    DSL45820
          GO TO 207                                                      DSL45830
    209   TLAST=C(KMEM+1)                                                DSL45840
C...      INITIALIZE DERIVATIVE SUBROUTINE                               DSL45850
          KEEP = 0                                                       DSL45860
          KOUNT=0                                                        DSL45870
          CALL UPDAT                                                     DSL45880
          CALL PLOTS(ISCOP)                                              DSL45882
          IF(ISCOP)206,206,205                                          DSL45884
    205   CALL DRGSC(2.,2.)                                              DSL45886
C...      TEST IF CONTINUATION OF PARAMETER STUDY                        DSL45890
    206   IF(KLOCK)220,221,220                                           DSL45900
    220   KEEP=-1                                                        DSL45910
          KLOCK=0                                                        DSL45915
          IF(DELTG)223,223,211                                          DSL45920
    211   CALL FPLOT(-2,SCAL1*C(INDX1),SCAL2*C(INDY))                    DSL45930
          GO TO 223                                                      DSL45940
    221   KEEP=1                                                         DSL45950
          TLAST=0.0                                                      DSL45960
          PLAST=0.0                                                      DSL45970
          IF(DELTG)223,223,219                                          DSL45975
    219   CALL FPLOT(2,0.,0.)                                           DSL45976
    223   CALL UPDAT                                                     DSL45980
          IF(KFINI)277,277,225                                          DSL45990
    225   DO 226  I=1,KFINI                                             DSL46000
          I2=IFINL(I)                                                    DSL46010
          IF(C(I2)- FINVA(I))227,227,228                                DSL46020
    227   JFINL(I)=1                                                     DSL46030
          GO TO 226                                                      DSL46040
    228   JFINL(I)=-1                                                    DSL46050
    226   CONTINUE                                                       DSL46060
    277   KTITL=IABS(KTITL)                                             DSL46070
          LINE1=KTITL/20+2                                               DSL46080
          KLINE=52                                                       DSL46090
          IF( KPRNT-10)46,46,66                                         DSL46100
     66   NLINS=(KPRNT+2)/4+1                                           DSL46110
          KLINE=54-NLINS                                                 DSL46120
          GO TO 46                                                       DSL46130
C...      LOOP FOR CENTRALIZED INTEGRATION                               DSL46140
C...      (INTEGRATION TYPE  -  RKS,RKS,RKSFX,ADAMS,TRAPZ,NON-CENTRAL)   DSL46150
     31   GO TO (60,60,92,94,95,98),INTYP                               DSL46160
C...      ADAMS-MOULTON INTEGRATION                                      DSL46170
     94   IF(LINK)377,375,377                                           DSL46180
    375   DO 376  I=1,NINTG                                             DSL46190
          Y(I) =Y(I)+0.5*DELT*(3.0*PO(I)-PD(I))                         DSL46200
    376   PD(I)=PO(I)                                                    DSL46210
          TIME=TIME+DELT                                                 DSL46220
          GO TO 390                                                      DSL46230
    377   LINK=0                                                         DSL46240
C...      TRAPEZOIDAL INTEGRATION                                        DSL46250
C...      STORE HISTORY                                                  DSL46260
     95   DO 360  I=1,NINTG                                             DSL46270
          PD(I)=PO(I)                                                    DSL46280
    360   YS(I)=Y(I)                                                     DSL46290
C...      PREDICT Y AT T + DELT                                          DSL46300
          TIME=TIME+DELT                                                 DSL46310
          DO 380  I=1,NINTG                                             DSL46320
    380   Y(I)=Y(I)+DELT*PO(I)                                          DSL46330
C...      UPDATE INPUTS                                                  DSL46340
          CALL UPDAT                                                     DSL46350
C...      COMPUTE OUTPUTS OF INTEGRATORS, CORRECT PREVIOUS COMPUTATION   DSL46360
          DO 385  I=1,NINTG                                             DSL46370
    385   Y(I)=YS(I)+0.5*DELT*(PO(I)+PD(I))                             DSL46380
          GO TO 390                                                      DSL46390
C...      NON-CENTRALIZED INTEGRATION                                    DSL46400
     98   TIME=TIME+DELT                                                DSL46410
C...      KEEP IS SET = 1 FOR DERIV AND DELAY TO IDENTIFY POINT TO STORE DSL46420
```

294

```
      390 KEEP=1                                                                DSL46430
          CALL UPDAT                                                            DSL46440
          KEEP=0                                                                DSL46450
          GO TO 46                                                              DSL46460
       60 IF(KPRNT)92,92,350                                                    DSL46470
      350 IF(TIME+1.5*DELT-TPRNT)92,92,61                                       DSL46480
       61 DELT=TPRNT-TIME                                                       DSL46490
          DELTA=1.                                                              DSL46500
C... RUNGE-KUTTA INTEGRATION     (RKS)                                          DSL46510
       92 KEEP=0                                                                DSL46520
          TSTRT=TIME                                                            DSL46530
          GO TO(190,190,21),INTYP                                              DSL46540
      190 IF(DELTA)192,192,22                                                   DSL46550
      192 IF(AMAX-.75) 260,22,265                                               DSL46560
      260 IF(AMAX-.075) 262,22,22                                               DSL46570
       21 DEL=DELT                                                              DSL46580
          GO TO 24                                                              DSL46590
      265 DELT=DELT/RL                                                          DSL46600
       25 IF(DELT-XINT)150,22,22                                                DSL46610
      150 NALRM=28                                                              DSL46614
          CALL ERR                                                             DSL46615
          XINT=0.5*XINT                                                         DSL46616
          GO TO 22                                                              DSL46617
      262 DELT=DELT*RL                                                          DSL46620
          IF(DELT-H)22,22,263                                                   DSL46624
      263 DELT=H                                                                DSL46626
       22 DEL=0.5*DELT                                                          DSL46630
          DO 23 I=1,N                                                           DSL46640
       23 SD(I)=0.0                                                             DSL46650
          IFLAG=1                                                               DSL46660
          S=1.                                                                  DSL46670
       24 DO 26 I=1,N                                                           DSL46680
          YST(I)=Y(I)                                                           DSL46690
       26 DYST(I)=PO(I)                                                         DSL46700
      264 DO 27 I=1,N                                                           DSL46710
          DELY(I)=DEL*PO(I)                                                     DSL46720
       27 PD(I)=DELY(I)                                                         DSL46730
          GO TO(70,70,80),INTYP                                                DSL46740
       70 DO 71 I=1,N                                                           DSL46750
       71 SD(I)=SD(I)+S*PO(I)                                                   DSL46760
       80 TIME=TIME+0.5*DEL                                                     DSL46770
          DO 68 I=1,N                                                           DSL46780
          YS(I)=Y(I)                                                           DSL46790
       68 Y(I)=YS(I)+0.5*DELY(I)                                                DSL46800
          CALL UPDAT                                                            DSL46810
          DO 63 I=1,N                                                           DSL46820
          DELY(I)=DEL*PO(I)                                                     DSL46830
          PD(I)=PD(I)+2.*DELY(I)                                                DSL46840
       63 Y(I)=YS(I)+0.5*DELY(I)                                                DSL46850
          CALL UPDAT                                                            DSL46860
          DO 62 I=1,N                                                           DSL46870
          DELY(I)=DEL*PO(I)                                                     DSL46880
          PD(I)=PD(I)+2.*DELY(I)                                                DSL46890
       62 Y(I)=YS(I)+DELY(I)                                                    DSL46900
          TIME=TIME+0.5*DEL                                                     DSL46910
          CALL UPDAT                                                            DSL46920
          DO 280 I=1,N                                                          DSL46930
          DELY(I)=DEL*PO(I)                                                     DSL46940
          PD(I)=PD(I)+DELY(I)                                                   DSL46950
      280 Y(I)=YS(I)+PD(I)/6.                                                   DSL46960
          GO TO(120,120,270),INTYP                                            DSL46970
      120 CALL UPDAT                                                            DSL46980
          GO TO (130,285),IFLAG                                                DSL46990
      130 S=4.                                                                  DSL47000
          IFLAG=2                                                               DSL47010
          GO TO 264                                                            DSL47020
      285 AMAX =0.0                                                             DSL47030
          DO 180 I=1,N                                                          DSL47040
          SD(I)=SD(I)+PO(I)                                                     DSL47050
          E=ABS(Y(I)-(YST(I)+DEL*SD(I)/3.))/(XABS(I)+R(I)*ABS(Y(I)))           DSL47060
          IF(AMAX-E)175,180,180                                                DSL47070
      175 AMAX=E                                                                DSL47080
      180 CONTINUE                                                             DSL47090
          IF(AMAX-1.) 270,270,290                                              DSL47100
      290 J=1                                                                   DSL47110
      251 AM=AMAX/10.**J                                                        DSL47120
          IF(1.-AM) 255,257,257                                                DSL47130
      255 J=J+1                                                                 DSL47140
```

```
          GO TO 251                                           DSL47150
  257 TIME=TSTRT                                              DSL47160
      DELT=DELT/(RL**J)                                       DSL47170
      DO 295 I=1,N                                            DSL47180
      PO(I)=DYST(I)                                           DSL47190
  295 Y(I)=YST(I)                                             DSL47200
      GO TO 25                                                DSL47210
  270 KEEP=1                                                  DSL47240
      CALL UPDAT                                              DSL47250
      DELTA=0.                                                DSL47260
   46 TIMEX=TIME+0.5*DELT                                     DSL47270
      IF(KRANG)42,42,167                                      DSL47280
  167 DO 168   I=1,KRANG                                      DSL47290
      I2=IRANG(I)                                             DSL47300
      IF(RMIN(I)-C(I2))170,170,169                            DSL47310
  169 RMIN(I)=C(I2)                                           DSL47320
      IF(TIME)170,170,168                                     DSL47330
  170 IF(RMAX(I)-C(I2))171,168,168                            DSL47340
  171 RMAX(I)=C(I2)                                           DSL47350
  168 CONTINUE                                                DSL47360
   42 IF( KPRNT)44,44,45                                      DSL47370
   45 IF(TIMEX-TPRNT)44,34,34                                 DSL47380
   34 KCDE=1                                                  DSL47390
      CALL PRINT                                              DSL47400
      DPT=DPT+1.                                              DSL47410
      TPRNT =DPT*DELTP+PLAST                                  DSL47420
   44 IF(DELTG)48,48,33                                       DSL47430
   33 IF(TIMEX-TPLOT)48,36,36                                 DSL47440
   36 CALL FPLOT(0,SCAL1*C(INDX1),SCAL2*C(INDY))              DSL47450
      DPL=DPL+1.                                              DSL47460
      TPLOT=DPL*DELTG+TLAST                                   DSL47470
   48 IF(TIMEX-FINTI)229,229,3                                DSL47480
  229 IF(KFINI)224,224,222                                    DSL47490
  222 DO 235   I=1,KFINI                                      DSL47500
      I2=IFINL(I)                                             DSL47510
      IF( JFINL(I))231,232,232                                DSL47520
  231 IF(C(I2)- FINVA(I))240,240,235                          DSL47530
  232 IF(C(I2)- FINVA(I))235,240,240                          DSL47540
  235 CONTINUE                                                DSL47550
C...  TEST IF INTERRUPT FOR TYPEWRITER INPUT                  DSL47560
  224 CALL REMSW(1,IRD)                                       DSL47570
      GO TO(3,31),IRD                                         DSL47580
  240 KCDE=2                                                  DSL47590
      GO TO 215                                               DSL47600
    3 KCDE=3                                                  DSL47605
  215 IF(TPRNT)213,213,212                                    DSL47610
  213 C(KMEM)=TIME                                            DSL47620
      GO TO 214                                               DSL47630
  212 C(KMEM)=TPRNT-DELTP                                     DSL47640
  214 C(KMEM+1)=TIME                                          DSL47650
      CALL PRINT                                              DSL47655
      IF(DELTG)37,37,35                                       DSL47660
   35 CALL FPLOT(0,SCAL1*C(INDX1),SCAL2*C(INDY))              DSL47670
      CALL FPLOT(1,0.,0.)                                     DSL47680
   37 CALL REMSW(3,IRD)                                       DSL47705
      GO TO(236,237),IRD                                      DSL47706
  236 CALL TYPIN                                              DSL47707
      GO TO 200                                               DSL47708
  237 RETURN                                                  DSL47710
      END                                                     DSL47710
*STORE        UA       RKS                                    DSL47720
// ASM                                                        DSL47740
*LIST                                                         DSL47750
                                                              DSL47760
                      ENT      SETUP                          DSL47770
              SETUP   DC       0                              DSL47780
                      STX   1  SAVEX+1                        DSL47790
                      STX   2  SAVEX+3                        DSL47800
                      LDX   1  -93                            DSL47810
                      LDX   L2 IRES                           DSL47820
              LOOP1   LD    L1 DATA+93                        DSL47830
                      STO   2  0                              DSL47840
                      LD    L1 DATA+94                        DSL47850
                      STO   2  -31                            DSL47860
                      LD    L1 DATA+95                        DSL47870
                      STO   2  -62                            DSL47880
                      MDX   2  -1                             DSL47890
                      MDX   1  3                              DSL47900
                      MDX      LOOP1                          DSL47910
```

```
                    LDX     1   -8                          DSL47920
                    LDX     L2  IRES-93                      DSL47930
            LOOP2   LD      L1  DATA1+8                      DSL47940
                    STO     2   0                           DSL47950
                    MDX     2   -1                          DSL47960
                    MDX     1   1                           DSL47970
                    MDX         LOOP2                       DSL47980
                    LD          DEOF                        DSL47990
                    STO     L   IRES-55                     DSL48000
            SAVEX   LDX     L1  *-*                         DSL48010
                    LDX     L2  *-*                         DSL48020
                    BSC     I   SETUP                       DSL48030
            DATA    EBC         .PARAM INCON CONST PRINT SCOPE .    DSL48040
                    EBC         .GRAPH RANGE FINISHRELERRABSERR.    DSL48050
                    EBC         .LABEL CHART TITLE CONTRLCONTIN.    DSL48060
                    EBC         .INTEG AFGEN /NLFGENSTOP   END .    DSL48070
                    EBC         .SCALE TABLE / RESET TYPE   EN** .  DSL48080
                    EBC         .******MILNE RKS   RKSFX ADAMS .    DSL48090
                    EBC         .TRAPZ .                    DSL48100
            DATA1   EBC         .DSL/1800OVRLAY  .          DSL48110
            DEOF    DC          /C415                       DSL48120
            IRES    EQU         /7ED0                       DSL48130
                    END                                     DSL48140
*STORE          UA      SETUP                               DSL48160
// FOR SIM                                                  DSL48170
*ONE WORD INTEGERS                                          DSL48200
*NCNPROCESS PROGRAM                                         DSL48210
*LIST SYMBCL TABLE                                          DSL11581
*LIST SUBPROGRAM NAMES                                      DSL11582
*NONPROCESS PROGRAM                                         DSL11584
*LIST ALL
*IOCS(KEYBOARD,TYPEWRITER,CARD,1132 PRINTER)               DSL48220
*NAME SIMX                                                  DSL48230
**SIM       MAIN ROUTINE FOR ANALOG SIMULATION PROGRAM     DSL48240
            COMMON NALRM,NINTG,NSYMB,KPT,KPRNT,KRANG,KFINI,KTITL,KREL,KABS,  DSL48250
          1 KMEM,IDUMY(393),DELTG,JDUMY(428),C(500)         DSL48260
            IF(KPT-INPT(3))1,3,1                            DSL48270
          1 KPT=INPT(3)                                     DSL48280
            CALL INITL                                      DSL48290
          3 CALL INTRA                                      DSL48300
            IF(CELTG)5,5,4                                  DSL48310
          4 CALL LINK(PLOTG)                                DSL48320
          5 CALL LINK(SIMUL)                                DSL48330
            END                                             DSL48340
// DUP
*DELETE             SIM
*STORE      WS  UA  SIM

// ASM                                                      DSL48370
*LIST                                                       DSL48380
                    ENT     SPLIT                           DSL48390
                    ENT     BUILD                           DSL48400
            SPLIT   DC      0                               DSL48410
                    STX     1   SAVE+1                      DSL48420
                    STX     2   SAVE+3                      DSL48430
                    LDX     L1  LETRS-35                    DSL48440
                    LDX     2   -72                         DSL48450
            LOOP    SLT     8                               DSL48460
                    LD      1   0                           DSL48470
                    RTE     8                               DSL48480
                    STO     L2  LETRS+2                     DSL48490
                    SRA     8                               DSL48500
                    SLT     8                               DSL48510
                    STO     L2  LETRS+1                     DSL48520
                    MDX     1   1                           DSL48530
                    MDX     2   2                           DSL48540
                    MDX         LOOP                        DSL48550
            SAVE    LDX     L1  *-*                         DSL48560
                    LDX     L2  *-*                         DSL48570
                    BSC     I   SPLIT                       DSL48580
            BUILD   DC      0                               DSL48590
                    STX     1   SAVEI+1                     DSL48600
                    LDX     L1  KA-5                        DSL48610
                    LD      1   5                           DSL48620
                    SLA     8                               DSL48630
                    OR      1   4                           DSL48640
                    STO     1   8                           DSL48650
                    LD      1   3                           DSL48660
```

```
                         SLA      8                           DSL48670
                         OR       1 2                         DSL48680
                         STO      1 7                         DSL48690
                         LD       1 1                         DSL48700
                         SLA      8                           DSL48710
                         OR       1 0                         DSL48720
                         STO      1 6                         DSL48730
                   SAVEI LDX      L1 *-*                      DSL48740
                         BSC      I  BUILD                    DSL48750
                   KA    EQU      /7F1E                       DSL48760
                   LETRS EQU      /7F18                       DSL48770
                         END                                  DSL48780
*STORE        UA      SPLIT                                   DSL48800
// FOR TYPIN                                                  DSL48810
**DSL/1800  (DIGITAL SIMULATION LANGUAGE) - KEYBOARD TRANSLATOR  DSL48820
*LIST SYMBCL TABLE                                           DSL48830
*LIST SUBPROGRAM NAMES                                       DSL48840
*LIST ALL
*ONE WORD INTEGERS                                           DSL48850
      SUBROUTINE TYPIN                                       DSL48860
      INTEGER SYMB(420),VAR(3)                               DSL48870
      COMMON NALRM,NINTG,NSYMB,KD(9),KLOCK,ID(290),IRES(101),JD(10),  DSL48880
     1 SYMB,C(1)                                             DSL48885
      EQUIVALENCE (C(1),TIME)                                DSL48890
      KLOCK=1                                                DSL48900
      M=INPT(6)                                              DSL48910
  227 READ(M,820) VAR(1),VAR(2),VAR(3),VALUE                 DSL48920
  820 FORMAT(2A2,A1,1X,E12.6)                                DSL48930
      IF(VAR(1)-IRES(2C))4,10,4                              DSL48934
   10 IF(VAR(2)-IRES(51))11,228,11                           DSL48935
   11 IF(VAR(2)-IRES(56))4,228,4                             DSL48936
    4 DO 100 J = 1,NSYMB                                     DSL48940
      IF(VAR(1)-SYMB(J))100,5,100                            DSL48950
    5 IF(VAR(2)-SYMB(J+140))100,6,100                        DSL48960
    6 IF(VAR(3)-SYMB(J+280))100,7,100                        DSL48970
    7 C(J)=VALUE                                             DSL48980
      GO TO 227                                              DSL48990
  100 CONTINUE                                               DSL49000
      N=INPT(1)                                              DSL49010
      WRITE(N,50) VAR(1),VAR(2),VAR(3)                       DSL49020
   50 FORMAT(22H SYMBOL NOT IN TABLE  3A2)                   DSL49030
      GO TO 227                                              DSL49040
  228 IF(TIME)230,230,236                                    DSL49050
  230 KLOCK=0                                                DSL49060
  236 RETURN                                                 DSL49070
      END                                                    DSL49080
// DUP                                                       DSL49090
*STORE        UA      TYPIN                                  DSL49100
// FOR                                                       DSL60000
**DSL/1800  (DIGITAL SIMULATION LANGUAGE)                    DSL60010
*LIST SYMBCL TABLE                                           DSL60020
*LIST SUBPROGRAM NAMES                                       DSL60030
*LIST ALL
*ONE WORD INTEGERS                                           DSL60040
      FUNCTION AFGEN(LOCN,ARG)                               DSL60050
      DIMENSION KFCT(2)                                      DSL60060
      COMMON IDUMY(293),JDUMY(541),FCT(2)                    DSL60070
      EQUIVALENCE (KFCT(1),FCT(1)),(JDUMY(1),KOUNT),(NALRM,IDUMY(1)),  DSL60080
     1 (ISTRT,IDUMY(222))                                    DSL60090
C...  FCT IS EQUIVALENT TO C IN INTRAN AND SIMUL             DSL60100
      LOCT=LOCN                                              DSL60110
      IF(LOCT)3,3,4                                          DSL60120
    3 AFGEN=0.                                               DSL60130
      GO TO 8                                                DSL60140
    4 VAR=ARG                                                DSL60150
      IF(VAR-FCT(LOCT))10,9,11                               DSL60160
   10 IF(KOUNT)12,12,9                                       DSL60170
   12 NALRM=33                                               DSL60180
      ISTRT=LOCT                                             DSL60190
      CALL ERR                                               DSL60200
      KOUNT=KOUNT+1                                          DSL60210
      GO TO 9                                                DSL60220
   11 LOCT2=KFCT(2*LOCT-3)                                   DSL60230
      DO 5 I=LOCT,LOCT2,2                                    DSL60240
      IF(VAR-FCT(I))6,7,5                                    DSL60250
    5 CONTINUE                                               DSL60260
      I=LOCT2                                                DSL60270
      IF(KOUNT)13,13,7                                       DSL60280
```

298

```
     13 NALRM=34                                                         DSL60290
        ISTRT=LOCT                                                       DSL60300
        CALL ERR                                                         DSL60310
        KCUNT=KOUNT+1                                                    DSL60320
        GO TO 7                                                          DSL60330
      6 AFGEN=(VAR-FCT(I-2))*(FCT(I+1)-FCT(I-1))/(FCT(I)-FCT(I-2))       DSL60340
      1 +FCT(I-1)                                                        DSL60350
        GO TO 8                                                          DSL60360
      9 I=LOCT                                                           DSL60370
      7 AFGEN=FCT(I+1)                                                   DSL60380
      8 RETURN                                                           DSL60390
        END                                                             DSL60400
// DUP                                                                   DSL60410
*STORE        UA        AFGEN                                            DSL60420
// FOR                                                                   DSL60430
**DSL/1800  (DIGITAL SIMULATION LANGUAGE)                                DSL60440
*LIST SYMBOL TABLE                                                       DSL60450
*LIST SUBPROGRAM NAMES                                                   DSL60460
*LIST ALL
*ONE WORD INTEGERS                                                      DSL60470
        FUNCTION ALOGX(A)                                               DSL60480
        ALOGX=0.4342945*ALOG(A)                                         DSL60490
        RETURN                                                          DSL60500
        END                                                             DSL60510
// DUP                                                                   DSL60520
*STORE        UA        ALOGX                                           DSL60530
// FOR                                                                   DSL60540
**DSL/1800  (DIGITAL SIMULATION LANGUAGE)                                DSL60550
*LIST SYMBOL TABLE                                                       DSL60560
*LIST SUBPROGRAM NAMES                                                   DSL60570
*LIST ALL
*ONE WORD INTEGERS                                                      DSL60580
        FUNCTION AMAX1(A,B)                                             DSL60590
        IF(A-B)4,5,5                                                    DSL60600
      4 AMAX1=B                                                          DSL60610
        GO TO 6                                                          DSL60620
      5 AMAX1=A                                                          DSL60630
      6 RETURN                                                           DSL60640
        END                                                             DSL60650
// DUP                                                                   DSL60660
*STORE        UA        AMAX1                                           DSL60670
// FOR                                                                   DSL60680
**DSL/1800  (DIGITAL SIMULATION LANGUAGE)                                DSL60690
*LIST SYMBOL TABLE                                                       DSL60700
*LIST SUBPROGRAM NAMES                                                   DSL60710
*LIST ALL
*ONE WORD INTEGERS                                                      DSL60720
        FUNCTION AMIN1(A,B)                                             DSL60730
        IF(A-B)5,5,4                                                    DSL60740
      4 AMIN1=B                                                          DSL60750
        GO TO 6                                                          DSL60760
      5 AMIN1=A                                                          DSL60770
      6 RETURN                                                           DSL60780
        END                                                             DSL60790
// DUP                                                                   DSL60800
*STORE        UA        AMIN1                                           DSL60810
// FOR                                                                   DSL60820
**DSL/1800  (DIGITAL SIMULATION LANGUAGE)                                DSL60830
*LIST SYMBOL TABLE                                                       DSL60840
*LIST SUBPROGRAM NAMES                                                   DSL60850
*ONE WORD INTEGERS                                                      DSL60860
        FUNCTION AMOD(A,B)                                              DSL60870
        K=A/B                                                            DSL60880
        X=K                                                              DSL60890
        AMOD=A-X*B                                                       DSL60900
        RETURN                                                          DSL60910
        END                                                             DSL60920
// DUP                                                                   DSL60930
*STORE        UA        AMOD                                            DSL60940
// ASM                                                                   DSL60950
*LIST                                                                    DSL60960
                              ENT     AND                               DSL60970
                              ENT     OR                                DSL60980
                              ENT     EOR                               DSL60990
                     *        LOGICAL AND, OR, AND EXCLUSIVE OR ROUTINES DSL61000
                     AND      DC      0                                 DSL61010
                              STX   1 SAVE+1                            DSL61020
                              LDX  I1 AND                               DSL61030
```

```
                    LD      I1  0                          DSL61040
                    AND     I1  1                          DSL61050
           COMN     MDX     1   2                          DSL61060
                    STX     1   RET+1                      DSL61070
           SAVE     LDX     L1  *-*                        DSL61080
           RET      BSC     L   *-*                        DSL61090
           OR       DC          0                          DSL61100
                    STX     1   SAVE+1                     DSL61110
                    LDX     I1  OR                         DSL61120
                    LD      I1  0                          DSL61130
                    OR      I1  1                          DSL61140
                    MDX         COMN                       DSL61150
           EOR      DC          0                          DSL61160
                    STX     1   SAVE+1                     DSL61170
                    LDX     I1  EOR                        DSL61180
                    LD      I1  0                          DSL61190
                    EOR     I1  1                          DSL61200
                    MDX         COMN                       DSL61210
                    END                                    DSL61220
// DUP                                                     DSL61230
*STORE          UA        AND                              DSL61240
// FOR                                                     DSL61250
**DSL/1800   (DIGITAL  SIMULATION  LANGUAGE)               DSL61260
*LIST SYMBOL TABLE                                         DSL61270
*LIST SUBPROGRAM NAMES                                     DSL61280
*LIST ALL
*ONE WORD INTEGERS
        SUBROUTINE CMPXP(IC1,IC2,P1,P2,X,Y,YDOT)           DSL61290
        REAL INTGR,IC1,IC2                                 DSL61300
        COMMON IDUMY(302),KOUNT,KDUMY(531),C(2)            DSL61310
        YDOT=INTGR(IC2,X-2.*P1*P2*C(KOUNT+7)-P2*P2*C(KOUNT+8))  DSL61320
        Y=INTGR(IC1,YDOT)                                  DSL61330
        RETURN                                             DSL61340
        END                                                DSL61350
// DUP                                                     DSL61360
*STORE          UA        CMPXP                            DSL61370
// FOR                                                     DSL61380
**DSL/1800   (DIGITAL  SIMULATION  LANGUAGE)               DSL61390
*LIST SYMBOL TABLE                                         DSL61400
*LIST SUBPROGRAM NAMES                                     DSL61410
*LIST ALL                                                  DSL61420
*ONE WORD INTEGERS
        FUNCTION COMPR(EXPR1,EXPR2)                        DSL61430
        IF(EXPR1-EXPR2)4,5,5                               DSL61440
      4 COMPR=0.0                                          DSL61450
        GO TO 6                                            DSL61460
      5 COMPR=1.                                           DSL61470
      6 RETURN                                             DSL61480
        END                                                DSL61490
// DUP                                                     DSL61500
*STORE          UA        COMPR                            DSL61510
// FOR                                                     DSL61520
**DSL/1800   (DIGITAL  SIMULATION  LANGUAGE)               DSL61530
*LIST SYMBOL TABLE                                         DSL61540
*LIST SUBPROGRAM NAMES                                     DSL61550
*LIST ALL                                                  DSL61560
*ONE WORD INTEGERS
        FUNCTION DEADS(LOBRK,HIBRK,EXPR)                   DSL61570
        REAL LOBRK                                         DSL61580
        IF(EXPR-HIBRK)8,8,10                               DSL61590
      8 IF(EXPR-LOBRK)20,9,9                               DSL61600
      9 DEADS=0.0                                          DSL61610
        GO TO 30                                           DSL61620
     10 DEADS=EXPR-HIBRK                                   DSL61630
        GO TO 30                                           DSL61640
     20 DEADS=EXPR-LOBRK                                   DSL61650
     30 RETURN                                             DSL61660
        END                                                DSL61670
// DUP                                                     DSL61680
*STORE          UA        DEADS                            DSL61690
// FOR                                                     DSL61700
**DSL/1800   (DIGITAL  SIMULATION  LANGUAGE)               DSL61710
*LIST SYMBOL TABLE                                         DSL61720
*LIST SUBPROGRAM NAMES                                     DSL61730
*LIST ALL                                                  DSL61740
*ONE WORD INTEGERS
        FUNCTION DEBUG(M,N,START)                          DSL61750
        INTEGER SYMB(420)                                  DSL61760
                                                           DSL61770
```

300

```
      DIMENSION KX(2)                                          DSL61780
      COMMON NALRM,NINTG,NSYMB,JPT,IDUMY(410),SYMB,C(2)        DSL61790
      EQUIVALENCE (C(1),TIME,KX(1))                            DSL61800
      K=2*M-1                                                  DSL61810
      IF(TIME)2,2,4                                            DSL61820
    2 KX(K)=N                                                  DSL61830
    4 IF(KX(K))12,12,5                                         DSL61840
    5 IF(TIME-START)12,13,13                                   DSL61850
   13 KX(K)=KX(K)-1                                            DSL61860
      KPT = JPT                                                DSL61870
      WRITE(KPT,8)(SYMB(I),SYMB(I+140),SYMB(I+280),C(I),I=1,NSYMB) DSL61880
    8 FORMAT(1H0 /(1X,3A2,  E12.4,2X,3A2,E12.4,2X,3A2,E12.4,2X,3A2, DSL61890
    1 E12.4,2X,3A2,E12.4))                                     DSL61900
   12 DEBUG=START                                              DSL61910
      RETURN                                                   DSL61920
      END                                                      DSL61930
// DUP                                                         DSL61940
*STORE      UA      DEBUG                                      DSL61950
// FOR                                                         DSL61960
**DSL/1800  (DIGITAL SIMULATION LANGUAGE)                      DSL61970
*LIST SYMBOL TABLE                                             DSL61980
*LIST SUBPROGRAM NAMES                                         DSL61990
*LIST ALL                                                      DSL62000
*ONE WORD INTEGERS                                             DSL62010
      FUNCTION DELAY(L,M,TD,XIN)                               DSL62020
      DIMENSION X(1),KX(1)                                     DSL62030
      COMMON IDUMY(15),KEEP,H,KDUMY(816),R(2)                  DSL62040
      EQUIVALENCE (X(1),KX(1),TIME,R(1)),(R(2),DELT)           DSL62050
      K=L+5                                                    DSL62060
      J=K+K-11                                                 DSL62070
      IF(TIME)3,3,10                                           DSL62080
    3 X(K-4)=TD/FLOAT(M-2)                                     DSL62090
      X(K-3)=X(K-4)                                            DSL62100
      KX(J)=K                                                  DSL62110
      KX(J-1)=K+2*(M-1)                                        DSL62120
      X(K-2)=-1000.                                            DSL62130
      X(K-1)=XIN                                               DSL62140
      X(K)=0.                                                  DSL62150
      X(K+1)=XIN                                               DSL62160
      DELAY=0.                                                 DSL62170
      GO TO 30                                                 DSL62180
C...  TEST IF TIME FOR OUTPUT
   10 TIMEX=TIME-TD                                            DSL62190
      IF(TIMEX+.5*DELT)9,11,11                                 DSL62200
    9 DELAY=0.0                                                DSL62210
      GO TO 20                                                 DSL62220
   11 INDEX=KX(J)                                              DSL62230
      IF(X(K))15,15,8                                          DSL62240
    8 KEND=KX(J-1)                                             DSL62250
    7 IF(INDEX-KEND)6,15,6                                     DSL62260
    6 I2=INDEX+2                                               DSL62270
      DO 12  I=I2,KEND,2                                       DSL62280
      IF(TIMEX-X(I))16,14,12                                   DSL62290
   12 CONTINUE                                                 DSL62300
   15 DO 13  I=K,INDEX,2                                       DSL62310
      IF(TIMEX-X(I))16,14,13                                   DSL62320
   13 CONTINUE                                                 DSL62330
      NALRM=31                                                 DSL62340
      CALL ERR                                                 DSL62350
      GO TO 30                                                 DSL62360
   14 DELAY=X(I+1)                                             DSL62370
      GO TO 20                                                 DSL62380
   16 DELAY=X(I-1)+(X(I+1)-X(I-1))*(TIMEX-X(I-2))/(X(I)-X(I-2)) DSL62390
C...  TEST IF TIME TO STORE INPUT
   20 IF(KEEP)30,30,25                                         DSL62400
   25 IF(X(K-3)-TIME+.5*DELT)5,5,30                            DSL62410
    5 INDEX=KX(J)+2                                            DSL62420
      IF(INDEX-KX(J-1))28,28,27                                DSL62430
   27 X(K-2)=X(INDEX-2)                                        DSL62440
      X(K-1)=X(INDEX-1)                                        DSL62450
      INDEX=K                                                  DSL62460
   28 X(K-3)=X(K-3)+X(K-4)                                     DSL62470
      IF(X(K-3)-TIME)4,4,2                                     DSL62480
    4 X(K-3)=TIME+X(K-4)                                       DSL62490
    2 KX(J)=INDEX                                              DSL62500
      X(INDEX)=TIME                                            DSL62510
      X(INDEX+1)=XIN                                           DSL62520
   30 RETURN                                                   DSL62530
```

```
        END                                                 DSL62550
// DUP                                                      DSL62560
*STORE        UA       DELAY                                DSL62570
// FOR                                                      DSL62580
**DSL/1800  (DIGITAL SIMULATION LANGUAGE)                   DSL62590
*LIST SYMBOL TABLE                                          DSL62600
*LIST SUBPROGRAM NAMES                                      DSL62610
*LIST ALL
*ONE WORD INTEGERS                                          DSL62620
      FUNCTION DERIV(L,XIC,XIN)                             DSL62630
      DIMENSION C(1)                                        DSL62640
      COMMON IDUMY(15),KEEP,H,KDUMY(816),R(2)               DSL62650
      EQUIVALENCE (C(1),TIME,R(1)),(R(2),DELT)              DSL62660
      K=L                                                   DSL62670
      IF(TIME)3,3,4                                         DSL62680
    3 C(K+4)=-1.                                            DSL62690
      DERIV=XIC                                             DSL62700
      C(K+1)=XIN                                            DSL62710
      C(K)=0.                                               DSL62720
      GO TO 9                                               DSL62730
    4 IF(C(K+4))5,2,6                                       DSL62740
    2 IF(KEEP)1,1,7                                         DSL62750
    5 C(K+4)=0.                                             DSL62760
      C(K+3)=C(K+1)                                         DSL62770
      C(K+2)=0.                                             DSL62780
    1 C(K+1)=XIN                                            DSL62790
      C(K)=TIME                                             DSL62800
      DERIV = (XIN-C(K+3))/TIME                             DSL62810
      GO TO 9                                               DSL62820
    6 IF(KEEP)8,10,7                                        DSL62830
    7 C(K+5)=C(K+3)                                         DSL62840
      C(K+4)=C(K+2)                                         DSL62850
      C(K+3)=C(K+1)                                         DSL62860
      C(K+2)=C(K)                                           DSL62870
    8 C(K+1)=XIN                                            DSL62880
      C(K)=TIME                                             DSL62890
   10 A = TIME-C(K+2)                                       DSL62900
      B = TIME-C(K+4)                                       DSL62910
      D = C(K+2)-C(K+4)                                     DSL62920
      DERIV = C(K+5)*A/(B*D)-C(K+3)*B/(A*D)+XIN/A+XIN/B     DSL62930
    9 RETURN                                                DSL62940
      END                                                   DSL62950
// DUP                                                      DSL62960
*STORE        UA       DERIV                                DSL62970
// FOR                                                      DSL62980
**DSL/1800  (DIGITAL SIMULATION LANGUAGE)                   DSL62990
*LIST SYMBOL TABLE                                          DSL63000
*LIST SUBPROGRAM NAMES                                      DSL63010
*LIST ALL
*ONE WORD INTEGERS                                          DSL63020
      FUNCTION  FCNSW(EXPR0,EXPR1,EXPR2,EXPR3)              DSL63030
      IF(EXPR0) 10,20,30                                    DSL63040
   10 FCNSW=EXPR1                                           DSL63050
      GO TO 40                                              DSL63060
   20 FCNSW=EXPR2                                           DSL63070
      GO TO 40                                              DSL63080
   30 FCNSW=EXPR3                                           DSL63090
   40 RETURN                                                DSL63100
      END                                                   DSL63110
// DUP                                                      DSL63120
*STORE        UA       FCNSW                                DSL63130
// FOR                                                      DSL63140
**DSL/1800  (DIGITAL SIMULATION LANGUAGE)                   DSL63150
*LIST SYMBOL TABLE                                          DSL63160
*LIST SUBPROGRAM NAMES                                      DSL63170
*LIST ALL
*ONE WORD INTEGERS                                          DSL63180
      FUNCTION HSTRS(L,IC,XOFS1,XOFS2,Y)                    DSL63190
      REAL IC                                               DSL63200
      DIMENSION A(1)                                        DSL63210
      COMMON IDUMY(834),R(2)                                DSL63220
      EQUIVALENCE (A(1),TIME,R(1)),(R(2),DELT)              DSL63230
      K=L                                                   DSL63240
      X=Y                                                   DSL63250
      IF(TIME) 1,1,5                                        DSL63260
    1 A(K+1)=IC                                             DSL63270
      GO TO 40                                              DSL63280
    5 IF(X-A(K)) 15,40,10                                   DSL63290
```

```
      10 IF(X-XOFS2 -A(K+1)) 40,20,20                              DSL63300
      15 IF(A(K+1)-X+XOFS1 ) 40,30,30                              DSL63310
      20 A(K+1)=X-XOFS2                                            DSL63320
         GO TO 40                                                  DSL63330
      30 A(K+1)=X-XOFS1                                            DSL63340
      40 A(K)=X                                                    DSL63350
         HSTRS =A(K+1)                                             DSL63360
         RETURN                                                    DSL63370
         END                                                       DSL63380
                                                                   DSL63390
// DUP                                                             DSL63400
*STORE          UA         HSTRS
// FOR                                                             DSL63410
**DSL/1800  (DIGITAL SIMULATION LANGUAGE)                          DSL63420
*LIST SYMBOL TABLE                                                 DSL63430
*LIST SUBPROGRAM NAMES                                             DSL63440
*LIST ALL
*ONE WORD INTEGERS                                                 DSL63450
      REAL FUNCTION IMPL(N,FIRST,ERROR,FUNCT)                      DSL63460
      DIMENSION C(2)                                               DSL63470
      COMMON NALR ,IDUMY(14),KEEP,H,KDUMY(816),R(2)                DSL63480
      EQUIVALENCE (R(1),C(1),TIME),(NALR,NALRM)                    DSL63490
      K = N                                                        DSL63500
      XNP1=FUNCT                                                   DSL63510
      IF(NALRM)25,1,6                                              DSL63520
    1 IF(TIME)5,5,11                                               DSL63530
    6 GO TO(11,2,3),LINK                                           DSL63540
    5 C(K) = FIRST                                                 DSL63550
   11 NALRM=1                                                      DSL63560
      KOUNT = -100                                                 DSL63570
      LINK = 2                                                     DSL63580
      GO TO 20                                                     DSL63590
    2 C(K+1) = C(K)                                                DSL63600
      C(K) = XNP1                                                  DSL63610
      LINK = 3                                                     DSL63620
      GO TO 20                                                     DSL63630
    3 DIFR=C(K)-C(K+1)                                             DSL63634
      IF(DIFR)12,18,12                                             DSL63636
   12 A = (XNP1-C(K))/DIFR                                         DSL63640
      IF(A-1.0)4,2,4                                               DSL63650
    4 Q = A/(A-1.0)                                                DSL63660
      C(K+1) = C(K)                                                DSL63670
      C(K) = Q*C(K) + (1.0-Q)*XNP1                                 DSL63680
      IF(ABS(C(K))-1.0)17,17,8                                     DSL63690
    8 IF(ABS((C(K)-C(K+1))/C(K))-ERROR)18,18,15                    DSL63700
   15 KCUNT = KOUNT+1                                              DSL63710
      IF(KCUNT)20,20,7                                             DSL63720
    7 IF(KEEP)16,10,16                                             DSL63730
   10 IF(TIME)18,18,16                                             DSL63740
   16 NALRM=30                                                     DSL63750
      CALL ERR                                                     DSL63760
      NALRM=-1                                                     DSL63770
      GO TO 19                                                     DSL63780
   17 IF(ABS(C(K)-C(K+1))-ERROR)18,18,15                           DSL63790
   18 NALRM=0                                                      DSL63800
   19 LINK = 1                                                     DSL63810
   20 IMPL = C(K)                                                  DSL63820
   25 RETURN                                                       DSL63830
      END                                                          DSL63840
                                                                   DSL63850
// DUP                                                             DSL63860
*STORE          UA         IMPL
// FOR                                                             DSL63870
**DSL/1800  (DIGITAL SIMULATION LANGUAGE)                          DSL63880
*LIST SYMBOL TABLE                                                 DSL63890
*PUNCH
*LIST SUBPROGRAM NAMES                                             DSL63900
*LIST ALL
*ONE WORD INTEGERS                                                 DSL63910
      REAL FUNCTION IMPUL(L,DELAY,PERID)                           DSL63920
      DIMENSION A(1)                                               DSL63930
      COMMON IDUMY(13),INTYP,KDUMY(820),R(2)                       DSL63940
      EQUIVALENCE (A(1),TIME,R(1)),(R(2),DELT)                     DSL63950
      K = L                                                        DSL63960
      IF(TIME) 10,10,11                                            DSL63970
   10 A(K) = 1.0E+37                                               DSL63980
      A(K+1) = DELAY                                               DSL63990
      A(K+2) = 0.0                                                 DSL64000
C... MILNE,RKS,RKSFX,ADAMS,TRAPZ,NON-CENTRAL                       DSL64010
   11 GO TO(15,12,14,15,15,15),INTYP                               DSL64020
```

```
    12 CFACT=0.125                                                DSL64030
       GO TO 20                                                   DSL64040
    14 DFACT=0.25                                                 DSL64050
       GO TO 20                                                   DSL64060
    15 DFACT=0.5                                                  DSL64070
    20 DEL=DFACT*DELT                                             DSL64080
       IF(ABS(A(K)-TIME) - DEL) 50, 30, 30                        DSL64090
    30 IF(ABS(A(K+1)-TIME) - DEL) 40, 90, 90                      DSL64100
    40 A(K+2) = A(K+2) + 1.0                                      DSL64110
       A(K) = A(K+1)                                              DSL64120
       A(K+1) = DELAY + A(K+2)*PERID                              DSL64130
    50 IMPUL=1.0                                                  DSL64140
       GO TO 100                                                  DSL64150
    90 IMPUL=0.0                                                  DSL64160
   100 RETURN                                                     DSL64170
       END                                                        DSL64180
*STORE        UA        IMPUL                                     DSL64200
// FOR                                                            DSL64210
**DSL/1800  (DIGITAL SIMULATION LANGUAGE)                         DSL64220
*LIST SYMBOL TABLE                                                DSL64230
*LIST SUBPROGRAM NAMES                                            DSL64240
*LIST ALL
*ONE WORD INTEGERS
       REAL FUNCTION   INSW(EXPR0,EXPR1,EXPR2)                    DSL64250
       IF(EXPR0) 10,20,20                                         DSL64260
    10 INSW=EXPR1                                                 DSL64270
       GO TO 30                                                   DSL64280
    20 INSW=EXPR2                                                 DSL64290
    30 RETURN                                                     DSL64300
       END                                                        DSL64310
// DUP                                                            DSL64320
*STORE        UA        INSW                                      DSL64330
// FOR                                                            DSL64340
**DSL/1800  (DIGITAL SIMULATION LANGUAGE)                         DSL64350
*LIST SYMBOL TABLE                                                DSL64360
*LIST SUBPROGRAM NAMES                                            DSL64370
*LIST ALL                                                         DSL64380
*ONE WORD INTEGERS
       REAL FUNCTION INTGR(A,X)                                   DSL64390
       DIMENSION C(2)                                             DSL64400
       COMMON NALRM,NINTG,IDUMY(8),KMEM,KDUMY(291),JDUMY(532),R(2)  DSL64410
       EQUIVALENCE (R(1),TIME,C(1)),(JDUMY(1),KOUNT)              DSL64420
       J=KMEM+KOUNT                                               DSL64430
       KOUNT=KOUNT+1                                              DSL64440
       IF(TIME)3,3,4                                              DSL64450
     3 C(KOUNT+6)=A                                               DSL64460
     4 C(J)=X                                                     DSL64470
       INTGR=C(KOUNT+6)                                           DSL64480
       IF(KCUNT-NINTG)8,7,7                                       DSL64490
     7 KCUNT=0                                                    DSL64500
     8 RETURN                                                     DSL64510
       END                                                        DSL64520
// DUP                                                            DSL64530
*STORE        UA        INTGR                                     DSL64540
// FOR                                                            DSL64550
**DSL/1800  (DIGITAL SIMULATION LANGUAGE)                         DSL64560
*LIST SYMBOL TABLE                                                DSL64570
*LIST SUBPROGRAM NAMES                                            DSL64580
*LIST ALL                                                         DSL64590
*ONE WORD INTEGERS
       SUBROUTINE LEDLG(IC,P1,P2,X,A,Y)                           DSL64600
       REAL INTGR,IC                                              DSL64610
       COMMON IDUMY(302),KOUNT,KDUMY(531),C(2)                    DSL64620
       YCOT=(X-C(KOUNT+7))/P2                                     DSL64630
       Y=INTGR(IC,YCOT)                                           DSL64640
       A=P1*YDOT+Y                                                DSL64650
       RETURN                                                     DSL64660
       END                                                        DSL64670
// DUP                                                            DSL64680
*STORE        UA        LEDLG                                     DSL64690
// FOR                                                            DSL64700
**DSL/1800  (DIGITAL SIMULATION LANGUAGE)                         DSL64710
*LIST SYMBOL TABLE                                                DSL64720
*LIST SUBPROGRAM NAMES                                            DSL64730
*LIST ALL                                                         DSL64740
*ONE WORD INTEGERS
       REAL FUNCTION LIMIT(P1,P2,X)                               DSL64750
       IF(X-P1)5,5,7                                              DSL64760
                                                                  DSL64770
```

304

```
      5 LIMIT=P1                                                      DSL64780
        GO TO 9                                                       DSL64790
      7 IF(X-P2)6,8,8                                                 DSL64800
      8 LIMIT=P2                                                      DSL64810
        GO TO 9                                                       DSL64820
      6 LIMIT=X                                                       DSL64830
      9 RETURN                                                        DSL64840
        END                                                           DSL64850
// DUP                                                                DSL64860
*STORE          UA        LIMIT                                       DSL64870
// FOR                                                                DSL64880
**DSL/1800  (DIGITAL SIMULATION LANGUAGE)                             DSL64890
*NONPROCESS PROGRAM                                                   DSL64900
*LIST SYMBOL TABLE                                                    DSL64910
*LIST SUBPROGRAM NAMES                                                DSL64920
*LIST ALL
*ONE WORD INTEGERS                                                    DSL64930
        FUNCTION LOCK(VAR)                                            DSL64940
C...    SUBROUTINE TO SEARCH FOR INDEX OF VARIABLE IN SYMB TABLE      DSL64950
        INTEGER SYMB(420),VAR(3)                                      DSL64960
        COMMON NALRM,NINTG,NSYMB,IDUMY(411),SYMB                      DSL64970
        EQUIVALENCE (NALR,NALRM)                                      DSL64980
        DO 100 J = 1,NSYMB                                            DSL64990
        IF(VAR(1)-SYMB(J))100,5,100                                   DSL65000
      5 IF(VAR(2)-SYMB(J+140))100,6,100                               DSL65010
      6 IF(VAR(3)-SYMB(J+280))100,7,100                               DSL65020
      7 LOOK = J                                                      DSL65030
        GO TO 200                                                     DSL65040
    100 CONTINUE                                                      DSL65050
        NALR=32                                                       DSL65060
        CALL ERR                                                      DSL65070
        STOP                                                          DSL65080
    200 RETURN                                                        DSL65090
        END                                                           DSL65100
// DUP                                                                DSL65110
*STORE          UA        LOOK                                        DSL65120
// FOR                                                                DSL65130
**DSL/1800  (DIGITAL SIMULATION LANGUAGE)                             DSL65140
*LIST SYMBOL TABLE                                                    DSL65150
*LIST SUBPROGRAM NAMES                                                DSL65160
*LIST ALL
*ONE WORD INTEGERS                                                    DSL65170
        REAL FUNCTION MODIN(IC,P1,P2,X)                               DSL65180
        REAL INTGR,IC                                                 DSL55190
        IF(P1)3,3,2                                                   DSL65200
      2 Y=INTGR(IC,X)                                                 DSL65210
        GO TO 8                                                       DSL65220
      3 Y=INTGR(IC,0.)                                                DSL65230
        IF(P2)8,8,4                                                   DSL65240
      4 Y=IC                                                          DSL65250
      8 MODIN=Y                                                       DSL65260
        RETURN                                                        DSL65270
        END                                                           DSL65280
// DUP                                                                DSL65290
*STORE          UA        MODIN                                       DSL65300
// FOR                                                                DSL65310
**DSL/1800  (DIGITAL SIMULATION LANGUAGE)                             DSL65320
*LIST SYMBOL TABLE                                                    DSL65330
*LIST SUBPROGRAM NAMES                                                DSL65340
*ONE WORD INTEGERS                                                    DSL65350
        REAL FUNCTION NLFGN(LOCN,ARG)                                 DSL65360
        DIMENSION KFCT(2)                                             DSL65370
        COMMON IDUMY(293),JDUMY(541),FCT(2)                           DSL65380
        EQUIVALENCE (KFCT(1),FCT(1)),(JDUMY(1),KOUNT),(NALRM,IDUMY(1)),DSL65390
      1 (ISTRT,IDUMY(222))                                            DSL65400
C...    FCT IS EQUIVALENT TO C AND DYNAM IN INTRAN AND SIMUL          DSL65410
        LCCT=LOCN                                                     DSL65420
        IF(LOCT)3,3,4                                                 DSL65430
      3 NLFGN=0.0                                                     DSL65440
        GO TO 8                                                       DSL65450
      4 VAR=ARG                                                       DSL65460
        IF(VAR-FCT(LOCT))10,9,11                                      DSL65470
     10 IF(KOUNT)12,12,9                                              DSL65480
     12 NALRM=33                                                      DSL65490
        ISTRT=LOCT                                                    DSL65500
        CALL ERR                                                      DSL65510
        KCUNT=KOUNT+1                                                 DSL65520
        GO TO 9                                                       DSL65530
```

305

```
   11 LOCT2=KFCT(2*LOCT-3)                                                      DSL65540
      DO 5  I=LOCT,LOCT2,2                                                      DSL65550
      IF(VAR-FCT(I))6,7,5                                                       DSL65560
    5 CONTINUE                                                                  DSL65570
      I=LOCT2                                                                   DSL65580
      IF(KCUNT)13,13,7                                                          DSL65590
   13 NALRM=34                                                                  DSL65600
      ISTRT=LOCT                                                                DSL65610
      CALL ERR                                                                  DSL65620
      KCUNT=KOUNT+1                                                             DSL65630
      GO TO 7                                                                   DSL65640
    6 IF(I-LOCT+2)14,14,15                                                      DSL65650
   14 NLFGN =(VAR-FCT(I-2))*(FCT(I+1)-FCT(I-1))/(FCT(I)-FCT(I-2))+              DSL65660
     1 FCT(I-1)                                                                 DSL65670
      GO TO 8                                                                   DSL65680
   15 A=FCT(I-2)-FCT(I-4)                                                       DSL65690
      B=FCT(I)-FCT(I-2)                                                         DSL65700
      C=FCT(I)-FCT(I-4)                                                         DSL65710
      D=VAR-FCT(I-2)                                                            DSL65720
      E=VAR-FCT(I)                                                              DSL65730
      F=VAR-FCT(I-4)                                                            DSL65740
      NLFGN =FCT(I-3)*D*E/(A*C)-FCT(I-1)*F*E/(A*B)+FCT(I+1)*F*D/(C*B)           DSL65750
      GO TO 8                                                                   DSL65760
    9 I=LOCT                                                                    DSL65770
    7 NLFGN =FCT(I+1)                                                           DSL65780
    8 RETURN                                                                    DSL65790
      END                                                                       DSL65800
// DUP                                                                          DSL65810
*STORE         UA        NLFGN                                                  DSL65820
// FOR                                                                          DSL65830
**DSL/1800  (DIGITAL SIMULATION LANGUAGE)                                       DSL65840
*LIST SYMBOL TABLE                                                              DSL65850
*LIST SUBPROGRAM NAMES                                                          DSL65860
*LIST ALL
*ONE WORD INTEGERS                                                              DSL65870
      REAL FUNCTION NORML(ODD1,AVG,SIGMA)                                       DSL65880
      U1=UZRP1(ODD1)                                                            DSL65890
      U2=UZRP1(ODD1)*6.2831968                                                  DSL65900
      NORML=SIGMA*SQRT(-2.0*ALOG(U1))*COS(U2)+AVG                               DSL65910
      RETURN                                                                    DSL65920
      END                                                                       DSL65930
// DUP                                                                          DSL65940
*STORE         UA        NORML                                                  DSL65950
// FOR                                                                          DSL65960
**DSL/1800  (DIGITAL SIMULATION LANGUAGE)                                       DSL65970
*LIST SYMBOL TABLE                                                              DSL65980
*LIST SUBPROGRAM NAMES                                                          DSL65990
*LIST ALL
*ONE WORD INTEGERS                                                              DSL66000
      SUBROUTINE  OUTSW(EXPRO,EXPR1,A,B)                                        DSL66010
      IF(EXPRO) 10,20,20                                                        DSL66020
   20 B=EXPR1                                                                   DSL66030
      A=0.0                                                                     DSL66040
      GO TO 30                                                                  DSL66050
   10 A=EXPR1                                                                   DSL66060
      B=0.0                                                                     DSL66070
   30 RETURN                                                                    DSL66080
      END                                                                       DSL66090
// DUP                                                                          DSL66100
*STORE         UA        OUTSW                                                  DSL66110
// FOR                                                                          DSL66120
**DSL/1800  (DIGITAL SIMULATION LANGUAGE)                                       DSL66130
*LIST SYMBOL TABLE                                                              DSL66140
*LIST SUBPROGRAM NAMES                                                          DSL66150
*LIST ALL
*ONE WORD INTEGERS                                                              DSL66160
      FUNCTION PULSE(L,TRIGR,WIDTH)                                             DSL66170
C...  NON-ZERO TRIGR INITIATES PULSE WIDTH WIDE                                 DSL66180
      DIMENSION A(1)                                                            DSL66190
      COMMON IDUMY(13),INTYP,JDUMY(820),R(2)                                    DSL66200
      EQUIVALENCE (A(1),TIME,R(1)),(R(2),DELT)                                  DSL66210
      K=L                                                                       DSL66220
      IF(TIME) 1,1,3                                                            DSL66230
    1 A(K)=-1.E37                                                               DSL66240
C...  MILNE,RKS,RKSFX,ADAMS,TRAPZ,NON-CENTRAL                                   DSL66260
    3 GO TO(16,12,14,16,16,16),INTYP                                            DSL66270
   12 DFACT=0.125                                                               DSL66280
      GO TO 5                                                                   DSL66290
```

306

```
   14 DFACT=0.25                                                    DSL66300
      GO TO 5                                                       DSL66310
   16 DFACT=0.5                                                     DSL66315
    5 DEL=DFACT*DELT                                                DSL66320
      IF(TIME-A(K))250,205,205                                      DSL66330
  205 IF(TIME-(A(K)+WIDTH+DEL))220,220,210                          DSL66340
  210 IF(TRIGR)310,310,212                                          DSL66350
  212 A(K+1)=A(K)                                                   DSL66360
      A(K)=TIME                                                     DSL66370
  220 PULSE=1.0                                                     DSL66380
      GO TO 300                                                     DSL66390
  250 IF(TIME-(A(K+1)+WIDTH+DEL))220,220,310                        DSL66400
  310 PULSE=0.                                                      DSL66410
  300 RETURN                                                        DSL66420
      END                                                           DSL66430
// DUP                                                              DSL66440
*STORE        UA        PULSE                                       DSL66450
// FOR                                                              DSL66460
**DSL/1800  (DIGITAL SIMULATION LANGUAGE)                           DSL66470
*LIST SYMBOL TABLE                                                  DSL66480
*LIST SUBPROGRAM NAMES                                              DSL66490
*LIST ALL
*ONE WORD INTEGERS                                                  DSL66500
      FUNCTION QNTZR(QUANT,EXPR)                                    DSL66510
      QOUT=IFIX((0.5*QUANT+ABS(EXPR))/QUANT)                        DSL66520
      QNTZR=SIGN(QUANT*QOUT,EXPR)                                   DSL66530
      RETURN                                                        DSL66540
      END                                                           DSL66550
// DUP                                                              DSL66560
*STORE        UA        QNTZR                                       DSL66570
// FOR                                                              DSL66580
**DSL/1800  (DIGITAL SIMULATION LANGUAGE)                           DSL66590
*LIST SYMBOL TABLE                                                  DSL66600
*LIST SUBPROGRAM NAMES                                              DSL66610
*LIST ALL
*ONE WORD INTEGERS                                                  DSL66620
      FUNCTION RAMP(DELAY)                                          DSL66630
      COMMON IDUMY(834),TIME                                        DSL66640
      TREF=TIME-DELAY                                               DSL66650
      IF(TREF) 1,2,2                                                DSL66660
    1 RAMP=0.0                                                      DSL66670
      GO TO 5                                                       DSL66680
    2 RAMP=TREF                                                     DSL66690
    5 RETURN                                                        DSL66700
      END                                                           DSL66710
// DUP                                                              DSL66720
*STORE        UA        RAMP                                        DSL66730
// FOR                                                              DSL66740
**DSL/1800  (DIGITAL SIMULATION LANGUAGE)                           DSL66750
*LIST SYMBOL TABLE                                                  DSL66760
*LIST SUBPROGRAM NAMES                                              DSL66770
*LIST ALL
*ONE WORD INTEGERS                                                  DSL66780
      FUNCTION REALP(IC,P,X)                                        DSL66790
      REAL INTGR,IC                                                 DSL66800
      COMMON IDUMY(302),KOUNT,KDUMY(531),C(2)                       DSL66810
      REALP=INTGR(IC,(X-C(KOUNT+7))/P)                              DSL66820
      RETURN                                                        DSL66830
      END                                                           DSL66840
// DUP                                                              DSL66850
*STORE        UA        REALP                                       DSL66860
// FOR                                                              DSL66870
**DSL/1800  (DIGITAL SIMULATION LANGUAGE)                           DSL66880
*LIST SYMBOL TABLE                                                  DSL66890
*LIST SUBPROGRAM NAMES                                              DSL66900
*LIST ALL
*ONE WORD INTEGERS                                                  DSL66910
      FUNCTION  RST(L,R,S,T)                                        DSL66920
      DIMENSION A(1)                                                DSL66930
      COMMON IDUMY(834),B(2)                                        DSL66940
      EQUIVALENCE (A(1),TIME,B(1))                                  DSL66950
      K = L                                                         DSL66960
      IF(TIME) 1, 1, 2                                              DSL66970
    1 A(K) = 0.0                                                    DSL66980
    2 IF(R) 10, 10, 5                                               DSL66990
    5 A(K) = 0.0                                                    DSL67000
      GO TO 50                                                      DSL67010
   10 IF(S) 20, 20, 35                                              DSL67020
```

307

```
      20 IF(T) 50, 50, 25                                    DSL67030
      25 IF(A(K)) 35, 35, 5                                  DSL67040
      35 A(K) = 1.0                                          DSL67050
      50 RST  = A(K)                                         DSL67060
         RETURN                                              DSL67070
         END                                                 DSL67080
// DUP                                                       DSL67090
*STORE         UA       RST                                  DSL67100
// FOR                                                       DSL67110
**DSL/1800   (DIGITAL SIMULATION LANGUAGE)                   DSL67120
*LIST SYMBOL TABLE                                           DSL67130
*LIST SUBPROGRAM NAMES                                       DSL67140
*LIST ALL
*ONE WORD INTEGERS                                           DSL67150
         FUNCTION SINE(DELAY,OMEGA,PHASE)                    DSL67160
         COMMON IDUMY(834),TIME,DELT                         DSL67170
         TIMER=TIME-DELAY                                    DSL67180
         IF(-TIMER-0.5*DELT)  3,3,2                          DSL67190
       2 SINE=0.0                                            DSL67200
         GO TO 10                                            DSL67210
       3 SINE=SIN(OMEGA*TIMER+PHASE)                         DSL67220
      10 RETURN                                              DSL67230
         END                                                 DSL67240
// DUP                                                       DSL67250
*STORE         UA       SINE                                 DSL67260
// FOR                                                       DSL67270
**DSL/1800   (DIGITAL SIMULATION LANGUAGE)                   DSL67280
*LIST SYMBOL TABLE                                           DSL67290
*LIST SUBPROGRAM NAMES                                       DSL67300
*LIST ALL
*ONE WORD INTEGERS                                           DSL67310
         FUNCTION STEP(DELAY)                                DSL67320
         COMMON IDUMY(834),TIME                              DSL67330
         IF(TIME-DELAY) 1,2,2                                DSL67340
       1 STEP=0.0                                            DSL67350
         GO TO 5                                             DSL67360
       2 STEP=1.0                                            DSL67370
       5 RETURN                                              DSL67380
         END                                                 DSL67390
// DUP                                                       DSL67400
*STORE         UA       STEP                                 DSL67410
// JOB
// FOR                                                       DSL67420
*LIST ALL
**DSL/1800   (DIGITAL SIMULATION LANGUAGE)                   DSL67430
*LIST SYMBOL TABLE                                           DSL67440
*LIST SUBPROGRAM NAMES                                       DSL67450
*ONE WORD INTEGERS                                           DSL67460
         FUNCTION TRNFR(N,K,A,B,X)                           DSL67470
         REAL INTGR                                          DSL67480
         DIMENSION A(2),B(5)
         COMMON KDUMY(834),C(1)                              DSL67491
         EQUIVALENCE (TIME,C(1)),(KOUNT,KDUMY(303))          DSL67492
         M=K                                                 DSL67493
         I1=KOUNT+7                                          DSL67494
         IF(TIME)1,1,2                                       DSL67495
       1 IF(B(M+1)-1.)11,2,11                                DSL67496
      11 EX=B(M+1)                                           DSL67498
         CALL TRNFX(M,B)                                     DSL67497
         CALL TRNFY(N,A,BX)
       2 DO 10   I=1,M                                       DSL67500
         J=M-I                                               DSL67510
         IF(J-N)3,3,5                                        DSL67520
       3 IF(A(J+1))4,5,4                                     DSL67530
       4 R=A(J+1)*X                                          DSL67540
         GO TO 6                                             DSL67550
       5 R=0.                                                DSL67560
       6 IF(B(J+1))7,8,7                                     DSL67570
       7 R=R-B(J+1)*C(I1)                                    DSL67580
       8 IF(I-M)9,10,10                                      DSL67590
       9 R=R+C(KOUNT+8)                                      DSL67600
      10 R=INTGR(0.,R)                                       DSL67610
         TRNFR=C(I1)                                         DSL67620
         RETURN                                              DSL67630
         END                                                 DSL67640
// DUP
*DELETE                 TRNFR
*STORE         WS  UA   TRNFR
```

308

```
// FOR                                                          DSL67655
*LIST SYMBCL TABLE                                              DSL67656
*LIST SUBPRCGRAM NAMES                                          DSL67657
*ONE WORD INTEGERS                                              DSL67658
      SUBROUTINE TRNFX(M,B)                                     DSL67659
      DIMENSION B(1)                                            DSL67660
      DC 12  I=1,M                                              DSL67661
   12 B(I)=B(I)/B(M+1)                                          DSL67662
      B(M+1)=1.0                                                DSL67663
      RETURN                                                    DSL67664
      END                                                       DSL67665
// CUP
*DELETE               TRNFX
*STORE        WS   UA TRNFX
// FOR TRNFY
*LIST ALL
*ONE WORD INTEGERS
      SUBROUTINE TRNFY(N,A,BX)
      DIMENSION A(1)
      NX=N+1
      DO 10 I=1,NX
   10 A(I)=A(I)/BX
      RETURN
      END
// DUP
*STORE        WS   UA TRNFY
// ASM                                                          DSL67670
*LIST                                                           DSL67680
                  ENT     UM1P1                                 DSL67690
                  ENT     UATOB                                 DSL67700
                  ENT     UZRP1                                 DSL67710
          * RANDOM NUMBER GENERATOR                             DSL67720
          * UNIFORM(0,1) *** UZRP1(NODD)                        DSL67730
          * UNIFORM(-1,+1) *** UM1P1(NODD)                      DSL67740
          * UNIFORM(A,B) *** UATOB(NODD,Y,Z) WHERE Y=A, Z=B-A   DSL67750
          UM1P1 DC      0                                       DSL67760
                STX   1 RETRN+1                                 DSL67770
                LDX  I1 UM1P1                                   DSL67780
                STX   1 LOCN+1                                  DSL67790
                MDX   1 1                                       DSL67800
                STX   1 RETRN+3                                 DSL67810
                LD      K1                                      DSL67820
                MDX     COMN                                    DSL67830
          UATOB DC      0                                       DSL67840
                STX   1 RETRN+1                                 DSL67850
                LDX  I1 UATOB                                   DSL67860
                STX   1 LOCN+1                                  DSL67870
                MDX   1 1                                       DSL67880
                STX   1 STRT+1                                  DSL67890
                MDX   1 1                                       DSL67900
                STX   1 STRT+4                                  DSL67910
                MDX   1 1                                       DSL67920
                STX   1 RETRN+3                                 DSL67930
                LD      K2                                      DSL67940
                MDX     COMN                                    DSL67950
          UZRP1 DC      0                                       DSL67960
                STX   1 RETRN+1                                 DSL67970
                LDX  I1 UZRP1                                   DSL67980
                STX   1 LOCN+1                                  DSL67990
                MDX   1 1                                       DSL68000
                STX   1 RETRN+3                                 DSL68010
                LD      K3                                      DSL68020
          COMN  STO     TRA                                     DSL68030
          LOCN  LDX  I1 *-*                                     DSL68040
                LD      X                                       DSL68050
                M     1 0                                       DSL68060
                STO     SAVEA                                   DSL68070
                SLT     16                                      DSL68080
                BSC     +-                                      DSL68090
                LD      SAVEA                                   DSL68100
                STO   1 0                                       DSL68110
                LDX   1 /20                                     DSL68120
                SLCA  1 0                                       DSL68130
                MDX   1 /60                                     DSL68140
                SRT     1                                       DSL68150
                AND     MASK                                    DSL68160
                STD   3 126                                     DSL68170
                STX   1 SAVEA                                   DSL68180
```

309

```
                        LD         SAVEA                         DSL68190
                        STO      3 125                           DSL68200
               TRA      MDX        *                             DSL68210
                        LIBF       FSUB                          DSL68220
                        DC         HALF                          DSL68230
                        LIBF       FMPY                          DSL68240
                        DC         TWO                           DSL68250
                        MDX        RETRN                         DSL68260
               STRT     LDX    I1  *-*                           DSL68270
                        STX     1  A                             DSL68280
                        LDX    I1  *-*                           DSL68290
                        STX     1  B                             DSL68300
                        LIBF       FMPY                          DSL68310
               B        DC         *-*                           DSL68320
                        LIBF       FADD                          DSL68330
               A        DC         *-*                           DSL68340
               RETRN    LDX    L1  *-*                           DSL68350
                        BSC    L   *-*                           DSL68360
               X        DC         /0103                         DSL68370
               K1       MDX        *                             DSL68380
               K2       MDX        *+5                           DSL68390
               K3       MDX        *+15                          DSL68400
               SAVEA    DC         0                             DSL68410
               MASK     DC         /7FFF                         DSL68420
                        BSS     E  0                             DSL68430
               HALF     DEC        .5                            DSL68440
               TWO      DEC        2.                            DSL68450
                        END                                      DSL68460
// DUP                                                           DSL68470
*DELETE                 UM1P1
*STORE           WS  UA  UM1P1
// FOR                                                           DSL68490
**DSL/1800  (DIGITAL SIMULATION LANGUAGE)                        DSL68500
*LIST SYMBOL TABLE                                               DSL68510
*LIST SUBPROGRAM NAMES                                           DSL68520
*LIST ALL
*ONE WORD INTEGERS                                               DSL68530
        FUNCTION ZHOLD(K,TRIGR,FCN)                              DSL68540
        DIMENSION A(1)                                           DSL68550
        COMMON IDUMY(834),R(2)                                   DSL68560
        EQUIVALENCE (A(1),TIME,R(1))                             DSL68570
        IF(TIME)10,10,20                                         DSL68580
   10   A(K) = 0.0                                               DSL68590
   20   IF(TRIGR)40,40,30                                        DSL68600
   30   A(K) = FCN                                               DSL68610
   40   ZHOLD= A(K)                                              DSL68620
        RETURN                                                   DSL68630
        END                                                      DSL68640
// DUP                                                           DSL68650
*STORE           UA      ZHOLD                                   DSL68660
// PAUSE
// FOR PLOTR
*ONE WORD INTEGERS                                               DSL43850
*LIST ALL                                                        DSL43880
*NONPROCESS PROGRAM                                              DSL43870
**ROUTINE TO DRAW AND LABEL GRAPH OR SCOPE AXES                  DSL43890
*IOCS(KEYBOARD,TYPEWRITER,PLOTTER)                               DSL43900
*NAME PLOTR                                                      DSL43910
        INTEGER TITLE(60),HDNG(75),SYMB(420)                     DSL43920
        COMMON NALRM,NINTG,NSYMB,JPT,KPRNT,KRANG,KFINI,KTITL,KREL,KABS,  DSL43930
       1 KMEM,KPOIN,KLOCK,INTYP,INDX1,KEEP,DELTA,STEPC,NX,NY,RLAST,ALAST, DSL43940
       2 FINVA(5),TITLE,HDNG,INDXP(25),IRANG(15),IFINL(5),JFINL(5),      DSL43950
       3 ISTRT,RMIN(15),RMAX(15),XOUT(9),LOCY,LOCX,KOUNT,IRES(101),      DSL43960
       4 DELTG,SCAL1,SCAL2,YPOS,INDX2,ISCOP,SYMB,C(300)                  DSL43970
        EQUIVALENCE (IRES(26),INDX),(IRES(57),INDY)               DSL43980
C...    SET LOGICAL UNIT OF PLOTTER AND TYPEWRITER                DSL43990
        M=INPT(7)                                                DSL43994
        M2=INPT(1)                                               DSL43995
        IF(DELTG)353,353,348                                     DSL43996
  348   CALL FPLOT(1,0.,0.)                                      DSL44000
        CALL PLOTS(ISCOP)                                        DSL44010
        IF(ISCOP)349,351,349                                     DSL44015
  349   SCALR=1.0                                                DSL44020
C...    SET N, WHERE 10*N/32768 CM. IS ORIGIN ON SCOPE           DSL44030
        CALL ORGSC(2.,2.)                                        DSL44035
        GO TO 330                                                DSL44040
  351   SCALR=0.5                                                DSL44060
  330   TW=.2*SCALR                                              DSL44070
                                                                 DSL44100
```

310

```
        TH=.3*SCALR                                               DSL44110
        FO=.4*SCALR                                               DSL44120
        IF(YPOS)356,354,354                                       DSL44130
  356   CALL REMSW(5,L)                                           DSL44140
        IF(L-1)371,372,371                                        DSL44150
  372   IF(ISCOP)370,359,370                                      DSL44160
  371   IF(INDX-INDX1)359,357,359                                 DSL44170
  357   IF(INDX2)352,352,358                                      DSL44180
  358   IF(INDY-INDX2)360,380,360                                 DSL44190
  380   IF(KLOCK)352,370,352                                      DSL44200
  359   W=NX+3                                                    DSL44210
        CALL FPLOT(1,W,0.)                                        DSL44220
        CALL SCALF(1.,1.,0.,0.)                                   DSL44230
  370   YPOS=0.                                                   DSL44240
  354   RX=NX                                                     DSL44250
        CALL ERASE                                                DSL44260
C...    DELAY FOR CLEAN STORE AFTER ERASE                         DSL44262
        DO 382  I=1,2000                                          DSL44264
  382   A=I                                                       DSL44266
        INDX=INDX1                                                DSL44270
        CALL FGRID(0,0.,0.,1.,NX)                                 DSL44280
        CALL FCHAR(1.,-1.5*SCALR,FO,FO,0.)                        DSL44290
        WRITE(M,100)(TITLE(I),I=41,60)                            DSL44300
  100   FORMAT(20A2)                                              DSL44310
        AA=.5*RX-.375                                             DSL44320
        CALL FCHAR(AA,-.8*SCALR,TH,TH,0.)                         DSL44330
        WRITE(M,100) SYMB(INDX1),SYMB(INDX1+140),SYMB(INDX1+280)  DSL44340
        K=NX/2                                                    DSL44350
        DO 55  I=1,K                                              DSL44360
        W=I+I                                                     DSL44370
   55   XOUT(I)=W/SCAL1                                           DSL44380
        CALL FCHAR(0.,-TH,TW,TW,0.)                               DSL44390
        IF(ISCOP)341,340,341                                      DSL44400
  340   WRITE(M,102)(XOUT(I),I=1,K)                               DSL44410
  102   FORMAT(F22.2,7F20.2)                                      DSL44420
        GO TO 360                                                 DSL44430
  341   WRITE(M,101)(XOUT(I),I=1,K)                               DSL44440
  101   FORMAT(F12.2,7F10.2)                                      DSL44450
  360   INDY=IABS(INDX2)                                          DSL44460
        RY=NY                                                     DSL44470
        CALL FGRID(1,YPOS,0.,1.,NY)                               DSL44480
        AA=.5*RY-.375                                             DSL44490
        BB = YPOS-.6*SCALR                                        DSL44500
        CALL FCHAR(BB,AA,TH,TH,1.5708)                            DSL44510
        WRITE(M,100) SYMB(INDY),SYMB(INDY+140),SYMB(INDY+280)     DSL44520
        INDX2=-INDY                                               DSL44530
        K=NY/2                                                    DSL44540
        DO 56  I=1,K                                              DSL44550
        W=I+I                                                     DSL44560
   56   XOUT(I)=W/SCAL2                                           DSL44570
        CALL FCHAR(YPOS-.1,0.,TW,TW,1.5708)                       DSL44580
        IF(ISCOP)343,342,343                                      DSL44590
  342   WRITE(M,102)(XOUT(I),I=1,K)                               DSL44600
        GO TO 344                                                 DSL44610
  343   WRITE(M,101)(XOUT(I),I=1,K)                               DSL44620
  344   YPOS=YPOS-1.0                                             DSL44630
C...    TEST FOR CONTINUATION                                     DSL44640
  352   IF(KLOCK)345,333,345                                      DSL44650
  345   CALL FPLOT(1,0.,0.)                                       DSL44660
        GO TO 334                                                 DSL44670
  333   CALL FPLOT(-2,0.,0.)                                      DSL44680
  334   CALL REMSW(4,L)                                           DSL44690
        IF(L-1)353,363,353                                        DSL44700
  363   WRITE(M2,400)                                             DSL44710
  400   FORMAT(29H SET PLOT AT (0,0), SET 4 OFF)                  DSL44720
  364   CALL REMSW(4,L)                                           DSL44730
        GO TO(364,370),L                                          DSL44740
  353   CALL LINK(SIMUL)                                          DSL44750
        END                                                       DSL44760
*STORE          UA          PLOTR                                 DSL44780
// FOR RKS                                                        DSL45500
*LIST ALL                                                         DSL45520
*ONE WORD INTEGERS                                                DSL45530
**RUNGE-KUTTA 4TH ORDER VARIABLE STEP INTEGRATION ROUTINE         DSL45540
        SUBROUTINE    RKS(PO,XABS,R,YST,DYST)                     DSL45550
        INTEGER TITLE(60),HDNG(75),SYMB(420)                      DSL45560
        DIMENSION  Y(1),XABS(1),R(1),PO(1),KC(1),YST(1),DYST(1),  DSL45570
      3 PD(50),SD(50),YS(50),DELY(50)                             DSL45580
```

311

```
      COMMON NALRM,NINTG,NSYMB,JPT,KPRNT,KRANG,KFINI,KTITL,KREL,KABS,       DSL45590
     1 KMEM,KPOIN,KLOCK,INTYP,INDX1,KEEP,DELTA,STEPC,NX,NY,RLAST,ALAST,      DSL45600
     2 FINVA(5),TITLE,HDNG,INDXP(25),IRANG(15),IFINL(5),JFINL(5),            DSL45610
     3 ISTRT,RMIN(15),RMAX(15),KWORD(11),KAFGN,LINE,KLINE,LINE1,NLINS,       DSL45620
     4 KODE,I2,LOCY,LOCX,KOUNT,IRES(101),DELTG,SCAL1,SCAL2,YPCS,INDX2,       DSL45630
     5 ISCCF,SYMB,C(7)                                                       DSL45640
      EQUIVALENCE (C(1),TIME),(C(2),DELT),(C(3),XINT),(C(7),Y(1)),          DSL45650
     1 (DELTP,C(5)),(FINTI,C(4)),(C(1),KC(1)),(DELTA,H),(DELTC,C(6)),        DSL45660
     4 (IRES(57),INDY),(N,NINTG)                                            DSL45670
  200 NALRM=0                                                               DSL45680
      LOCY=0                                                                DSL45685
      LINK=1                                                                DSL45690
      LINE=60                                                               DSL45700
      TPRNT=0.0 +                                                           DSL45710
      DPT=0.                                                                DSL45720
      TFLOT=0.0                                                             DSL45730
      DPL=0.                                                                DSL45740
      KPT = JPT                                                             DSL45750
      KAFGN=-10                                                             DSL45760
      RL=10.**.2                                                            DSL45770
      DELTA=1.                                                              DSL45780
      PLAST=C(KMEM)                                                         DSL45790
      W=PLAST+DELTP-TIME                                                    DSL45800
      IF(DELTP)209,209,207                                                  DSL45805
  207 IF(W+DELTP*DPT)208,208,209                                            DSL45810
  208 DPT=DPT+1.0                                                           DSL45820
      GO TO 207                                                             DSL45830
  209 TLAST=C(KMEM+1)                                                       DSL45840
C...  INITIALIZE DERIVATIVE SUBROUTINE                                      DSL45850
      KEEP = 0                                                             DSL45860
      KCUNT=0                                                               DSL45870
      CALL UPDAT                                                            DSL45880
      CALL PLCTS(ISCOP)                                                     DSL45882
      IF(ISCOP)206,206,205                                                  DSL45884
  205 CALL CRGSC(2.,2.)                                                     DSL45886
C...  TEST IF CONTINUATION OF PARAMETER STUDY                              DSL45890
  206 IF(KLOCK)220,221,220                                                  DSL45900
  220 KEEP=-1                                                               DSL45910
      KLOCK=0                                                               DSL45915
      IF(DELTG)223,223,211                                                  DSL45920
  211 CALL FPLOT(-2,SCAL1*C(INDX1),SCAL2*C(INDY))                          DSL45930
      GO TO 223                                                             DSL45940
  221 KEEP=1                                                                DSL45950
      TLAST=0.0                                                             DSL45960
      PLAST=0.0                                                             DSL45970
      IF(DELTG)223,223,219                                                  DSL45975
  219 CALL FPLOT(2,0.,0.)                                                   DSL45976
  223 CALL UPDAT                                                            DSL45980
      IF(KFINI)277,277,225                                                  DSL45990
  225 DO 226  I=1,KFINI                                                     DSL46000
      I2=IFINL(I)                                                           DSL46010
      IF(C(I2)- FINVA(I))227,227,228                                       DSL46020
  227 JFINL(I)=1                                                            DSL46030
      GO TO 226                                                             DSL46040
  228 JFINL(I)=-1                                                           DSL46050
  226 CONTINUE                                                              DSL46060
  277 KTITL=IABS(KTITL)                                                     DSL46070
      LINE1=KTITL/20+2                                                      DSL46080
      KLINE=52                                                              DSL46090
      IF( KPRNT-10)46,46,66                                                 DSL46100
   66 NLINS=(KPRNT+2)/4+1                                                   DSL46110
      KLINE=54-NLINS                                                        DSL46120
      GO TO 46                                                              DSL46130
C...  LOOP FOR CENTRALIZED INTEGRATION                                     DSL46140
C...  (INTEGRATION TYPE -  RKS,RKS,RKSFX,ADAMS,TRAPZ,NON-CENTRAL)          DSL46150
   31 GO TO (60,60,92,94,95,98),INTYP                                      DSL46160
C...  ADAMS-MOULTON INTEGRATION                                            DSL46170
   94 IF(LINK)377,375,377                                                   DSL46180
  375 DO 376  I=1,NINTG                                                     DSL46190
      Y(I) =Y(I)+0.5*DELT*(3.0*P0(I)-PD(I))                                DSL46200
  376 PD(I)=P0(I)                                                           DSL46210
      TIME=TIME+DELT                                                        DSL46220
      GO TO 390                                                             DSL46230
  377 LINK=0                                                                DSL46240
C...  TRAPEZOIDAL INTEGRATION                                              DSL46250
C...  STORE HISTORY                                                        DSL46260
   95 DO 360  I=1,NINTG                                                     DSL46270
      PD(I)=P0(I)                                                           DSL46280
```

```
  360 YS(I)=Y(I)                                                         DSL46290
C...    PREDICT Y AT T + DELT                                            DSL46300
        TIME=TIME+DELT                                                   DSL46310
        DO 380   I=1,NINTG                                               DSL46320
  380 Y(I)=Y(I)+DELT*PO(I)                                               DSL46330
C...    UPDATE INPUTS                                                    DSL46340
        CALL UPDAT                                                       DSL46350
C...    COMPUTE OUTPUTS OF INTEGRATORS, CORRECT PREVIOUS COMPUTATION     DSL46360
        DC 385   I=1,NINTG                                               DSL46370
  385 Y(I)=YS(I)+0.5*DELT*(PO(I)+PD(I))                                  DSL46380
        GO TO 390                                                        DSL46390
C...    NON-CENTRALIZED INTEGRATION                                      DSL46400
   98 TIME=TIME+DELT                                                     DSL46410
C...    KEEP IS SET = 1 FOR DERIV AND DELAY TO IDENTIFY POINT TO STORE   DSL46420
  390 KEEP=1                                                             DSL46430
        CALL UPDAT                                                       DSL46440
        KEEP=0                                                           DSL46450
        GO TO 46                                                         DSL46460
   60 IF(KPRNT)92,92,350                                                 DSL46470
  350 IF(TIME+1.5*DELT-TPRNT)92,92,61                                    DSL46480
   61 DELT=TPRNT-TIME                                                    DSL46490
        DELTA=1.                                                         DSL46500
C...    RUNGE-KUTTA INTEGRATION    (RKS)                                 DSL46510
   92 KEEP=0                                                             DSL46520
        TSTRT=TIME                                                       DSL46530
        GO TO(190,190,21),INTYP                                          DSL46540
  190 IF(DELTA)192,192,22                                                DSL46550
  192 IF(AMAX-.75) 260,22,265                                           DSL46560
  260 IF(AMAX-.075) 262,22,22                                           DSL46570
   21 DEL=DELT                                                           DSL46580
        GO TO 24                                                         DSL46590
  265 DELT=DELT/RL                                                       DSL46600
   25 IF(DELT-XINT)150,22,22                                             DSL46610
  150 NALRM=28                                                           DSL46614
        CALL ERR                                                         DSL46615
        XINT=0.5*XINT                                                    DSL46616
        GC TO 22                                                         DSL46617
  262 DELT=DELT*RL                                                       DSL46620
   22 DEL=0.5*DELT                                                       DSL46630
        DO 23 I=1,N                                                      DSL46640
   23 SD(I)=0.0                                                          DSL46650
        IFLAG=1                                                          DSL46660
        S=1.                                                             DSL46670
   24 DO 26 I=1,N                                                        DSL46680
        YST(I)=Y(I)                                                      DSL46690
   26 DYST(I)=PO(I)                                                      DSL46700
  264 DC 27 I=1,N                                                        DSL46710
        DELY(I)=DEL*PO(I)                                                DSL46720
   27 PD(I)=DELY(I)                                                      DSL46730
        GO TO(70,70,80),INTYP                                           DSL46740
   70 DC 71 I=1,N                                                        DSL46750
   71 SD(I)=SD(I)+S*PO(I)                                                DSL46760
   80 TIME=TIME+0.5*DEL                                                  DSL46770
        DC 68 I=1,N                                                      DSL46780
        YS(I)=Y(I)                                                       DSL46790
   68 Y(I)=YS(I)+0.5*DELY(I)                                             DSL46800
        CALL UPDAT                                                       DSL46810
        DO 63 I=1,N                                                      DSL46820
        DELY(I)=DEL*PO(I)                                                DSL46830
        PD(I)=PD(I)+2.*DELY(I)                                           DSL46840
   63 Y(I)=YS(I)+0.5*DELY(I)                                             DSL46850
        CALL UPDAT                                                       DSL46860
        DC 62 I=1,N                                                      DSL46870
        DELY(I)=DEL*PO(I)                                                DSL46880
        PD(I)=PD(I)+2.*DELY(I)                                           DSL46890
   62 Y(I)=YS(I)+DELY(I)                                                 DSL46900
        TIME=TIME+0.5*DEL                                                DSL46910
        CALL UPDAT                                                       DSL46920
        DO 280 I=1,N                                                     DSL46930
        DELY(I)=DEL*PO(I)                                                DSL46940
        PD(I)=PD(I)+DELY(I)                                              DSL46950
  280 Y(I)=YS(I)+PD(I)/6.                                                DSL46960
        GC TO(120,120,270),INTYP                                        DSL46970
  120 CALL UPDAT                                                         DSL46980
        GO TO (130,285),IFLAG                                           DSL46990
  130 S=4.                                                               DSL47000
        IFLAG=2                                                          DSL47010
        GO TO 264                                                        DSL47020
```

313

```
  285 AMAX =0.0                                                      DSL47030
      DO 180 I=1,N                                                   DSL47040
      SD(I)=SD(I)+PO(I)                                              DSL47050
      E=ABS(Y(I)-(YST(I)+DEL*SD(I)/3.))/(XABS(I)+R(I)*ABS(Y(I))))   DSL47060
      IF(AMAX-E)175,180,180                                         DSL47070
  175 AMAX=E                                                         DSL47080
  180 CONTINUE                                                       DSL47090
      IF(AMAX-1.) 270,270,290                                       DSL47100
  290 J=1                                                            DSL47110
  251 AM=AMAX/10.**J                                                 DSL47120
      IF(1.-AM) 255,257,257                                         DSL47130
  255 J=J+1                                                          DSL47140
      GO TO 251                                                      DSL47150
  257 TIME=TSTRT                                                     DSL47160
      DELT=DELT/(RL**J)                                             DSL47170
      DO 295 I=1,N                                                   DSL47180
      PO(I)=DYST(I)                                                  DSL47190
  295 Y(I)=YST(I)                                                    DSL47200
      GO TO 25                                                       DSL47210
  270 KEEP=1                                                         DSL47240
      CALL UPDAT                                                     DSL47250
      DELTA=0.                                                       DSL47260
   46 TIMEX=TIME+0.5*DELT                                            DSL47270
      IF(KRANG)42,42,167                                            DSL47280
  167 DO 168  I=1,KRANG                                              DSL47290
      I2=IRANG(I)                                                    DSL47300
      IF(RMIN(I)-C(I2))170,170,169                                  DSL47310
  169 RMIN(I)=C(I2)                                                  DSL47320
      IF(TIME)170,170,168                                           DSL47330
  170 IF(RMAX(I)-C(I2))171,168,168                                  DSL47340
  171 RMAX(I)=C(I2)                                                  DSL47350
  168 CONTINUE                                                       DSL47360
   42 IF( KPRNT)44,44,45                                             DSL47370
   45 IF(TIMEX-TPRNT)44,34,34                                        DSL47380
   34 KODE=1                                                         DSL47390
      CALL PRINT                                                     DSL47400
      CPT=DPT+1.                                                     DSL47410
      TPRNT =DPT*DELTP+PLAST                                         DSL47420
   44 IF(DELTG)48,48,33                                              DSL47430
   33 IF(TIMEX-TPLOT)48,36,36                                        DSL47440
   36 CALL FPLOT(0,SCAL1*C(INDX1),SCAL2*C(INDY))                     DSL47450
      DFL=DPL+1.                                                     DSL47460
      TPLOT=DPL*DELTG+TLAST                                          DSL47470
   48 IF(TIMEX-FINTI)229,3,3                                         DSL47480
  229 IF(KFINI)224,224,222                                           DSL47490
  222 DO 235  I=1,KFINI                                              DSL47500
      I2=IFINL(I)                                                    DSL47510
      IF( JFINL(I))231,232,232                                      DSL47520
  231 IF(C(I2)- FINVA(I))240,240,235                                DSL47530
  232 IF(C(I2)- FINVA(I))235,240,240                                DSL47540
  235 CONTINUE                                                       DSL47550
C...  TEST IF INTERRUPT FOR TYPEWRITER INPUT                         DSL47560
  224 CALL REMSW(1,IRD)                                              DSL47570
      GO TO(3,31),IRD                                                DSL47580
  240 KODE=2                                                         DSL47590
      GO TO 215                                                      DSL47600
    3 KODE=3                                                         DSL47605
  215 IF(TPRNT)213,213,212                                          DSL47610
  213 C(KMEM)=TIME                                                   DSL47620
      GO TO 214                                                      DSL47630
  212 C(KMEM)=TPRNT-DELTP                                           DSL47640
  214 C(KMEM+1)=TIME                                                DSL47650
      CALL PRINT                                                     DSL47655
      IF(DELTG)37,37,35                                             DSL47660
   35 CALL FPLOT(0,SCAL1*C(INDX1),SCAL2*C(INDY))                     DSL47670
      CALL FPLOT(1,0.,0.)                                            DSL47680
   37 CALL REMSW(3,IRD)                                              DSL47705
      GO TO(236,237),IRD                                            DSL47706
  236 CALL TYPIN                                                     DSL47707
      GO TO 200                                                      DSL47708
  237 RETURN                                                         DSL47710
      END                                                            DSL47720
*STORE       UA       RKS                                            DSL47740
// FOR SIM                                                           DSL48170
*LIST ALL                                                            DSL48190
*ONE WORD INTEGERS                                                   DSL48200
*NONPROCESS PROGRAM                                                  DSL48210
*IOCS(KEYBOARD,TYPEWRITER,CARD,1443 PRINTER)                         DSL48220
```

314

```
*NAME SIM                                                          DSL48230
**SIM      MAIN ROUTINE FOR ANALOG SIMULATICN PROGRAM             DSL48240
      COMMON NALRM,NINTG,NSYMB,KPT,KPRNT,KRANG,KFINI,KTITL,KREL,KABS,  DSL48250
     1 KMEM,IDUMY(393),DELTG,JDUMY(428),C(300)                     DSL48260
      IF(KPT-INPT(3))1,3,1                                         DSL48270
    1 KPT=INPT(3)                                                  DSL48280
      CALL INITL                                                   DSL48290
    3 CALL INTRA                                                   DSL48300
      IF(CELTG)5,5,4                                               DSL48310
    4 CALL LINK(PLOTG)                                             DSL48320
    5 CALL LINK(SIMUL)                                             DSL48330
      END                                                          DSL48340
*STORE      UA      SIM                                            DSL48360
// JOB
// ASM
*LIST
                        *              BRUSH INCREMENTAL SUPPORT PACK    BRU00000
                        *              FEATURES SUPPORTED                BRU00010
                        *                  1. INCREMENTAL CHART DRIVE      00020
                        *                  2. COS ANALOG OUTPUT          BRU00030
                        *                  3. TIME MARKERS(EVENT MRKRS)  BRU00040
                        *                                                BRU00050
                        *              CALL PEN(N,KVAL)                  BRU00060
                        *                  N=0-15=PENNUMBER              BRU00070
                        *              CALL MOVE(MARK)                   BRU00080
                        *                  MARK=1 DEFLECT TIME MARKER    BRU00090
                        *                  MARK=0 STEP CHART ONLY        BRU00090
                        *                  MARK=-1 DEFLECT AND           BRU00100
                        *                         RESET TIME CTR         BRU00110
                        *           C3 UNTER DEFLECTS EVERY TENTH MARKER
                        *                         TO THE RIGHT
                        *              CALL SPEED(KRAT)
                        *                  KRAT =1 STEP/2
                        *                  KRAT=2 STEP/4
                        *                  KRAT=4  STEP/10
                        *                  KRAT=8 STEP/100
                        *
                             ENT     PEN
                             ENT     MOVE
                             ENT     SPEED
                        *              SYSTEM DEFINED LAVELS
                        DAO  EQU     /6000       DAO AREA
                        WRIT EQU     /0100       FUNCTION WRITE
                        CONT EQU     /0480       DAO CONTROL PULSE TIMER ON
                        SENS EQU     /0700       FUNCTION SENSE DSW
                        *              USER ASSIGNED ADDRESSES
                        SAMPL EQU    125         COS SAMPLE PULSE ADDRESS
                        ACAC EQU     /0         COS DAC ADDRESS
                        BCONT EQU    126         BRUSH PULSE CONTROL
                        BACV EQU     /8000       BIT 0 IS ADVANCE PULSE
                        BEVTL EQU    /0A00       EVENT LEFT DEFL BITS
                        BEVTR EQU    /0500       EVENT RIGHT DEFL BITS
                        BTRAN EQU    15          HIGH ORDER BIT FOR TRANS CONT
                        AECO EQU     111         ECO ADDRESS FOR TRANS CONT
                        *
                        PEN  DC      0
                             BSI     SAVE
                             LDX  I1 PEN
                             LD   I1 0           GET IPEN
                             AND     FOX
                             OR      SRA         MAKE SHIFT INSTRUN
                             STO     *+1
                             LD      H8000       LOAD SAMPLE BIT 0
                             DC      *-*
                             STO     BIT
                             LD   I1 1           GET KVAL
                             SLA     5
                             STO     DACV
                             XIO     STAT        TEST FOR PO TIMER ON
                             BSC     Z
                             MDX     *-3         STALL IF BUSY
                             LD      DLY         STALL.5 MS ANYWAY
                             S       ONE
                             BSC     Z-
                             MDX     *-3
                             XIO     DAC         SET DAC
                             XIO     PO          HIT SAMPLE
                             XIO     POT         START TIMER
```

315

```
         XIO    STAT        RESET DSW
         MDX  1 2           RETURN ADDRESS IN XR1
         STX  1 RET+1
RESTO LDX  L1 *-*
         LDD    AQ
         LDS    *-*
RET   BSC  L  *-*           RETURN ADDRESS
SAVE  DC     0
         STD    AQ
         STX  1 RESTO+1
         STS    RESTO+3
         BSC  I SAVE
*
SPEED DC     0
         BSI    SAVE
         LDX  I1 SPEED
         LD   I1 0          GET SPEED BIT PATTERN
         SLA    15-BTRAN    POSITION PROPERLY
         STO    BIT
         XIO    ECO
         MDX  1 1           RETURN ADDRESS
         STX  1 RET+1
         MDX    RESTO
*
MOVE  DC     0
         BSI    SAVE
         LDX  I1 MOVE
         LD   I1 0          GET MARK
         BSC  L  ADVAN,+-   IF ZERO, GO MOVE PAPER
         BSC    -           IF-, RESET COUNTER
         MDX    *+4         IF+ COUNT
         LD     TEN         DELECT RIGHT
         STO    CNTR        RESER COUNTR
         LD     RDEFL
         MDX    EVNT        GO MOVE MARKER
         LD     CNTR        COUNT MOVES
         S      ONE
         BSC    +
         MDX    *-8         IF 0 THROW RIGHR
         STO    CNTR
         LD     LDEFL
EVNT  STO    BIT           SAVE DEFL REGISTER
         MDX    *+2
ADVAN SLA    16            CLEAR  FOR NO MARKER
         STO    BIT
         LD     BIT         PREPAR TO MOVE CHART
         OR     ADBIT
         STO    BIT
         XIO    STAT        IS TIMER ON
         BSC    Z           NO-PROCEED
         MDX    *-3         YES-STALL OUT
         XIO    XMOVE
         LD     DLY2        WAIT 15-20 MS
         S      ONE         FOR PENS TO MOVE
         BSC    Z-
         MDX    *-3
         XIO    POT         START TIMER
         MDX  1 1           RETURN ADDRESS
         STX  1 RET+1
         MDX    RESTO
FOX   DC     /000F         HIGH ORDER BITS
SRA   DC     /1800         SRA OP CODE
H8000 DC     /8000         BIT 0
BIT   DC     0             DO BIT REGISTER STORAGE
DACV  DC     0             AO STORAGE
DLY   DC     100           .5 MS DLY CNST
ONE   DC     1
TEN   DC     10
CNTR  DC     0
RDEFL DC     BEVTR
LDEFL DC     BEVTL
DLY2  DC     3000
ADBIT DC     BADV
AQ    BSS  E 2             SAVE A AND Q
STAT  DC     0             IOCC FOR SENSE DSW
      DC     DAO+SENS
DAC   DC     DACV          IOCC FOR AO
      DC     DAO+WRIT+ADAC
```

```
                    PO      DC      BIT         IOCC FOR COS SAMPLE
                            DC      DAO+WRIT+SAMPL
                    POT     DC      0           IOCC TO START PO TIMER
                            DC      DAO+CONT
                    ECO     DC      BIT         IOCC FOR TRANS CONT
                            DC      DAO+WRIT+AECO
                    XMOVE   DC      BIT         IOCC FOR BRUSH PO CONT
                            DC      DAO+WRIT+BCONT
                            END
// JOB
// *PRIOR TO LOADING DSL/1800,SELECT DESIRED VERSION OF INTEG
// *AND INSET IN DECK FOLLOWING INTRA
// *PRESS START TO CONTINUE
// PAUSE
// * STORE DSL/1800 SUBROUTINES
// *UNLESS OTHERWISE NOTED, ALL SUBROUTINES AT AT LEVEL 0
// *REVISED SUBROUTINES HAVE DATE OF REVISION ON STORE CARD
// DUP
*DUMP       UA      ERR1
*DUMP       UA      INPT
*DUMP       UA      NUMRC
*DUMP       UA      OPEN
*DUMP       UA      OUTP
*DUMP       UA      READ1
*DUMP       UA      REMSW
*DUMP       UA      SCAN
*DUMP       UA      SORT
*DUMP       UA      START
*DUMP       UA      TRAN1
*DUMP       UA      TRANS
*DUMP       UA      TRDMP
*DUMP       UA      ALPHA
*DUMP       UA      ERASE
*DUMP       UA      ERR
*DUMP       UA      FXINT
*DUMP       UA      INITL
*DUMP       UA      INTRA
*DUMP       UA      INTRK
*DUMP       UA      MILNE
*DUMP       UA      NUMER
*DUMP       UA      PLOTI
*DUMP       UA      PLOTR
*DUMP       UA      PRINT
*DUMP       UA      READ
*DUMP       UA      RKS
*DUMP       UA      SETUP
*DUMP       UA      SIM
*DUMP       UA      SPLIT
*DUMP       UA      TYPIN
*DUMP       UA      AFGEN
*DUMP       UA      ALOGX
*DUMP       UA      AMAX1
*DUMP       UA      AMIN1
*DUMP       UA      AMOD
*DUMP       UA      AND
*DUMP       UA      CMPXP
*DUMP       UA      COMPR
*DUMP       UA      DEADS
*DUMP       UA      DEBUG
*DUMP       UA      DELAY
*DUMP       UA      DERIV
*DUMP       UA      FCNSW
*DUMP       UA      HSTRS
*DUMP       UA      IMPL
*DUMP       UA      INSW
*DUMP       UA      INTGR
*DUMP       UA      LEDLG
*DUMP       UA      LIMIT
*DUMP       UA      LOOK
*DUMP       UA      MODIN
*DUMP       UA      NLFGN
*DUMP       UA      NORML
*DUMP       UA      OUTSW
*DUMP       UA      PULSE
*DUMP       UA      QNTZR
*DUMP       UA      RAMP
*DUMP       UA      REALP
*DUMP       UA      RST
```

```
*DUMP          UA      SINE
*DUMP          UA      STEP
*DUMP          UA      TRNFR
*DUMP          UA      UM1P1
*DUMP          UA      ZHOLD
// JOB    X     X
// DUP
*DELETE                1
*DELETE                1  ERR1
*DELETE                1  INPT
*DELETE                1  NUMRC
*DELETE                1  OPEN
*DELETE                1  OUTP
*DELETE                1  READ1
*DELETE                1  REMSW
*DELETE                1  SCAN
*DELETE                1  SORT
*DELETE                1  START
*DELETE                1  TRAN1
*DELETE                1  TRANS
*DELETE                1  TRDMP
*DELETE                1  ALPHA
*DELETE                1  ERASE
*DELETE                1  ERR
*DELETE                1  FXINT
*DELETE                1  INITL
*DELETE                1  INTRA
*DELETE                1  INTRK
*DELETE                1  MILNE
*DELETE                1  NUMER
*DELETE                1  PLOTI
*DELETE                1  PLOTR
*DELETE                1  PRINT
*DELETE                1  READ
*DELETE                1  RKS
*DELETE                1  SETUP
*DELETE                1  SIM
*DELETE                1  SPLIT
*DELETE                1  TYPIN
*DELETE                1  AFGEN
*DELETE                1  ALOGX
*DELETE                1  AMAX1
*DELETE                1  AMIN1
*DELETE                1  AMOD
*DELETE                1  AND
*DELETE                1  CMPXP
*DELETE                1  COMPR
*DELETE                1  DEADS
*DELETE                1  DEBUG
*DELETE                1  DELAY
*DELETE                1  DERIV
*DELETE                1  FCNSW
*DELETE                1  HSTRS
*DELETE                1  IMPL
*DELETE                1  INSW
*DELETE                1  INTGR
*DELETE                1  LEDLG
*DELETE                1  LIMIT
*DELETE                1  LOOK
*DELETE                1  MODIN
*DELETE                1  NLFGN
*DELETE                1  NORML
*DELETE                1  OUTSW
*DELETE                1  PULSE
*DELETE                1  QNTZR
*DELETE                1  RAMP
*DELETE                1  REALP
*DELETE                1  RST
*DELETE                1  SINE
*DELETE                1  STEP
*DELETE                1  TRNFR
*DELETE                1  UM1P1
*DELETE                1  ZHOLD
// JOB
// DUP
*DEFINE PAKDK 1
// JOB    X     X
// END
```

```
// FOR INPT                                                            DSL00123
*LIST ALL                                                              DSL00125
*ONE WORD INTEGERS                                                     DSL00126
**ROUTINE TO SET LOGICAL UNIT NUMBERS OF I/O                           DSL00127
      FUNCTION INPT(K)                                                 DSL00128
      GO TO(1,2,3,4,20,6,7,8,20),K                                     DSL00129
C...   SET LOGICAL UNIT FOR TYPEWRITER(1),CARD PUNCH(2),PRIMARY OUTPUT  DSL00130
C...   DEVICE(3), KEYBOARD(6), PLOTTER(7), CARD READER(8)              DSL00131
    1 N=1                                                              DSL00132
      GO TO 20                                                         DSL00133
    2 N=2                                                              DSL00134
      GO TO 20                                                         DSL00135
    3 N=3                                                              DSL00136
      GO TO 20                                                         DSL00137
C...   SET NUMBER OF COMMON STORAGE LOCS. TO BE USED IN SIMULATION MODEL DSL00138
    4 N=300                                                            DSL00139
      GO TO 20                                                         DSL00140
    6 N=6                                                              DSL00141
      GO TO 20                                                         DSL00142
    7 N=7                                                              DSL00143
      GO TO 20                                                         DSL00144
    8 N=2                                                              DSL00145
   20 INPT=N                                                           DSL00151
      RETURN                                                           DSL00152
      END                                                              DSL00153
*STORE        UA        INPT                                           DSL00156
// FOR INTEG                                                           DSL33210
*LIST ALL                                                              DSL33220
*ONE WORD INTEGERS                                                     DSL33230
*NONPROCESS PROGRAM                                                    DSL33240
*IOCS(KEYBOARD,TYPEWRITER,1443 PRINTER,PLOTTER)                        DSL33250
**MAIN ROUTINE FOR INTEGRATION CONTROL                                 DSL33260
*NAME INTEG                                                            DSL33270
      COMMON NALRM,NINTG,NSYMB,KPT,KPRNT,KRANG,KFINI,KTITL,KREL,KABS,  DSL33280
     1 KMEM,KPOIN,IDUMY(822),C(300)                                    DSL33290
      K1=KPOIN                                                         DSL33300
      K2=KPOIN+NINTG                                                   DSL33310
      CALL   RKS(C(KMEM),C(KABS),C(KREL),C(K1),C(K2))                  DSL33320
      CALL LINK(SIMX)                                                  DSL33330
      END                                                              DSL33340
/*
SCOPE    .02,8,6,DISPL,LOAD                                            DSL90890
LABEL    DSL/1800                                                      DSL90900
SCALE    1.,  2.0                                                      DSL90910
// JOB   X      X            DSL/1800 SAMPLE PROBLEM
// XEQ DSL    FX  DIGITAL SIMULATION PROGRAM
SCOPE .05,8,6,TIME,VC
SCOPE .05,8,6,TIME,ERROR
// JOB                                                                 DSL90940
// FOR                                                                 DSL90950
*LIST ALL                                                             DSL90960
*ONE WORD INTEGERS                                                    DSL90970
**DSL/1800 (DIGITAL SIMULATION LANGUAGE)                              DSL90980
      SUBROUTINE UPDAT                                                DSL90990
      REAL  LOAD ,INPUT                                               DSL91000
      INTEGER    J                                                    DSL91010
      COMMON NALRM,IZZZ(833) ,DISPL,DELP ,DELMI,FINTI,DELTP,DELTC,LOAD ,DSL91020
     1ZZ001,ZZ002,ZZ003,ERROR,FGAIN,INPUT,INZZ0,J                     DSL91030
     1 ,ZZ990(  2) ,A   (  5) ,B    (  2)                             DSL91040
      GO TO(3,7),J                                                    DSL91050
    3 A(5)=1.0                                                        DSL91060
      A(4)=.864/.016                                                  DSL91070
      A(3)=3.268/.016                                                 DSL91080
      A(2)=3.42/.016                                                  DSL91090
      A(1)=1./.016                                                    DSL91100
      J=2                                                             DSL91110
    7 CONTINUE                                                        DSL91120
      ERROR=INPUT-FGAIN*LOAD                                          DSL91130
      LOAD=TRNFR(1,4,B,A,ERROR,LOAD)                                  DSL91140
      WW=DEBUG(5,0.)                                                  DSL91150
      RETURN                                                          DSL91160
      END                                                             DSL91170
// DUP                                                                 DSL91180
*DELETE                SIMUL                                           DSL91190
*STORECIX    UA      0 SIMUL INTRK                                     DSL91200
*LOCALPRINT,(FPLOT,XYPLT,PLOTI)                                        DSL91210
*CCEND                                                                 DSL91220
// JOB                                                                 DSL91230
```

319

```
// XEQ SIMX    FX                                                      DSL91240
    4  16  24                                                          DSL91250
DISPL DELP  DELMI FINTI DELTP DELTC LOAD  ZZ001 ZZ002 ZZ003 ERROR FGAIN DSL91260
INPUT J      A      B                                                  DSL91270
PARAM   A     = 17 ,B    = 22                                          DSL91280
TITLE   DYSAC TEST PROB. USING TRNFR                                   DSL91290
TITLE   DSL/1800      8/25/66                                          DSL91300
INCON   J=1                                                            DSL91310
PARAM   INPUT=20.4, FGAIN= 10.                                         DSL91320
TABLE    B(1-2)=312.5, 375.                                            DSL91330
INTEG   RKS                                                            DSL91340
RELERR  LOAD=1.E-4                                                     DSL91350
ABSERR  LOAD=1.E-3                                                     DSL91360
CONTRL DELP=.005,  FINTI=6.0                                           DSL91370
PRINT   0.2,ERROR,LOAD,DELP                                            DSL91380
GRAPH   .02,8,6,DISPL,LOAD                                             DSL91390
LABEL   DSL/1800                                                       DSL91400
SCALE   1., 2.0                                                        DSL91410
END                                                                    DSL91420
STOP                                                                   DSL91430
```

320

B
LISTINGS FOR DSL
SYSTEM/3 MODULES

```
// CALL FORTRN,F1
// RUN
*PROCESS OBJECT(R,LIB(R1))
       SUBROUTINE PACK
       IMPLICIT INTEGER*2(I-N)
       DIMENSION KX(6)
       GLOBAL IDM(6),KA(6),LETRS(72),IDMY(1808),KFORT(2)
       M=0
       DO 1 I=1,6
     1 KX(I)=KA(I)
       DO 2 I=1,5,2
     2 CALL SHIFT(KX(I),KX(I+1),M)
       KA(1)=KX(1)
       KA(2)=KX(3)
       KA(3)=KX(5)
       RETURN
       END
/*
// CALL FORTRN,F1
// RUN
*PROCESS OBJECT(R,LIB(R1))
       SUBROUTINE START
       IMPLICIT INTEGER*2(I-N)
       DIMENSION JDATA(105),JDAT2(53),JDAT1(66),JCON1(10),JCNST(12)
       DIMENSION KDATA(4)
       GLOBAL IDM(6),KA(6),LT(73),ICHAR(12),IRES(158),KSYMB(600),KX(962),
      1 KMEM(15),KDMY(269)
       DATA KDATA/'CA','LL','D ',ZFF00/
       DATA JDATA/'PA','RA','M ','CO','NS','T ','IN','CO','N ','AF','GE',
      1 'N ', 'NL','FG','EN','TA','BL','E ','PR','OC','ED','ME','MO','RY'
      2,'ST','OR','AG','IN','TG','ER','RE','NA','ME','IN','TG','RL','PR'
      3'IN','T ','CO','NT','RL','RE','LE','RR','AB','SE','RR','RA','NG',
      4'E ','FI','NI','SH','EN','D ','  ','ST','OP','  ','SO','RT','  ',
      5'NO','SO','RT','EN','DP','RO','TI','TL','E ','SC','OP','E ','GR',
      6'AP','H ','LA','BE','L ','SC','AL','E ','CO','NT','IN','IN','TE',
      7'G ','RE','SE','T ','CH','AR','T ','TY','PE','  ','DU','MP','  ',
      8'*L','OC','AL'/
       DATA JDAT2/'$$','  ','  ','  ','IF','(N','AL','RM',') ','92','00',
      1          '  ','92','00','  ','93','00','$$','93','00','  ','CO',
      2          'NT','IN','UE','$$','  ','  ','  ','GO','T','0 ','91',
      3          '00','$$','92','00','  ','CO','NT','IN','UE','$$','ZZ',
      4          '99','0 ','RE','AL','IN','TE','GE','R ','  1'/
       DATA JDAT1/'TI','ME','  ','DE','LT','  ','DE','LM','I ','FI','NT',
      1'I ','DE','LT','P ','DE','LT','C ','IN','TG','R ','RE','AL','P ',
      2'MO','DI','N ','TR','NF','R ','CM','PX','P ','LE','DL','G ','DE',
```

```
       3'BU','G ','DE','RI','V ','HS','TR','S ','IM','PU','L ','PU','LS',
       4'E ','RS','T ','   ','ZH','OL','D ','IM','PL','  ','IM','PL','C ',
       5'DE','LA','Y '/
       DATA JCON1/ 1,6,2,3,2,1,1,2,3,0/,JCNST/107,77,108,93,76,123,78,80,
      1        96,92,75,97/
C      INITIALIZE IRES **
       J=1
       DO 400 I=1,35
       IRES(I)=JDATA(J)
       IRES(I+35)=JDATA(J+1)
       IRES(I+70)=JDATA(J+2)
  400  J=J+3
       DO 401 I=1,53
  401  IRES(I+105)=JDAT2(I)
C      INITIALIZE KSYMB. NOTICE DIMENSION OF KSYMB INCREASED TO 600.
       J=1
       DO 402 I=1,22
       KSYMB(I     )=JDAT1(J)
       KSYMB(I+200)=JDAT1(J+1)
       KSYMB(I+400)=JDAT1(J+2)
  402  J=J+3
C      INITIALIZE ICHAR AND KMEM
       DO 403  I=1,12
  403  ICHAR(I)=JCNST(I)
       DO 404 I=1,10
  404  KMEM(I)=JCON1(I)
       DO 500 I=1,4
  500  KA(I)=KDATA(I)
       RETURN
       END
/*
// CALL FORTRN,F1
// RUN
*PROCESS OBJECT(R,LIB(R1))
       SUBROUTINE BUILD
       IMPLICIT INTEGER*2(I-N)
       GLOBAL IDM(222),IWRD(3),KA(6),LETRS(72)
       M=0
       CALL SHIFT(KA(1),KA(2),M)
       IWRD(1)=KA(1)
       CALL SHIFT(KA(3),KA(4),M)
       KA(2)=KA(3)
       IWRD(2)=KA(3)
       CALL SHIFT(KA(5),KA(6),M)
       IWRD(3)=KA(5)
       KA(3)=KA(5)
       RETURN
       END
/*
// LOAD $ASSEM,F1
// FILE NAME-$SOURCE,UNIT-R1,PACK-R1R1R1,TRACKS-20,RETAIN-S
// FILE NAME-$WORK,UNIT-R1,PACK-R1R1R1,TRACKS-20,RETAIN-S
// FILE NAME-$WORK2,UNIT-R1,PACK-R1R1R1,TRACKS-15,RETAIN-S
// COMPILE OBJECT-R1
// RUN
 OPTIONS LIST,XREF,REL,OBJ(P)
*          FORTRAN CALLABLE ASSEMBLY ROUTINE REQUIRED FOR SYSTEM/3 DSL
*          ASSEMBLY LANGUAGE SUBROUTINE TO PACK TWO WORDS TO ONE OR UNPACK
*          ONE WORD TO TWO OR TRANSFER  RIGHT HALF OF SECOND WORD TO FIRST
@XR1       EQU     X'01'
@XR2       EQU     X'02'
@ARR       EQU     X'08'
```

322

```
@IAR       EQU    X'10'
           ENTRY  SHIFT
SHIFT      START  X'0000'
           ST     SAVAR1,@XR1
           LA     SAVA,@XR1
           USING  SAVA,@XR1
           ST     SAVAR2(,@XR1),@XR2
           ST     SAVART(,@XR1),@ARR
           L      SAVART(,@XR1),@XR2
           L      1(,@XR2),@XR2
           ALC    SAVART(,@XR1),NPAR(,@XR1)      RETURN POINT UPDATE
           L      3(,@XR2),@XR1         REG 1 HAS FIRST PARAM ADDRS
           MVC    WK1(2),1(,@XR1)       FIRST PARM NOW IN WK STRG 1
           L      5(,@XR2),@XR1         REG 1 HAS FIRST PARAM ADDRS
           MVC    WK2(2),1(,@XR1)       SECOND PARM IN WRK STRG 2
*          CHECK  IF THE THIRD WORD IS 0,1 OR 2. IF 0 PACK TWO WORDS TO ONE
*          IF IT  IS 2 OR MORE  SHIFT RIGHT HALF OF SECOND WORD TO RT HALF OF
*          FIRST  WORD. IF IT IS 1 UNPACK ONE WORD TO TWO.
           L      7(,@XR2),@XR1         GET ADDRESS OF THIRD PARM IN XR1
           CLC    1(,@XR1),COMPAR       CHECK IF THIRD PARM IS >,= OR< ONE
           JL     PACK                  LESS THAN ONE
           JH     TRNSFR                MORE THAN ONE
UNPACK     MVC    WK2,WK1(1)            MOVE RIGHT HALF OF 1ST PARM TO RT HF OF 2ND P
           MVC    WK1(1),WK1-1          MOVE LEFT HALF OF 1ST PARM TO RIGHT HALF
           MVC    WK1-1,ZERO(1)         ZERO LEFT HALF OF 1ST PARM WORD
           MVC    WK2-1,ZERO(1)         ZERO LEFT HALF OF SECOND PARM
           J      OUT
PACK       MVC    WK1-1,WK1(1)          MOVE RT HALF OF FIRST WORD TO LEFT
           MVC    WK1,WK2(1)            MOVE RT HALF OF 2ND PARM TO RT HALF OF 1ST PA
           J      OUT
TRNSFR     MVC    WK1,WK2(1)            MOVE RT HALF OF 2ND PARM TO RT HALF OF 1ST PA
OUT        L      3(,@XR2),@XR1         GET 1ST PARM ADDR IN REG 1
           MVC    1(,@XR1),WK1(2)       STORE NEW PARM IN LOCATION
           L      5(,@XR2),@XR1
           MVC    1(,@XR1),WK2(2)       STORE NEW PARM VALUE IN PARM 2 LOCATION
           LA     SAVA,@XR1
           L      SAVAR2(,@XR1),@XR2    RESTORE XR2
           L      SAVAR1(,@XR1),@XR1    RESTORE 1
           L      SAVART,@IAR          RETURN
SAVA       DC     XL1'30'
           DC     CL6'SHIFT'
SAVAR1     DC     XL2'00'
SAVAR2     DC     XL2'00'
SAVART     DC     AL2(00)
NPAR       DC     IL2'2'
ZERO       DC     XL1'00'
COMPAR     DC     XL2'0001'
WK1        DC     XL2'0000'
WK2        DC     XL2'0000'
           END
/*
// CALL FORTRN,F1
// RUN
*PROCESS OBJECT(R,LIB(R1))
      FUNCTION IABS1(I)
      INTEGER*2 IABS1,I
C     FUNCTION TO SUBSTITUTE S/3 IABS FUNCTION TO BE ABLE TO WORK WITH
C     ONE WORD INTEGERS IN DSL
      I1=I
      IX=IABS(I1)
      IABS1=IX
      RETURN
      END
```

```
/*
// CALL FORTRN,F1
// RUN
*PROCESS OBJECT(R,LIB(R1))
       SUBROUTINE ORGSC(A,B)
       WRITE(3,1)
     1 FORMAT(' NO PLOTTING IN DSL FOR SYSTEM/3')
       RETURN
       END
/*
// CALL FORTRN,F1
// RUN
*PROCESS OBJECT(R,LIB(R1))
       SUBROUTINE FPLOT(M,A,B)
       INTEGER*2 M
       WRITE(3,1)
     1 FORMAT(' NO PLOTTING IN DSL FOR SYSTEM / 3')
       RETURN
       END
/*
// CALL FORTRN,F1
// RUN
*PROCESS OBJECT(R,LIB(R1))
       SUBROUTINE PLOTS(M)
       INTEGER*2 M
       RETURN
       END
/*
// CALL FORTRN,F1
// RUN
// PRINT DEVICE-1403
*PROCESS OBJECT(R,LIB(R1)),LINK(R,LIB(R1))
       PROGRAM PLOTG
       WRITE(3,1)
     1 FORMAT(' PLOTTING NOT POSSIBLE WITH SYSTEM/3. REMOVE GRAPH AND
      1SCALE CARDS')
       INVOKE SIMUL
       END
/*
// CALL FORTRN,F1
// RUN
*PROCESS OBJECT(R,LIB(R1))
       SUBROUTINE SETUP
       IMPLICIT INTEGER*2(I-N)
       INTEGER*2 JDATA(93),JDAT1(8)
       GLOBAL IDM(303),IRES(101)
       DATA JDATA/'PA','RA','M ','IN','CO','N ','CO','NS','T ','PR','IN',
      1'T ','SC','OP','E ','GR','AP','H ','RA','NG','E ','FI','NI','SH ',
      2'RE','LE','RR','AB','SE','RR','LA','BE','L ','CH','AR','T ','TI',
      3'TL','E ','CO','NT','RL','CO','NT','IN','IN','TE','G ','AF','GE',
      4'N ','NL','FG','EN','ST','OP','   ','EN','D ','   ','SC','AL','E ',
      5'TA','BL','E ','RE','SE','T ','TY','PE','   ','EN','**','   ','**',
      6'**','**','MI','LN','E ','RK','S ','   ','RK','SF','X ','AD','AM',
      7'S ','TR','AP','Z '/
       DATA JDAT1/'DS','L/','18','00','OV','ER','LA','Y '/
       DATA IEOF/ZC415/
       J=1
       DO 2 I=1,31
       IRES(I)=JDATA(J)
       IRES(I+31)=JDATA(J+1)
       IRES(I+62)=JDATA(J+2)
     2 J=J+3
```

324

```
            DO 3 I=1,8
        3   IRES(I+93)=JDAT1(I)
            IRES(55)=IEOF
            RETURN
            END
/*
// CALL FORTRN,F1
// RUN
*PROCESS OBJECT(R,LIB(R1))
            SUBROUTINE UNPAK
            IMPLICIT INTEGER*2(I-N)
            RETURN
            END
/*
// CALL FORTRN,F1
// RUN
*PROCESS OBJECT(R,LIB(R1)),MAP
            SUBROUTINE OPEN
            IMPLICIT INTEGER*2(I-N)
            GLOBAL ID(6),KA(6),LETRS(72),IDUMY(1808),KFORT(2)
            DIMENSION LX(72)
            J=1
            DO 100 I=1,36
            M2=1
            M=LETRS(I)
            CALL SHIFT(M,M1,M2)
            LX(J)=M
            LX(J+1)=M1
      100  J=J+2
            DO 101 I=1,72
      101  LETRS(I)=LX(I)
            RETURN
            END
/*
// CALL FORTRN,F1
// RUN
*PROCESS OBJECT(R,LIB(R1))
            SUBROUTINE SPLIT
            IMPLICIT INTEGER*2(I-N)
            DIMENSION LX(80)
            GLOBAL IDM(225),KA(6),LETRS(72)
            M2=1
            J=1
            DO 100 I=1,36
            M=LETRS(I)
            CALL SHIFT(M,M1,M2)
            LX(J)=M
            LX(J+1)=M1
      100  J=J+2
            DO 101 I=1,72
      101  LETRS(I)=LX(I)
            RETURN
            END
/*
// CALL FORTRN,F1
// RUN
*PROCESS OBJECT(R,LIB(R1))
            SUBROUTINE BNDEC
            IMPLICIT INTEGER*2(I-N)
            GLOBAL ID(6),KA(6),LETRS(72),IDUMY(1808),KFORT(128)
            NB1=KA(6)/100
            NB2=KA(6)-NB1*100
```

```
      NB3=NB2/10
      NB4=NB2-NB3*10
      M=1
      IF(KA(5))5,9,6
    5 K=-KA(5)+1
      KA(1)=77
      GO TO 7
    9 KA(1)=64
      K=1
      GO TO 7
    6 K=KA(5)+1
      KA(4)=107
      M=0
    7 KA(M+1)=NB1+240
      KA(M+2)=NB3+240
      KA(M+3)=NB4+240
      CALL PACK
      KFORT(K)=KA(1)
      KFORT(K+1)=KA(2)
      RETURN
      END



// CALL FORTRN,F1
// RUN
*PROCESS OBJECT(R,LIB(R1))
      SUBROUTINE DOUTN(IA,IB,IC)
      IMPLICIT INTEGER*2(I-N)
      RETURN
      END
/*
// CALL FORTRN,F1
// RUN
*PROCESS OBJECT(R,LIB(R1))
      SUBROUTINE ERR1                                          DSL00060
      IMPLICIT INTEGER*2(I-N)
      INTEGER*4 L
      GLOBAL NALRM,IERR,ICONT,JPT,KEY,ISTRT,KA(6),LETRS(72)    DSL00070
      L=JPT                                                    DSL00075
      WRITE(L,8) ISTRT,NALRM                                   DSL00080
    8 FORMAT(12H ERR IN COL I2,9H   DSL ERR I4)                DSL00090
      NALRM=0                                                  DSL00100
      RETURN                                                   DSL00110
      END                                                      DSL00120
```

```
/*
// CALL FORTRN,F1
// RUN
*PROCESS OBJECT(R,LIB(R1))
      FUNCTION INPT(K)                                              DSL00128
      GO TO(1,2,3,4,20,6,7,8,20),K                                  DSL00129
C...  SET LOGICAL UNIT FOR TYPEWRITER(1),CARD PUNCH(2),PRIMARY OUTPUT DSL00130
C...  DEVICE(3), KEYBOARD(6), PLOTTER(7), CARD READER(8)            DSL00131
    1 N=6                                                           DSL00133
      GO TO 20                                                      DSL00134
    2 N=2                                                           DSL00135
      GO TO 20                                                      
    3 N=3                                                           DSL00137
      GO TO 20                                                      
C...  SET NUMBER OF COMMON STORAGE LOCS. TO BE USED IN SIMULATION MODEL DSL00138
    4 N=500                                                         DSL00139
      GO TO 20                                                      DSL00140
    6 N=5                                                           DSL00142
      GO TO 20                                                      
    7 N=3                                                           DSL00144
      GO TO 20                                                      
    8 N=1                                                           DSL00151
   20 INPT=N                                                        DSL00152
      RETURN                                                        DSL00153
      END
/*
// CALL FORTRN,F1
// RUN
*PROCESS OBJECT(R,LIB(R1))
      SUBROUTINE NUMRC                                              DSL00200
      IMPLICIT INTEGER*2(I-N)
      GLOBAL NALRM,IERR,ICONT,JPT,KEY,ISTRT,KA(6),LETRS(72)         DSL00210
      EQUIVALENCE (KA(6),NMBR)                                      DSL00220
      NMBR=0                                                        DSL00230
    3 IF(LETRS(ISTRT)-64)10,4,5                                     DSL00240
    5 IF(LETRS(ISTRT)-240)7,6,6                                     DSL00250
    6 NMBR=10*NMBR+LETRS(ISTRT)-240                                 DSL00260
    4 ISTRT=ISTRT+1                                                 DSL00270
      IF(ISTRT-72)3,3,10                                            DSL00280
```

```
    7 IF(LETRS(ISTRT)-107)8,15,10                                       DSL00290
    8 IF(LETRS(ISTRT)-93)9,15,10                                        DSL00300
    9 IF(LETRS(ISTRT)-76)10,15,10                                       DSL00310
   10 NALRM=4                                                           DSL00320
      CALL ERR1                                                         DSL00330
   15 ISTRT=ISTRT+1                                                     DSL00340
      RETURN                                                            DSL00350
      END                                                               DSL00360
/*
// CALL FORTRN,F1
// RUN
*PROCESS OBJECT(R,LIB(R1))
      SUBROUTINE REMSW(J,K)                                             DSL05630
      IMPLICIT INTEGER*2(I-N)
      INTEGER*4 J1,K1
      J1=J
      CALL DATSW(J1,K1)
      K=K1
      RETURN                                                            DSL05650
      END                                                               DSL05660
/*
// CALL FORTRN,F1
// RUN
*PROCESS OBJECT(R,LIB(R1))
      SUBROUTINE READR                                                  DSL05450
      IMPLICIT INTEGER*2(I-N)
      INTEGER*4 L,KPT
      GLOBAL NALRM,IERR,ICONT,JPT,KEY,ISTRT,KA(6),LETRS(72),ID(2010),   DSL05460
     1 KEYB,KARD,ISC(5)
    1 CALL REMSW(2,IRD)                                                 DSL05480
      GO TO(4,5),IRD                                                    DSL05490
    4 L=KEYB                                                            DSL05500
      GO TO 6                                                           DSL05504
    5 L=KARD                                                            DSL05506
    6 READ(   L,10)(LETRS(K),K=1,36)                                    DSL05510
   10 FORMAT(36A2)                                                      DSL05520
      CALL REMSW(6,K)                                                   DSL05530
      GO TO(12,13),K                                                    DSL05540
   13 KPT = JPT                                                         DSL05545
      WRITE (KPT,11) (LETRS(K),K=1,36)                                  DSL05550
   11 FORMAT(1H  36A2)                                                  DSL05560
   12 RETURN                                                            DSL05570
      END                                                               DSL05580
/*
// CALL FORTRN,F1
// RUN
*PROCESS OBJECT(R,LIB(R1))
      SUBROUTINE SCAN                                                   DSL05740
      IMPLICIT INTEGER*2(I-N)
      GLOBAL NALRM,IERR,ICONT,JPT,KEY,ISTRT,KA(6),LETRS(73),ICHAR(12),  DSL05750
     1 IRES(158),KSYMB(600),NSYMB,ITYPE,NOUT,NINP,NPAR,NINT,NDATA,NFOR, DSL05760
     2 NSORT,ISORT,NMEM,NING,NINGL,NSTOR,NTSEQ,NREAL,MFOR,              DSL05770
     3 KOUT(100),LFORT(100),NFORT(100),LIN(100),NIN(100),INPUT(300),    DSL05780
     4 KPAR( 60),KSTOR(10),LINT(50),INGER(10),MEMRY(15),KMEM(15),       DSL05790
     5 LSTOR(10),INGRL(10),KREAL(20),KSORT(20)                          DSL05800
      GLOBAL ID(204),KEY1,KPREV,KEXP,KPT,J                              DSL05805
C...  SUBROUTINE TO SELECT ALPHA-NUMERIC NAME FROM LETRS(ISTRT) WITH    DSL05810
C...  EQUALS, COMMA, LEFT OR RIGHT PARENTHESIS, END OF CARD, OR         DSL05820
C...  AN ARITHMETIC OPERATOR AS SEPARATOR                               DSL05830
C...   (... WILL FORCE READ + CONTINUATION)                             DSL05840
```

328

```
// CALL FORTRN.F1
// FILE NAME-$SOURCE,RETAIN-S,PACK-F2F2F2,UNIT-F2,TRACKS-50
// FILE NAME-$WORK,RETAIN-S,PACK-F2F2F2,UNIT-F2,TRACKS-50
// RUN
// DAD44 UNITNO-'11,12'
// PUNCH DEVICE-'MFCU2'
// PRINT DEVICE-'1403,5471'
// READ DEVICE-'MFCU1,5471'
*PROCESS OBJECT(R,LIB(R1)),NOLINK,NOSHRBUFF
        PROGRAM DSL
        IMPLICIT INTEGER*2(I-N)
        INTEGER*4 MAXN1,INPT
        DEFINE FILE 11(5000,1,U,MFOR)
        DEFINE FILE 12(1000,1,U,NDATA)
        GLOBAL NALRM,IERR,ICONT,JPT,KEY,ISTRT,KA(6),LETRS(73),ICHAR(12),    DSL11589
       1 IRES(158),KSYMB(600),NSYMB,ITYPE,NOUT,NINP,NPAR,NINT,NDATA,NFOR,   DSL11590
       2 NSORT,ISORT,NMEM,NING,NINGL,NSTOR,NTSEQ,NREAL,MFOR,                DSL11591
       3 KOUT(100),LFORT(100),NFORT(100),LIN(100),NIN(100),INPUT(300),      DSL11592
       4 KPAR( 60),KSTOR(10),LINT(50),INGER(10),MEMRY(15),KMEM(15),         DSL11593
       5 LSTOR(10),INGRL(10),KREAL(20),KSORT(20),                          DSL11594
       6      KFORT(128),KSPEC(36),LOCAL(36),NLOC,NSPEC,KEYB,KARD,ISC(5)    DSL11595
        EQUIVALENCE (IRES(106),IDOLR),(IRES(107),IBLNK),(KA(4),LOC),        DSL11596
       1 (KA(6),NMBR),(IRES(158),IONE)                                      DSL11597
        DATA MASK3/' ,'/
        DATA MASK2/'  '/
        MM2=2
        KEND=0                                                             DSL11598
        IPRO=0                                                             DSL11600
        NIMP=0                                                             DSL11601
        IMP1=600                                                          DSL11602
        KDUMP=0                                                            DSL11604
        MPL=0                                                             DSL11606
C...    INITIALIZE DATA AREAS, STORE SYMBOL TABLES IN COMMON              DSL11609
        CALL TRAN1                                                         DSL11610
        CALL START                                                        DSL11611
        MAXN1=INPT(4)
        MAXN=MAXN1
        ICA=KA(1)                                                          DSL11613
        LL=KA(2)                                                           DSL11614
        ID=KA(3)                                                           DSL11615
        MASK=KA(4)                                                         DSL11616
C...    START OF LOOP FOR EACH CARD,  READ AND PRINT INPUT                DSL11620
      4 ICONT=0                                                            DSL11622
        ITYPE=0                                                            DSL11624
        MEMR=0                                                             DSL11626
        JINTG=0                                                            DSL11627
      5 CALL READR                                                         DSL11628
        ISTRT=7                                                            DSL11630
        IF(ICONT)63,2,172                                                  DSL11640
      2 IF(NFOR-1)15,15,12                                                 DSL11642
     12 N2=0                                                               DSL11643
        IF(MPL)320,320,310                                                 DSL11644
    310 MPL=0                                                              DSL11645
C...    MULTIPLE OUTPUT BLOCK, INSERT OUTPUTS IN CALL STATEMENT           DSL11646
C...    TEST FOR LEADING BLANK OR ALPHA IN PAIR, OR RIGHT PAREN          DSL11647
        IF(KFORT(NFOR-1))323,323,322                                       DSL11649
C...    INSERT BLANK-COMMA(16491), RIGHT PAREN-BLANK(23872), + COMMA(107) DSL11651
    322 KFORT(NFOR-1)=16491                                                DSL11652
        GO TO 324                                                          DSL11653
    323 CALL SHIFT(KFORT(NFOR-1),MASK3,MM2)
```

```
 324 DO 325  NF= 94,NF8                                            DSL11657
     KFORT(NFOR)=KFORT(NF)                                         DSL11658
 325 NFOR=NFOR+1                                                   DSL11659
     KFORT(NFOR)=23872                                             DSL11660
     GO TO 450                                                     DSL11661
 320 NFOR=NFOR-1                                                   DSL11662
 450 WRITE(11'MFOR)(KFORT(K),K=1,NFOR)
     NFOR=1                                                        DSL11680
     IF(MFOR-5000)15,15,100                                        DSL11685
  15 IF(LETRS(1)-IBLNK)6,70,6                                      DSL11690
C... DETERMINE DATA STATEMENT TYPE   (1-35)                        DSL11700
C... PARAM,CONST,INCON,AFGEN,NLFGEN,TABLE,PROCED,MEMORY,STORAG,INTGER,  DSL11710
C... RENAME,INTGRL,PRINT,CONTRL,RELERR,ABSERR,RANGE,FINISH,END,STOP,    DSL11720
C... SORT,NOSORT,ENDPRO,TITLE,SCOPE,GRAPH,LABLE,SCALE,CONTIN,INTEG,     DSL11730
C... RESET,CHART,TYPE,DUMP,*LOCAL                                  DSL11740
   6 DO 7  I=1,35                                                  DSL11750
     IF(LETRS(1)-IRES(I))7,8,7                                     DSL11760
   8 IF(LETRS(2)-IRES(I+35))7,9,7                                  DSL11770
   9 IF(LETRS(3)-IRES(I+70))7,10,7                                 DSL11780
   7 CONTINUE                                                      DSL11790
C... TEST FOR D IN COL 1                                           DSL11800
     IF(LETRS(1)-ID)70,28,70                                       DSL11805
  10 ITYPE=I                                                       DSL11810
  63 IF(ITYPE-7)14,21,64                                           DSL11820
  64 IF(ITYPE-12)21,21,66                                          DSL11830
  66 IF(ITYPE-21)14,51,67                                          DSL11840
  67 IF(ITYPE-23)52,53,25                                          DSL11850
  25 IF(ITYPE-34)14,36,26                                          DSL11860
  36 KDUMP=1                                                       DSL11864
     GO TO 4                                                       DSL11866
  26 NLOC=1                                                        DSL11870
     DO 27  I=1,36                                                 DSL11880
  27 LOCAL(I)=LETRS(I)                                             DSL11890
     GO TO 4                                                       DSL11900
  28 DO 29  I=2,36                                                 DSL11910
  29 KSPEC(I)=LETRS(I)                                             DSL11920
     KSPEC(1)=IBLNK                                                DSL11930
     NSPEC=1                                                       DSL11940
     GO TO 4                                                       DSL11950
C... STORE DATA CARD AS VARIABLE WORD RECORD WITH $$ LAST WORD     DSL11960
  14 IEND=36                                                       DSL11970
  16 IF(LETRS(IEND)-IBLNK)18,17,18                                 DSL11980
  17 IEND=IEND-1                                                   DSL11990
     GO TO 16                                                      DSL12000
  18 IEND=IEND+1                                                   DSL12010
     LETRS(IEND)=IDOLR                                             DSL12020
     DO 1002 K=1,IEND
1002 WRITE(12'NDATA)LETRS(K)
     IF(NDATA-1000)20,20,100                                       DSL12040
  20 IF(KEND)61,11,24                                              DSL12050
C... NOTE - ADD 30 TO KEY NAME SEQUENCE NO. TO GET FORTRAN STATEMENT NODSL12060
  11 IF(ITYPE-19)21,49,24                                          DSL12070
  24 IF(ITYPE-20)4,50,4                                            DSL12080
  21 CALL OPEN                                                     DSL12090
     IF(ITYPE-12)22,42,43                                          DSL12100
  22 GO TO(31,31,31,34,34,43,37,38,39,40,41),ITYPE                 DSL12110
  31 CALL SCAN                                                     DSL12120
     L=1                                                           DSL12130
     GO TO(61,61,62,61,61,4,48),KEY                                DSL12140
  61 NALRM=6                                                       DSL12150
```

```
           GO TO 99                                          DSL12160
    62 DO 122   M=1,NPAR                                      DSL12170
       IF(LOC-KPAR(M))122,123,122                            DSL12180
   122 CONTINUE                                               DSL12190
       IF(NPAR-60)108,100,100                                DSL12200
   108 NPAR=NPAR+1                                            DSL12210
       KPAR(NPAR)=LOC                                         DSL12220
   123 IF(L-1)61,31,43                                        DSL12230
    34 IF(ICONT)43,65,61                                      DSL12240
    65 CALL SCAN                                              DSL12250
       L=2                                                    DSL12260
       IF(KEY-3)65,62,61                                      DSL12270
C...   PROCEDURAL BLOCK                                       DSL12280
    37 IPRO=1                                                 DSL12290
       NFOR1=MFOR                                             DSL12300
       GO TO 68                                               DSL12310
C...   MEMORY BLOCK SPECIFICATION                            DSL12320
    38 CALL SCAN                                              DSL12330
       IF(KEY-4)61,150,149                                    DSL12340
   149 IF(KEY-5)61,61,4                                       DSL12350
   150 IF(NMEM-15)151,100,100                                 DSL12360
   151 NMEM=NMEM+1                                            DSL12370
       MEMRY(NMEM)=LOC                                        DSL12380
       CALL NUMRC                                             DSL12390
       KMEM(NMEM)=NMBR                                        DSL12400
       GO TO 38                                               DSL12410
C...   STORAG BLOCK SPECIFICATION                            DSL12420
    39 CALL SCAN                                              DSL12430
       IF(KEY-4)61,160,159                                    DSL12440
   159 IF(KEY-5)61,61,4                                       DSL12450
   160 IF(NSTOR)162,162,163                                   DSL12460
   162 NSTOR=2                                                DSL12470
       GO TO 164                                              DSL12480
   163 IF(NSTOR-10)161,100,100                                DSL12490
   161 NSTOR=NSTOR+1                                          DSL12500
   164 KSTOR(NSTOR)=LOC                                       DSL12510
       CALL NUMRC                                             DSL12520
       LSTOR(NSTOR)=NMBR                                      DSL12530
       GO TO 39                                               DSL12540
    40 CALL SCAN                                              DSL12550
       IF(KEY-2)4,110,4                                       DSL12560
   110 IF(NING-10)111,100,100                                 DSL12570
   111 NING=NING+1                                            DSL12580
       INGER(NING)=LOC                                        DSL12590
       GO TO 40                                               DSL12600
C...   RENAME                                                 DSL12610
    41 CALL SCAN                                              DSL12620
       IF(KEY-3)61,112,61                                     DSL12630
   112 IF(LOC-6)113,113,61                                    DSL12640
   113 I=LOC                                                  DSL12650
       CALL SCAN                                              DSL12660
       KSYMB(I)=KSYMB(NSYMB)                                  DSL12670
       KSYMB(I+200)=KSYMB(NSYMB+200)                          DSL12680
       KSYMB(I+400)=KSYMB(NSYMB+400)
       NSYMB=NSYMB-1                                          DSL12690
       IF(ISTRT-72)41,4,4                                     DSL12700
C...   INTEGRAL OUTPUT VARIABLE SPECIFICATION                DSL12710
    42 CALL SCAN                                              DSL12720
       IF(KEY-2)61,170,61                                     DSL12730
   170 IF(NINGL-10)171,100,100                                DSL12740
```

```
    171  NINGL=NINGL+1                                                  DSL12750
         INGRL(NINGL)=LOC                                               DSL12760
         IF(ISTRT-72)72,4,4                                            DSL12770
C...     TEST FOR CONTINUATION OF DATA CARD                            DSL12780
     43  KPT=0                                                         DSL12790
         I=IEND+IEND-2                                                 DSL12800
     44  IF(LETRS(I)-64)4,45,46                                        DSL12810
     45  I=I-1                                                         DSL12820
         GO TO 44                                                      DSL12830
     46  IF(LETRS(I)-75)4,47,4                                         DSL12840
     47  KPT=KPT+1                                                     DSL12850
         IF(KPT-3)45,48,48                                            DSL12860
     48  ICONT=-1                                                      DSL12870
         GO TO 5                                                       DSL12880
C...     END AND STOP CARDS                                            DSL12890
     49  KEND=1                                                        DSL12900
         GO TO 4                                                       DSL12910
     51  IF(ISORT)264,264,4                                           DSL12950
    264  ISORT=1                                                       DSL12960
         IF(NOUT-1)4,4,175                                            DSL12970
    175  NSORT=NSORT+2                                                 DSL12980
         KSORT(NSORT-1)=NOUT                                          DSL12990
         GO TO 4                                                       DSL13000
     52  IF(ISORT)265,4,265                                           DSL13005
    265  ISORT=0                                                       DSL13010
         IF(NOUT-1)176,176,177                                        DSL13020
    176  NSORT=0                                                       DSL13030
         GO TO 4                                                       DSL13040
    177  KSORT(NSORT)=NOUT-1                                          DSL13050
         GO TO 4                                                       DSL13060
     53  IPRU=0                                                        DSL13070
         NFORT(NOUT1)=NFOR1-MFOR                                      DSL13080
         LFORT(NOUT1)=NFOR1                                           DSL13090
         LIN(NOUT1)=NINP1                                             DSL13100
         NIN(NOUT1)=NINP-NINP1+1                                      DSL13110
         GO TO 4                                                       DSL13120
C...     START SCAN OF DSL STRUCTURE STATEMENT                         DSL13130
C...     CONTINUATION OF STRUCTURE STATEMENT                           DSL13140
    172  NB1=NMBR/2                                                    DSL13150
         NFOR=NFORA+NB1+MEMR+1                                        DSL13160
         MEMR=0                                                        DSL13164
         KFORT(NFOR-1)=IDOLR                                          DSL13170
         IF(NMBR-NB1-NB1)94,174,70                                    DSL13180
    174  KFORT(NFOR-2)=KFORT(NFOR-2)-11                               DSL13190
     70  IEND=36                                                       DSL13200
     71  IF(LETRS(IEND)-IBLNK)73,72,73                                DSL13210
     72  IEND=IEND-1                                                   DSL13220
         GO TO 71                                                      DSL13230
     73  NFORA=NFOR                                                    DSL13240
         DO 74  K=1,IEND                                              DSL13250
         KFORT(NFOR)=LETRS(K)                                         DSL13260
     74  NFOR=NFOR+1                                                   DSL13270
         IF(NFOR- 200)116,116,100                                     DSL13280
    116  CALL OPEN                                                     DSL13290
C...     TEST FOR COMMENT (ASTERISK COL. 1), REFERENCE NUMBER, CONTINUATION DSL13300
         IF(LETRS(1)-92)92,91,92                                      DSL13310
     91  NFOR=NFOR-IEND                                               DSL13320
         GO TO 5                                                       DSL13330
     92  DO 95  K=1,5                                                 DSL13340
         IF(LETRS(K)-64)94,95,93                                      DSL13350
```

```
       93 IF(LETRS(K)-240)94,95,95                                      DSL13360
       95 CONTINUE                                                       DSL13370
          IF(ICONT)94,210,211                                           DSL13380
      210 IF(LETRS(6)-64)94,96,94                                       DSL13390
       94 NALRM=9                                                        DSL13400
          GO TO 99                                                       DSL13410
      211 KFORT(NFORA+2)=IONE                                            DSL13420
       96 IF(IPRO)120,121,121                                           DSL13430
      120 KFORT(NFOR)=IDOLR                                              DSL13440
          NFOR=NFOR+1                                                    DSL13450
          GO TO 4                                                        DSL13460
      121 IF(ICONT)94,101,102                                           DSL13470
      102 KFORT(NFOR)=IDOLR                                              DSL13480
          NFOR=NFOR+1                                                    DSL13490
          NFORT(NOUT1)=NFOR1-NFOR                                        DSL13500
          IF(L-2)77,94,82                                                DSL13510
C...   FIND ALL OUTPUT VARIABLES                                         DSL13520
      101 LFORT(NOUT)=MFOR                                               DSL13530
          NFORT(NOUT)=IEND                                               DSL13540
          NFOR1=NFORA                                                    DSL13550
       68 NOUT1=NOUT                                                     DSL13560
          L=1                                                            DSL13570
       77 CALL SCAN                                                      DSL13580
C...   BRANCH ON NUMERIC,OPERATOR,=,(,GT 5 CHAR.,EOR,CONTINUE           DSL13585
          GO TO(94,79,78,263,130,94,103),KEY                            DSL13590
C...   TEST IF END OF CARD BEFORE COMMA OR EQUALS                        DSL13600
       79 IF(ISTRT-72)350,190,190                                       DSL13610
      350 MPL=1                                                          DSL13614
          GO TO 355                                                      DSL13616
      263 L=3                                                            DSL13618
          GO TO 355                                                      DSL13619
      190 IF(NSYMB-LOC)94,191,94                                        DSL13620
      191 NSYMB=NSYMB-1                                                  DSL13630
      130 IF(ISORT)94,131,94                                            DSL13640
      131 KOUT(NOUT)=0                                                   DSL13642
          NOUT=NOUT+1                                                    DSL13643
          GO TO 4                                                        DSL13644
       78 L=2                                                            DSL13645
      355 KSTR=1                                                         DSL13647
C...   TEST IF OUTPUT IS STORAG VAR                                      DSL13649
          IF(NSTOR)353,353,351                                          DSL13650
      351 DO 352 I=2,NSTOR                                               DSL13651
          IF(LOC-KSTOR(I))352,354,352                                   DSL13652
      354 MPL=1                                                          DSL13653
          KSTR=LSTOR(I)                                                  DSL13654
          GO TO 353                                                      DSL13655
      352 CONTINUE                                                       DSL13656
C...   TEST IF OUTPUT VARIABLE IS AN INTEGRAL                            DSL13660
      353 IF(NINGL)196,196,192                                          DSL13661
      192 DO 193  I=1,NINGL                                              DSL13662
          IF(INGRL(I)-LOC)193,194,193                                   DSL13663
      194 IF(NINT- 50)195,100,100                                       DSL13664
      195 LINT(NINT+1)=NOUT                                              DSL13665
          NINT=NINT+KSTR                                                 DSL13666
          JINTG=1                                                        DSL13667
          GO TO 196                                                      DSL13668
      193 CONTINUE                                                       DSL13669
      196 GO TO(76,356,262),L                                           DSL13672
      356 IF(MPL)76,76,272                                              DSL13674
      262 MPL=0                                                          DSL13678
```

```
          GO TO 225                                         DSL13680
   272 NF1=(ISTRT-1)/2                                      DSL13685
       NF8=NF1+90                                           DSL13686
       IF(ISTRT-NF1-NF1-2)276,275,100                       DSL13687
C...   ADD BLANK-ZERO (16384)                               DSL13689
   275 KFORT(NF1+1)=LETRS(ISTRT)+16384                      DSL13690
       GO TO 277                                            DSL13691
   276 KFORT(NF1)=LETRS(ISTRT-2)
       MM0=0
       CALL SHIFT(KFORT(NF1),MASK2,MM0)
   277 DO 273  NF=4,NF1                                     DSL13698
   273 KFORT(NF+ 90)=KFORT(NF)                              DSL13699
       IF(NF1-5)271,279,278                                 DSL13700
   271 NALRM=8                                              DSL13701
       GO TO 99                                             DSL13702
   278 NF3=5                                                DSL13703
       DO 274  NF=NF1,NFOR                                  DSL13704
       KFORT(NF3)=KFORT(NF)                                 DSL13705
   274 NF3=NF3+1                                            DSL13706
       NF3=NF1-5                                            DSL13707
       MEMR=MEMR-NF3                                        DSL13709
       NFOR=NFOR-NF3                                        DSL13710
C...   INSERT CA-LL IN 4 AND 5                              DSL13711
   279 KFORT(4)=ICA                                         DSL13712
       KFORT(5)=LL                                          DSL13713
       NFORT(NOUT1)=NFORT(NOUT1)+3                          DSL13714
    76 IF(NOUT-100)109,109,100                              DSL13715
   109 IF(NIMP)94,225,220                                   DSL13718
   220 IF(LOC-IMP3)225,221,225                              DSL13720
   221 LOC=IMP2                                             DSL13730
       NIMP=-2                                              DSL13740
   225 KOUT(NOUT)=LOC                                       DSL13750
       NOUT=NOUT+1                                          DSL13850
       IF(L-1)94,77,81                                      DSL13860
C...   FIND ALL INPUT VARIABLES                             DSL13870
    81 NINP1=NINP+1                                         DSL13880
    82 CALL SCAN                                            DSL13890
       L=3                                                  DSL13900
C...   BRANCH ON NUMERIC,OPERATOR,=,(,GT 5 CHAR.,EOR,CONTINUE  DSL13905
       GO TO(94,84,94,85,94,89,103),KEY                    DSL13910
   103 ICONT=1                                              DSL13920
       GO TO 5                                              DSL13930
    84 IF(NINP1-NINP)138,138,106                            DSL13960
   138 DO 104  M=NINP1,NINP                                 DSL13970
       IF(INPUT(M)-LOC)104,82,104                           DSL13980
   104 CONTINUE                                             DSL13990
       IF(NINP-300)106,100,100                              DSL14000
   106 IF(NIMP)228,230,228                                  DSL14010
   228 IF(LOC-IMP2)230,229,230                              DSL14020
   229 LOC=IMP1                                             DSL14030
   230 NINP=NINP+1                                          DSL14040
       INPUT(NINP)=LOC                                      DSL14050
       GO TO 82                                             DSL14060
C...   TEST IF INTEGRAL BLOCK (INTGR,REALP,MODIN,TRNFR,CMPXP,LEDLG)  DSL14070
    85 IF(LOC-7)82,141,148                                  DSL14090
   148 IF(LOC-12)141,145,180                                DSL14100
   141 IF(JINTG)266,266,82                                  DSL14103
   266 IF(NINT)143,143,147                                  DSL14105
   147 IF(LINT(NINT)-NOUT1)240,142,143                      DSL14110
   240 IF(LINT(NINT)+NOUT1)143,244,143                      DSL14120
```

334

```
  143 ICH=ISTRT-6                                            DSL14130
  246 ICH=ICH-1                                              DSL14140
      IF(LETRS(ICH)-64)103,246,248                           DSL14150
  248 IF(LETRS(ICH)-126)249,247,246                          DSL14155
  249 IF(LETRS(ICH)-123)241,247,241                          DSL14160
  241 NIN(NOUT1)=NINP-NINP1+1                                DSL14170
      LIN(NOUT1)=NINP1                                       DSL14180
      NOUT1=NOUT                                             DSL14190
      NOUT=NOUT+1                                            DSL14200
      NINP1=NINP+1                                           DSL14210
      GO TO 245                                              DSL14220
  244 LINT(NINT)=0                                           DSL14230
  245 LINT(NINT+1)=-NOUT1                                    DSL14240
      GO TO 107                                              DSL14250
  142 LINT(NINT)=0                                           DSL14260
  247 LINT(NINT+1)=NOUT1                                     DSL14270
  107 NINT=NINT+1                                            DSL14280
      IF(LOC-10)146,144,145                                  DSL14290
  144 CALL NUMRC                                             DSL14300
      CALL NUMRC                                             DSL14310
      NINT=NINT+NMBR-1                                       DSL14320
  146 IF(NINT-50)82,82,100                                   DSL14330
  145 NINT=NINT+1                                            DSL14340
      LINT(NINT)=NOUT1+1                                     DSL14350
      GO TO 146                                              DSL14360
C... TEST IF SYSTEM MEMORY BLOCK, DELAY BLOCK,               DSL14370
C... OR IMPLICIT FUNCTION (IMPL, IMPLC)                      DSL14380
  180 DO 181  I=1,NMEM                                       DSL14390
      IF(MEMRY(I)-LOC)181,183,181                            DSL14400
  181 CONTINUE                                               DSL14410
      GO TO 82                                               DSL14420
  183 IF(NFORT(NOUT1))255,256,256                            DSL14422
  255 NFORT(NOUT1)=NFORT(NOUT1)-2                            DSL14424
      GO TO 257                                              DSL14426
  256 NFORT(NOUT1)=NFORT(NOUT1)+2                            DSL14430
  257 NF1=(ISTRT-1)/2                                        DSL14440
      NF2=NFORA+NF1+MEMR-1                                   DSL14450
      MEMR=MEMR+2                                            DSL14455
      NF3=NFOR-1                                             DSL14460
      NFOR=NFOR+2                                            DSL14470
  184 KFORT(NF3+2)=KFORT(NF3)                                DSL14480
      NF3=NF3-1                                              DSL14490
      IF(NF3-NF2)185,185,184                                 DSL14500
  185 IF(ISTRT-NF1-NF1-2)179,178,100                         DSL14510
  178 KA(5)=-NF2                                             DSL14520
C... TEST IF PAREN FROM 026 OR 029                           DSL14530
      IF(KFORT(NF2+3)-20000)281,280,280                      DSL14540
  280 KFORT(NF2+3)=KFORT(NF2+3)-256                          DSL14550
      GO TO 188                                              DSL14560
  281 KFORT(NF2+3)=KFORT(NF2+3)+7680                         DSL14570
      GO TO 188                                              DSL14580
  179 KA(5)=NF2                                              DSL14590
C... TEST IF DELAY BLOCK                                     DSL14598
  188 IF(I-10)187,186,187                                    DSL14600
  186 CALL NUMRC                                             DSL14610
      KMEM(I)=NMBR+NMBR+5                                    DSL14620
  187 MAXN=MAXN-KMEM(I)                                      DSL14630
      NMBR=MAXN+1                                            DSL14640
      CALL BNDEC                                             DSL14650
C... TEST IF IMPLICIT FUNCTION BLOCK                         DSL14658
```

335

```
      IF(I-8)82,88,189                                    DSL14660
189   IF(I-9)82,88,82                                     DSL14670
 88   KFORT(NFORA)=IRES(138)                              DSL14680
      KFORT(NFORA+1)=IRES(139)                            DSL14690
      DO 250  KR=106,131                                  DSL14700
      KFORT(NFOR)=IRES(KR)                                DSL14710
250   NFOR=NFOR+1                                         DSL14720
      NFORT(NOUT1)=-NFORT(NOUT1)-26                       DSL14730
      IMP1=IMP1+1                                         DSL14740
      IMP2=KOUT(NOUT1)                                    DSL14750
      KOUT(NOUT1)=IMP1                                    DSL14760
      NIMP=-1                                             DSL14770
      GO TO 82                                            DSL14780
 89   IF(NIMP)267,270,270                                 DSL14790
267   IF(NIMP+2)94,268,269                                DSL14800
268   NIMP=0                                              DSL14810
      IF(NFORT(NOUT1))223,224,224                         DSL14820
223   NFOR=NFOR-1                                         DSL14830
      NFORT(NOUT1)=NFORT(NOUT1)-17                        DSL14840
      GO TO 226                                           DSL14850
224   NFORT(NOUT1)=-NFORT(NOUT1)-18                       DSL14860
226   DO 222  KR=131,148                                  DSL14870
      KFORT(NFOR)=IRES(KR)                                DSL14880
222   NFOR=NFOR+1                                         DSL14890
      IRES(125)=IRES(125)+1                               DSL14900
      IRES(139)=IRES(139)+1                               DSL14910
      IRES(142)=IRES(142)+1                               DSL14920
      IRES(116)=IRES(142)                                 DSL14930
      IRES(119)=IRES(142)                                 DSL14940
      IRES(122)=IRES(125)                                 DSL14950
      GO TO 270                                           DSL14960
269   NIMP=1                                              DSL14970
      IMP3=INPUT(NINP)                                    DSL14980
      INPUT(NINP)=0                                       DSL14990
270   IF(IPRO)4,97,98                                     DSL15000
 97   NIN(NOUT1)=NINP-NINP1+1                             DSL15010
      LIN(NOUT1)=NINP1                                    DSL15020
      GO TO 4                                             DSL15030
 98   IPRO=-1                                             DSL15040
      GO TO 4                                             DSL15050
100   NALRM=10                                            DSL15060
      NSORT=0                                             DSL15070
      IERR=1                                              DSL15080
      CALL ERR1                                           DSL15090
      GO TO 50                                            DSL15100
 99   CALL ERR1                                           DSL15110
      GO TO 4                                             DSL15120
 50   IF(KDUMP)55,56,55                                   DSL15130
 55   INVOKE TRDMP
 56   CONTINUE
      INVOKE SORTX
      END                                                 DSL15170
/*
```

```
// CALL FORTRN,F1
// RUN
// READ DEVICE-'MFCU1,5471'
// PRINT DEVICE-'1403,5471'
// PUNCH DEVICE-'MFCU2'
*PROCESS OBJECT(R,LIB(R1)),NOLINK
      PROGRAM SORTX
      IMPLICIT INTEGER*2(I-N)
      INTEGER*4 KPT
      GLOBAL NALRM,IERR,ICONT,JPT,KEY,ISTRT,KA(6),LETRS(73),ICHAR(12),      DSL07550
     1 IRES(158),KSYMB(600),NSYMB,ITYPE,NOUT,NINP,NPAR,NINT,NDATA,NFOR,    DSL07560
     2 NSOR1,ISORT,NMEM,NING,NINGL,NSTOR,NTSEQ,NREAL,MFOR,                 DSL07570
     3 KOUT(100),LFORT(100),NFORT(100),LIN(100),NIN(100),INPUT(300),       DSL07580
     4 KPAR( 60),KSTOR(10),LINT(50),INGER(10),MEMRY(15),KMEM(15),          DSL07590
     5 LSTOR(10),INGRL(10),KREAL(20),KSORT(20),KFORT(202),                 DSL07600
     6       KSEQ(120),KSEQ2(30),LIN2(30),NIN2(30),KSEQ3(10)               DSL07602
      KPT=JPT                                                             DSL07604
      JINT=1                                                             DSL07606
      J2=0                                                               DSL07608
      KGO=1                                                              DSL07610
      NSOR1=1                                                            DSL07612
      NOUT1=1                                                            DSL07614
      MFOR=MFOR-1                                                        DSL07616
      NOUT=NOUT-1                                                        DSL07618
      NDATA=NDATA-1                                                      DSL07620
      IF(NSORT)101,101,18                                                DSL07622
  101 NTSEQ=NOUT                                                         DSL07624
      DO 102  I=1,NOUT                                                   DSL07626
  102 KSEQ(I)=I                                                          DSL07628
C...  TEST IF ALL PARAMETERS INPUT                                       DSL07630
      DO 104  I=1,NINP                                                   DSL07632
      KVAR=INPUT(I)                                                      DSL07634
      DO 105  J=1,NOUT                                                   DSL07636
      IF(KOUT(J)-KVAR)105,104,105                                        DSL07638
  105 CONTINUE                                                           DSL07640
C...  TEST IF PARAMETER OR STORAG VARIABLE                              DSL07642
      DO 106  J=1,NPAR                                                   DSL07644
      IF(KPAR(J)-KVAR)106,104,106                                        DSL07646
  106 CONTINUE                                                           DSL07648
      IF(NSTOR)117,117,108                                               DSL07650
  108 DO 109  J=1,NSTOR                                                  DSL07652
      IF(KSTOR(J)-KVAR)109,104,109                                       DSL07654
  109 CONTINUE                                                           DSL07656
  117 IF(KVAR)104,104,118                                                DSL07658
  118 KA(1)=KSYMB(KVAR)                                                  DSL07660
      KA(2)=KSYMB(KVAR+200)                                              DSL07662
      KA(3)=KSYMB(KVAR+400)                                              DSL07663
      CALL UNPAK                                                         DSL07664
      WRITE(KPT,200) KA(1),KA(2),KA(3)                                   DSL07666
      IF(NPAR-60)107,210,210                                            DSL07668
  107 NPAR=NPAR+1                                                        DSL07672
      KPAR(NPAR)=KVAR                                                    DSL07674
  104 CONTINUE                                                           DSL07676
      GO TO 400                                                         DSL07678
C...  ESTABLISH CORRECT SEQUENCE OF OUTPUTS                             DSL07680
   18 IF(ISORT)28,28,19                                                  DSL07687
   19 KSORT(NSORT)=NOUT                                                  DSL07688
C...  SET LOCATIONS OF 1ST AND LAST OUTPUT WITHIN NEXT SORT SECTION      DSL07690
   28 NSOR2=KSORT(NSOR1)                                                 DSL07700
      NSOR3=KSORT(NSOR1+1)                                               DSL07710
```

```
      NTIN1=LIN(NSOR2)                                           DSL07720
      IF(NSOR3-NOUT)30,29,30                                     DSL07730
   29 NTIN2=NINP                                                 DSL07740
      GO TO 31                                                   DSL07750
   30 NSOR4=NSOR3                                                DSL07753
  170 IF(LIN(NSOR4))171,171,172                                  DSL07754
  171 NSOR4=NSOR4-1                                              DSL07755
      GO TO 170                                                  DSL07756
  172 NTIN2=LIN(NSOR4)+NIN(NSOR4)-1                              DSL07760
   31 IF(NSOR2-NOUT1)32,35,32                                    DSL07770
   32 NOUT2=NSOR2-1                                              DSL07780
   36 DO 33   I=NOUT1,NOUT2                                      DSL07790
      IF(LFORT(I))34,33,34                                       DSL07800
   34 J2=J2+1                                                    DSL07810
      KSEQ(J2)=I                                                 DSL07820
   33 CONTINUE                                                   DSL07830
      GO TO(35,130),KGO                                          DSL07840
...   REMOVE VARIABLES FROM INPUT TABLE WHICH ARE NOT COMPUTED   DSL07850
   35 DO 4   I=NTIN1,NTIN2                                       DSL07860
      KVAR=INPUT(I)                                              DSL07870
      IF(KVAR)65,4,16                                            DSL07875
   16 DO 5   J=NSOR2,NSOR3                                       DSL07880
      IF(KOUT(J)-KVAR)5,3,5                                      DSL07890
...   REMOVE FROM INPUT TABLE IF INTEGRAL                        DSL07900
    3 IF(NINT)2,2,9                                              DSL07910
    9 DO 10   K=1,NINT                                           DSL07920
      IF(LINT(K)-J)10,8,10                                       DSL07930
   10 CONTINUE                                                   DSL07940
    2 INPUT(I)=J                                                 DSL07950
      GO TO 4                                                    DSL07960
    5 CONTINUE                                                   DSL07970
...   TEST IF PARAMETER OR STORAG VARIABLE OR OUTPUT IN PREV. SECTION DSL07980
      DO 6   J=1,NPAR                                            DSL07990
      IF(KPAR(J)-KVAR)6,8,6                                      DSL08000
    6 CONTINUE                                                   DSL08010
      IF(NSOR2-1)14,14,13                                        DSL08020
   13 DO 17   J=1,NSOR2                                          DSL08030
      IF(KOUT(J)-KVAR)17,8,17                                    DSL08040
   17 CONTINUE                                                   DSL08050
   14 IF(NSTOR)15,15,11                                          DSL08060
   11 DO 12   J=1,NSTOR                                          DSL08070
      IF(KSTOR(J)-KVAR)12,8,12                                   DSL08080
   12 CONTINUE                                                   DSL08090
   15 KA(1)=KSYMB(KVAR)                                          DSL08110
      KA(2)=KSYMB(KVAR+200)                                      DSL08120
      KA(3)=KSYMB(KVAR+400)                                      DSL08121
      CALL UNPAK                                                 DSL08130
      WRITE(KPT,200) KA(1),KA(2),KA(3)                           DSL08140
  200 FORMAT(25H PARAM NOT INPUT, SET=0   3A2)                   DSL08150
      IF(NPAR-60)7,300,300                                       DSL08160
  300 NALRM=10                                                   DSL08170
  301 CALL ERR1                                                  DSL08180
      IERR=1                                                     DSL08190
      GO TO 8                                                    DSL08195
    7 NPAR=NPAR+1                                                DSL08200
      KPAR(NPAR)=KVAR                                            DSL08210
    8 INPUT(I)=0                                                 DSL08220
    4 CONTINUE                                                   DSL08230
      IF(NINT)100,100,40                                         DSL08240
   40 KSW=1                                                      DSL08250
```

338

```
          NSEQ=J2+1                                               DSL08260
C...   BEGIN EACH SEQUENCE WITH INTEGRATOR OUTPUT AND WORK BACK    DSL08270
   41  JLOC=LINT(JINT)                                            DSL08280
       IF(JLOC)95,95,46                                           DSL08290
C...   TEST IF INTEGRAL IS WITHIN SORT SECTION                    DSL08300
   46  IF(JLOC-NSOR3)42,42,100                                    DSL08310
   42  IF(JLOC-NSOR2)95,43,43                                     DSL08320
   43  KSEQ2(1)=JLOC                                              DSL08330
   44  J=J2                                                       DSL08340
       J1=J+1                                                     DSL08350
       J3=J1                                                      DSL08360
       LEV=1                                                      DSL08370
       NSEQ3=0                                                    DSL08380
C...   TEST FOR MULTIPLE OUTPUTS, PLACE ALL IN SAME LOOP AT FIRST USE  DSL08390
C...   TEST IF DUMMY ENTRY (INTEGRATOR WITHIN PROCEDURE)          DSL08400
   50  IF(LFORT(JLOC))52,52,51                                    DSL08410
   51  IF(LFORT(JLOC+1))58,53,58                                  DSL08420
   52  JLOC=JLOC-1                                                DSL08430
       IF(LFORT(JLOC))52,52,53                                    DSL08440
   53  IF(JLOC-NOUT)153,58,58                                     DSL08450
  153  IF(KOUT(JLOC+1))154,154,45                                 DSL08460
  154  NSEQ3=NSEQ3+1                                              DSL08470
       KSEQ3(NSEQ3)=JLOC+1                                        DSL08480
       IF(NSEQ3-10)58,58,210                                      DSL08490
   45  JL1=JLOC                                                   DSL08500
   54  J=J+1                                                      DSL08510
       IF(J-120)56,56,55                                          DSL08520
   55  WRITE(KPT,201)                                             DSL08530
  201  FORMAT(30H0MORE THAN 120 OUTPUTS IN SORT)                  DSL08540
       GO TO 65                                                   DSL08550
   56  KSEQ(J)=JLOC                                               DSL08560
       JLOC=JLOC+1                                                DSL08570
       IF(LFORT(JLOC))57,155,57                                   DSL08580
  155  IF(JLOC-NOUT)54,54,57                                      DSL08585
   57  JLOC=JL1                                                   DSL08590
       GO TO 60                                                   DSL08600
   58  J=J+1                                                      DSL08610
       IF(J-120)59,59,55                                          DSL08620
   59  KSEQ(J)=KSEQ2(LEV)                                         DSL08630
   60  LIN2(LEV)=LIN(JLOC)                                        DSL08640
       NIN2(LEV)=NIN(JLOC)                                        DSL08650
C...   TEST IF MORE INPUTS TO BLOCK ON PRESENT LEVEL              DSL08660
   61  IF(NIN2(LEV))65,78,66                                      DSL08670
C...   TEST IF INTEGRATOR, PARAMETER, OR OUTPUT NEXT              DSL08730
   66  K=LIN2(LEV)                                                DSL08740
       IF(INPUT(K))73,80,72                                       DSL08750
   72  JLOC=INPUT(K)                                              DSL08760
C...   TEST IF WITHIN SORT SECTION                                DSL08770
   73  IF(JLOC-NSOR2)80,75,74                                     DSL08780
   74  IF(JLOC-NSOR3)75,75,80                                     DSL08790
C...   INCREASE LEVEL                                             DSL08800
   75  LEV=LEV+1                                                  DSL08810
       IF(LEV-30)77,77,76                                         DSL08820
   76  WRITE(KPT,203)                                             DSL08830
  203  FORMAT(36H0MORE THAN 30 OUTPUTS IN SINGLE LOOP)            DSL08840
       GO TO 65                                                   DSL08850
   77  KSEQ2(LEV)=INPUT(K)                                        DSL08860
       GO TO 50                                                   DSL08870
C...   DECREASE LEVEL                                             DSL08880
   78  LEV=LEV-1                                                  DSL08890
```

```
      IF(LEV)180,180,80                                    DSL08900
   80 NIN2(LEV)=NIN2(LEV)-1                                 DSL08910
      LIN2(LEV)=LIN2(LEV)+1                                 DSL08920
      GO TO 61                                              DSL08930
  180 IF(NSEQ3)81,81,181                                    DSL08940
  181 JLOC=KSEQ3(NSEQ3)                                     DSL08950
      NSEQ3=NSEQ3-1                                         DSL08960
      KSEQ2(1)=JLOC                                         DSL08970
      LEV=1                                                 DSL08980
C...  REMOVE COMPLETED ELEMENTS FROM INPUT FOR PARTIAL LOOP DSL08990
      DO 182  I=J3,J                                        DSL09000
      KVAR=KSEQ(I)                                          DSL09010
      DO 182  K=NTIN1,NTIN2                                 DSL09020
      IF(KVAR-INPUT(K))182,183,182                          DSL09030
  183 INPUT(K)=0                                            DSL09040
  182 CONTINUE                                              DSL09050
      J3=J+1                                                DSL09060
      GO TO 58                                              DSL09070
C...  END OF SEQUENCE FOR ONE LOOP, ELIMINATE DUPLICATES    DSL09080
   81 IF(J-J1)82,82,83                                      DSL09090
   82 J2=J2+1                                               DSL09100
      GO TO 91                                              DSL09110
   83 J5=J-1                                                DSL09120
      DO 86  I=J1,J5                                        DSL09130
      KVAR=KSEQ(I)                                          DSL09140
      I2=I+1                                                DSL09150
      DO 84  K=I2,J                                         DSL09160
      IF(KVAR-KSEQ(K))84,85,84                              DSL09170
   84 CONTINUE                                              DSL09180
      GO TO 86                                              DSL09190
   85 KSEQ(I)=0                                             DSL09200
   86 CONTINUE                                              DSL09210
C...  CONDENSE, THEN REVERSE SEQUENCE                       DSL09220
      DO 88  I=J1,J                                         DSL09230
      IF(KSEQ(I))87,88,87                                   DSL09240
   87 J2=J2+1                                               DSL09250
      KSEQ(J2)=KSEQ(I)                                      DSL09260
   88 CONTINUE                                              DSL09270
      J3=(J2-J1+1)/2+J1-1                                   DSL09280
      J4=J2                                                 DSL09290
      DO 90  I=J1,J3                                        DSL09300
      KSAVE=KSEQ(J4)                                        DSL09310
      KSEQ(J4)=KSEQ(I)                                      DSL09320
      KSEQ(I)=KSAVE                                         DSL09330
   90 J4=J4-1                                               DSL09340
C...  REMOVE COMPLETED ELEMENTS FROM INPUT                  DSL09350
   91 DO 92  I=J1,J2                                        DSL09360
      KVAR=KSEQ(I)                                          DSL09370
      DO 92  K=NTIN1,NTIN2                                  DSL09380
      IF(KVAR-INPUT(K))92,93,92                             DSL09390
   93 INPUT(K)=0                                            DSL09400
   92 CONTINUE                                              DSL09410
      GO TO(95,121),KSW                                     DSL09420
C...  TEST IF LAST INTEGRATOR                               DSL09430
   95 JINT=JINT+1                                           DSL09440
      IF(JINT-NINT)160,160,100                              DSL09450
C...  TEST IF ALREADY IN SEQUENCE                           DSL09453
  160 KVAR=LINT(JINT)                                       DSL09454
      DO 162  K=1,J2                                        DSL09455
      IF(KVAR-KSEQ(K))162,95,162                            DSL09456
```

340

```
 162 CONTINUE                                                         DSL09457
     GO TO 41                                                         DSL09458
 100 KSW=2                                                            DSL09460
C... TRACE BRANCHES NOT WITHIN INTEGRATION OR MEMORY LOOPS            DSL09470
 121 LTEST=0                                                          DSL09480
     DO 126  I=NSOR2,NSOR3                                            DSL09490
     IF(LFORT(I))126,126,122                                          DSL09500
 122 DO 123  K=NSEQ,J2                                                DSL09510
     IF(KSEQ(K)-I)123,126,123                                         DSL09520
 123 CONTINUE                                                         DSL09530
C... TEST IF INPUT TO ANOTHER BLOCK                                   DSL09540
     DO 124  K=NTIN1,NTIN2                                            DSL09550
     IF(INPUT(K)-I)124,125,124                                        DSL09560
 124 CONTINUE                                                         DSL09570
     KSEQ2(1)=I                                                       DSL09580
     JLOC=I                                                           DSL09590
     GO TO 44                                                         DSL09600
 125 LTEST=1                                                          DSL09610
 126 CONTINUE                                                         DSL09620
     IF(LTEST)127,127,128                                             DSL09630
 127 NSOR1=NSOR1+2                                                    DSL09640
     NOUT1=NSOR3+1                                                    DSL09650
     IF(NSOR1-NSORT)28,28,129                                         DSL09660
 129 IF(NOUT1-NOUT)131,131,130                                        DSL09670
 131 NOUT2=NOUT                                                       DSL09680
     KGO=2                                                            DSL09690
     GO TO 36                                                         DSL09700
 210 NALRM=10                                                         DSL09702
     CALL ERR1                                                        DSL09704
     IERR=1                                                           DSL09706
     GO TO 400                                                        DSL09708
 128 WRITE(KPT,204)                                                   DSL09710
 204 FORMAT(25H0UNDEFINED IMPLICIT LOOP )                             DSL09720
  65 IERR=1                                                           DSL09730
     J2=J                                                             DSL09740
 130 NTSEQ=J2                                                         DSL09750
     WRITE(KPT,205) (KSEQ(I),I=1,J2)                                  DSL09760
 205 FORMAT(25H0OUTPUT VARIABLE SEQUENCE /(20I4))                     DSL09770
 400 WRITE(KPT,900) NINT,NINP,NOUT,NPAR,NSYMB,MFOR,NDATA              DSL09772
 900 FORMAT(21H0STORAGE USED/MAXIMUM /6H INTGR I4,12H/50, IN VARS I4, DSL09773
   1 14H/300,  OUT VARS I4,12H/100, PARAMS I3,12H/60, SYMBOLS I4,     DSL09774
   2 14H/200, FORT WDS I5,15H/5000, DATA WDS I4,5H/1000)              DSL09775
     IF(IERR)150,151,150                                              DSL09777
 150 STOP                                                             DSL09778
 151 ISTRT=INPT(2)                                                    DSL09779
     INVOKE OUTPX
     END                                                              DSL09790
/*
```

341

```
// CALL FORTRN,F1
// RUN
// READ DEVICE-'MFCU1,5471'
// PRINT DEVICE-'1403,5471'
// DAD44 UNITNO-'11,12'
// PUNCH DEVICE-'MFCU2'
*PROCESS OBJECT(R,LIE(R1)),NOLINK,NOSHRBUFF
      PROGRAM OUTPX
      IMPLICIT INTEGER*2(I-N)
      INTEGER*4 L,KPT
      DEFINE FILE 11(5000,1,U,MFOR),12(1000,1,U,NDATA)
      GLOBAL NALRM,IERR,ICONT,JPT,KEY,ISTRT,KA(6),LETRS(73),ICHAR(12),   DSL01910
     1 IRES(158),KSYMB(600),NSYMB,ITYPE,NOUT,NINP,NPAR,NINT,NDATA,NFOR,  DSL01920
     2 NSORT,ISORT,NMEM,NING,NINGL,NSTOR,NTSEQ,NREAL,MFOR,               DSL01930
     3 KOUT(100),LFORT(100),NFORT(100),LIN(100),NIN(100),INPUT(300),     DSL01940
     4 KPAR( 60),KSTOR(10),LINT(50),INGER(10),MEMRY(15),KMEM(15),        DSL01950
     5 LSTOR(10),INGRL(10),KREAL(20),KSORT(20) ,                         DSL01960
     6      KFORT(128),KSPEC(36),LOCAL(36),NLOC,NSPEC,KSEQ(120),         DSL01970
     7 KOMN(110),KBUFR(36)                                               DSL01980
      EQUIVALENCE (IRES(106),IDOLR),(IRES(107),IBLNK),(IRES(158),IONE)   DSL01990
      DATA IAST2/'*2'/
      KPT=JPT                                                            DSL02000
      L=ISTRT                                                            DSL02005
 1111 CONTINUE
      IF(NSORT)101,101,103                                              DSL02020
  101 NTSEQ=NOUT                                                        DSL02030
      DO 102  I=1,NOUT                                                  DSL02040
  102 KSEQ(I)=I                                                         DSL02050
C...  TEST IF ALL PARAMETERS INPUT                                      DSL02060
      DO 104  I=1,NINP                                                  DSL02070
      KVAR=INPUT(I)                                                     DSL02080
      DO 105  J=1,NOUT                                                  DSL02090
      IF(KOUT(J)-KVAR)105,104,105                                       DSL02100
  105 CONTINUE                                                          DSL02110
C...  TEST IF PARAMETER OR STORAG VARIABLE                             DSL02120
      DO 106  J=1,NPAR                                                  DSL02130
      IF(KPAR(J)-KVAR)106,104,106                                       DSL02140
  106 CONTINUE                                                          DSL02150
      IF(NSTOR)117,117,108                                              DSL02160
  108 DO 109  J=1,NSTOR                                                 DSL02170
      IF(KSTOR(J)-KVAR)109,104,109                                      DSL02180
  109 CONTINUE                                                          DSL02190
  117 IF(KVAR)104,104,118                                               DSL02200
  118 KA(1)=KSYMB(KVAR)                                                 DSL02210
      KA(2)=KSYMB(KVAR+200)                                             DSL02220
      KA(3)=KSYMB(KVAR+400)                                             DSL02221
      WRITE(KPT,200) KA(1),KA(2),KA(3)                                  DSL02240
  200 FORMAT(25H PARAM NOT INPUT, SET=0   3A2)                          DSL02250
      IF(NPAR-60)107,300,300                                            DSL02260
  300 WRITE(KPT,210) NPAR,NCOM                                          DSL02270
  210 FORMAT(33H CHECK MAX VS. PARAM, COMMON VARS 2I4)                  DSL02280
      GO TO 150                                                         DSL02290
  107 NPAR=NPAR+1                                                       DSL02300
      KPAR(NPAR)=KVAR                                                   DSL02310
  104 CONTINUE                                                          DSL02320
C...  FORM COMMON VARIABLE LIST (ALSO SYMBOL TABLE)                     DSL02330
C...  SIX SYSTEMS VARS, INTEGRALS, OUTPUT VARS, PARAMS, STORAG VARS     DSL02340
  103 KA(4)=240
      KA(5)=240                                                         DSL02370
      KA(6)=64
```

```
         DO 5  I=1,6                                              DSL02380
       5 KOMN(I)=I                                                DSL02390
         IF(NINT)15,15,6                                          DSL02400
       6 NKS=1                                                    DSL02402
         M=1                                                      DSL02404
     222 NEXT=KSEQ(NKS)                                           DSL02406
C...   USE OUTPUT SEQUENCE TO SET UP INTEGRATORS IN COMMON        DSL02408
         DO 10  I=1,NINT                                          DSL02410
         J=LINT(I)                                                DSL02420
         IF(J)203,7,223                                           DSL02432
     223 IF(J-NEXT)10,8,10                                        DSL02432
     203 IF(J+NEXT)10,7,10                                        DSL02434
       7 KA(1)=233                                                DSL02440
         KA(2)=233                                                DSL02450
         KA(3)=240                                                
         IF(KA(5)-249)3,2,2                                       DSL02460
       2 KA(4)=KA(4)+1                                            DSL02470
         KA(5)=240                                                DSL02480
         GO TO 4                                                  DSL02490
       3 KA(5)=KA(5)+1                                            DSL02500
       4 CALL PACK                                                DSL02510
         NSYMB=NSYMB+1                                            DSL02520
         KSYMB(NSYMB)=KA(1)                                       DSL02530
         KSYMB(NSYMB+200)=KA(2)                                   DSL02540
         KSYMB(NSYMB+400)=KA(3)                                   DSL02541
         KOMN(M+6)=NSYMB                                          DSL02550
         GO TO 204                                                DSL02560
       8 KOMN(M+6)=KOUT(J)                                        DSL02570
         KOUT(J)=-KOUT(J)                                         DSL02580
     204 IF(M-NINT)206,15,15                                      DSL02581
     206 M=M+1                                                    DSL02582
         I=I+1                                                    DSL02583
         IF(I-NINT)224,224,205                                    DSL02584
     224 J=LINT(I)                                                DSL02587
         IF(J)205,7,205                                           DSL02588
      10 CONTINUE                                                 DSL02590
     205 NKS=NKS+1                                                DSL02592
         IF(NKS-NTSEQ)222,222,15                                  DSL02594
      15 NCOM=NINT+6                                              DSL02600
         IF(NING)115,115,110                                      DSL02610
C...   ADD DUMMY WORDS IN COMMON FOR INTEGERS TO MATCH SYMBOL TABLE DSL02620
     110 KOUNT=IRES(151)                                          DSL02630
         DO 112  I=1,NING                                         DSL02640
         DO 112  J=7,NPAR                                         DSL02650
         IF(INGER(I)-KPAR(J))112,111,112                          DSL02660
     111 KPAR(J)=KPAR(NPAR)                                       DSL02670
         NPAR=NPAR-1                                              DSL02680
     112 CONTINUE                                                 DSL02690
         DO 113  I=1,NING                                         DSL02700
         NPAR=NPAR+2                                              DSL02710
         KPAR(NPAR-1)=-1                                          DSL02720
     113 KPAR(NPAR)=INGER(I)                                      DSL02730
     115 DO 17  I=1,NOUT                                          DSL02740
         KVAR=KOUT(I)                                             DSL02750
         IF(KVAR)17,17,16                                         DSL02760
C...   SKIP IF ALSO A PARAMETER                                   DSL02762
      16 DO 13  K=7,NPAR                                          DSL02764
         IF(KVAR-KPAR(K))13,17,13                                 DSL02766
      13 CONTINUE                                                 DSL02768
         IF(KVAR-500)116,116,17                                   DSL02770
```

343

```
  116 DO 12   K=1,NCOM                                        DSL02780
      IF(KVAR-KOMN(K))12,17,12                                DSL02790
   12 CONTINUE                                                DSL02800
      IF(NSTOR)215,215,214                                    DSL02833
  214 DO 211   K=2,NSTOR                                      DSL02834
      IF(KVAR-KSTOR(K))211,17,211                             DSL02835
  211 CONTINUE                                                DSL02836
  215 NCOM=NCOM+1                                             DSL02840
      KOMN(NCOM)=KVAR                                         DSL02850
      IF(NCOM-110)17,17,300                                   DSL02860
   17 CONTINUE                                                DSL02870
  264 IF(KPAR(NPAR))263,263,262                               DSL02880
  263 NPAR=NPAR-1                                             DSL02882
      IF(NPAR)262,262,264                                     DSL02884
C... OUTPUT CONTROL CARDS FOR UPDATE SUBROUTINE               DSL02888
  262 WRITE(L,250)                                            DSL02890
  250 FORMAT('// CALL FORTRN,F1'/'// RUN'/'*PROCESS OBJECT(R,LIB(R1))'/
     1 '        SUBROUTINE UPDAT'/'        IMPLICIT INTEGER*2(I-N)')
C... OUTPUT COMMON VARIABLE LIST                              DSL02920
      KGO=0                                                   DSL02930
      NB1=6                                                   DSL02940
      KGO1=0                                                  DSL02950
      NWDS=10                                                 DSL02960
      KOUN=0                                                  DSL02965
      KBUFR(1)=IBLNK                                          DSL02970
      KBUFR(2)=IBLNK                                          DSL02980
      IF(NREAL)51,51,9                                        DSL02990
C... REMOVE INTEGERS FROM REAL VARIABLE TABLE                 DSL03000
    9 IF(NING)11,11,14                                        DSL03010
   14 DO 29   I=1,NING                                        DSL03020
      DO 27   J=1,NREAL                                       DSL03030
      IF(INGER(I)-KREAL(J))27,20,27                           DSL03040
   20 KREAL(J)=KREAL(NREAL)                                   DSL03050
      NREAL=NREAL-1                                           DSL03060
      IF(NREAL)51,51,29                                       DSL03070
   27 CONTINUE                                                DSL03080
   29 CONTINUE                                                DSL03090
   11 KBUFR(4)=IRES(152)                                      DSL03100
      KBUFR(5)=IRES(153)                                      DSL03110
      KBUFR(6)=IBLNK                                          DSL03120
      KSW=1                                                   DSL03130
      N3=NREAL                                                DSL03140
   18 N1=1                                                    DSL03150
      KBUFR(3)=IBLNK                                          DSL03160
   19 NBUF=NB1                                                DSL03170
      N2=N1+NWDS-1                                            DSL03180
   26 IF(N2-N3)22,22,21                                       DSL03190
   21 N4=N3-N1+1                                              DSL03200
      N2=N3                                                   DSL03210
   22 DO 30   I=N1,N2                                         DSL03220
      IF(KSW-3)23,24,80                                       DSL03230
   80 IF(KSW-7)81,25,25                                       DSL03240
   23 J=KREAL(I)                                              DSL03250
      GO TO 28                                                DSL03260
   24 J=INGER(I)                                              DSL03270
      GO TO 28                                                DSL03280
   81 J=KPAR(I)                                               DSL03290
      IF(J)128,128,28                                         DSL03300
  128 NBUF=NBUF+3                                             DSL03310
      KBUFR(NBUF-2)=IRES(3)                                   DSL03320
```

344

```
      KBUFR(NBUF-1)=IRES(149)                                       DSL03330
      KBUFR(NBUF)=KOUNT                                             DSL03340
      KOUNT=KOUNT+256                                               DSL03350
      GO TO 30                                                      DSL03360
   25 J=KOMN(I)                                                     DSL03370
   28 KA(1)=KSYMB(J)                                                DSL03380
      KA(2)=KSYMB(J+200)                                            DSL03390
      KA(3)=KSYMB(J+400)                                            DSL03391
      NBUF=NBUF+3                                                   DSL03410
      KBUFR(NBUF-2)=KA(1)                                           DSL03420
      KBUFR(NBUF-1)=KA(2)                                           DSL03430
      KBUFR(NBUF)=KA(3)                                             DSL03440
   30 CONTINUE                                                      DSL03450
      IF(N2-N3)31,35,35                                             DSL03460
   31 KEND=36                                                       DSL03470
   32 IF(KGO)33,33,83                                               DSL03480
   33 K1=NB1+3                                                      DSL03490
      IF(K1-KEND)75,75,40                                           DSL03500
   75 DO 34  K=K1,KEND,3                                            DSL03510
   34 KBUFR(K)=KBUFR(K)+43                                          DSL03520
      GO TO 40                                                      DSL03530
   35 KEND=NBUF-3                                                   DSL03540
      KSW=KSW+1                                                     DSL03550
      IF(KSW-7)32,36,36                                             DSL03560
   36 IF(NPAR-6)38,38,37                                            DSL03570
   37 IF(NBUF-36)39,31,31                                           DSL03580
   38 KSW=6                                                         DSL03590
      GO TO 32                                                      DSL03600
   39 N2=NWDS-N4+6                                                  DSL03610
      KSW=5                                                         DSL03620
      N3=NPAR                                                       DSL03630
      N1=7                                                          DSL03640
      GO TO 26                                                      DSL03650
   40 IF(KGO1)59,52,54                                              DSL03660
   54 WRITE(L,254) (KBUFR(I),I=16,NBUF)                             DSL03670
  254 FORMAT(30H       GLOBAL NALRM,IZZZ(833) , 21A2)
      KGO1=0                                                        DSL03690
      KOUN=7                                                        DSL03700
      GO TO 53                                                      DSL03710
   59 WRITE(L,255)                                                  DSL03720
  255 FORMAT(12H       COMMON)                                      DSL03730
      KGO1=0                                                        DSL03740
      KOUN=0                                                        DSL03750
      GO TO 57                                                      DSL03760
   52 IF(KOUN-33)57,56,56                                           DSL03770
   56 IF(KEND-36)83,84,84                                           DSL03775
   84 KBUFR(KEND)=KBUFR(KEND)-43                                    DSL03780
      KGO1=-1                                                       DSL03790
   57 KOUN=KOUN+11                                                  DSL03800
   83 WRITE(L,251) (KBUFR(I),I=1,NBUF)                              DSL03810
  251 FORMAT(36A2)                                                  DSL03820
   53 GO TO(41,51,41,42,41,44,41,43),KSW                           DSL03830
   41 N1=N2+1                                                       DSL03840
      IF(KGO)49,49,19                                               DSL03850
   49 KBUFR(3)=IONE                                                 DSL03860
      NB1=3                                                         DSL03870
      NWDS=11                                                       DSL03880
      GO TO 19                                                      DSL03890
C...  WRITE DIMENSION STMNT FOR ALL STORAG + INTGRL VARS            DSL03892
   42 IF(NSTOR)134,134,130                                          DSL03893
```

```
130 IF(NINGL)134,134,244                                            DSL03895
244 J=2                                                              DSL03896
249 DO 245   I=1,NINGL                                              DSL03897
    IF(KSTOR(J)-INGRL(I))245,246,245                               DSL03898
246 NSTOR=NSTOR-1                                                   DSL03900
    IJ=INGRL(I)                                                     DSL03902
    KA(1)=KSYMB(IJ)                                                 DSL03903
    KA(2)=KSYMB(IJ+200)                                            DSL03904
    KA(3)=KSYMB(IJ+400)
    WRITE(L,270) KA(1),KA(2),KA(3)                                 DSL03906
270 FORMAT(16H        DIMENSION 2A2,A1,3H(1) )                     DSL03907
    DO 248   J2=J,NSTOR                                            DSL03910
    KSTOR(J2)=KSTOR(J2+1)                                          DSL03911
248 LSTOR(J2)=LSTOR(J2+1)                                          DSL03912
    GO TO 247                                                      DSL03913
245 CONTINUE                                                       DSL03914
    J=J+1                                                          DSL03915
247 IF(J-NSTOR)249,249,134                                        DSL03916
134 IF(NSPEC)141,141,140                                          DSL03918
140 WRITE(L,251)(KSPEC(I),I=1,36)                                 DSL03919
141 KGO1=1                                                        DSL03920
    NB1=15                                                        DSL03930
    NWDS=7                                                        DSL03940
 46 KSW=7                                                         DSL03950
    N3=NCOM                                                       DSL03960
    GO TO 18                                                      DSL03970
 43 KSW=5                                                         DSL03980
    N3=NPAR                                                       DSL03990
    N1=7                                                          DSL04000
    GO TO 19                                                      DSL04010
 51 IF(NING)42,42,55                                             DSL04020
 55 NB1=9                                                         DSL04030
    NWDS=9                                                        DSL04040
    KSW=3                                                         DSL04050
    N3=NING                                                       DSL04060
    KBUFR(4)=IRES(154)                                           DSL04070
    KBUFR(5)=IRES(155)                                           DSL04080
    KBUFR(6)=IRES(156)                                           DSL04090
    KBUFR(7)=IRES(157)                                           DSL04100
    KBUFR(8)=IAST2
    KBUFR(9)=IBLNK                                               DSL04120
    GO TO 18                                                     DSL04130
 44 IF(KGO)45,45,67                                              DSL04140
 45 KPOIN=NCOM+NPAR-NING-5                                       DSL04150
    KPOIT=KPOIN                                                  DSL04160
C... MOVE STORAG VARIABLES TO PARAMETER TABLE                    DSL04170
    IF(NSTOR)66,66,61                                            DSL04180
C... WRITE STORAG VARIABLES   (ADD TO COMMON)                    DSL04190
 61 DO 62   J=2,NSTOR                                            DSL04200
    NPAR=NPAR+1                                                  DSL04210
 62 KPAR(NPAR)=KSTOR(J)                                          DSL04220
    LSTOR(1)=NSTOR-1                                             DSL04230
    KPOIT=KPOIT+LSTOR(1)                                         DSL04240
    KBUFR(1)=IRES(149)                                          DSL04250
    KBUFR(13)=IRES(150)                                         DSL04260
    KBUFR(25)=IRES(151)                                         DSL04270
    DO 98   I=2,NSTOR                                            DSL04280
    KPOIT=KPOIT+LSTOR(I)                                        DSL04290
    J=KSTOR(I)                                                  DSL04300
    KA(1)=KSYMB(J)                                              DSL04310
```

346

```
      KA(2)=KSYMB(J+200)                                          DSL04320
      KA(3)=KSYMB(J+400)                                          DSL04321
      KBUFR(I)=KA(1)                                              DSL04340
      KBUFR(I+12)=KA(2)                                           DSL04350
   98 KBUFR(I+24)=KA(3)                                           DSL04360
      N1=1                                                        DSL04370
      ICOMA=IBLNK+43                                              DSL04380
      IF(NSTOR-5)91,91,90                                         DSL04390
   90 N2=5                                                        DSL04400
      GO TO 92                                                    DSL04410
   91 N2=NSTOR                                                    DSL04420
   92 WRITE(L,260) (ICOMA,KBUFR(I),KBUFR(I+12),KBUFR(I+24),       DSL04430
     1 LSTOR(I),I=N1,N2)                                          DSL04440
  260 FORMAT(6H       1 5(    3A2,A1,1H( I3,1H)))                 DSL04450
      IF(N2-NSTOR)93,66,66                                        DSL04460
   93 N1=6                                                        DSL04470
      GO TO 91                                                    DSL04480
C... WRITE FORTRAN STATEMENTS                                     DSL04490
   66 DO 60  I=1,NTSEQ                                            DSL04500
      J=KSEQ(I)                                                   DSL04510
      N1=LFORT(J)                                                 DSL04520
      N2=NFORT(J)                                                 DSL04530
      N3=IABS1(N2)
      IF(N2)74,60,74                                              DSL04550
   74 READ(11'N1)(KFORT(M),M=1,N3)
      IF(N2)47,60,58                                              DSL04570
   47 N4=1                                                        DSL04580
      DO 50  M=1,N3                                               DSL04590
      IF(KFORT(M)-IDOLR)50,48,50                                  DSL04600
   48 N5=M-1                                                      DSL04610
      WRITE(L,251) (KFORT(K),K=N4,N5)                             DSL04620
      N4=M+1                                                      DSL04630
   50 CONTINUE                                                    DSL04640
      GO TO 60                                                    DSL04650
   58 WRITE(L,251) (KFORT(K),K=1,N3)                              DSL04660
   60 CONTINUE                                                    DSL04670
C... END OF UPDAT SUBROUTINE                                      DSL04680
      WRITE(L,252)                                                DSL04690
C... WRITE SYMBOL TABLE                                           DSL04770
  252 FORMAT('        RETURN'/'       END'/'/*'/'// CALL FORTL,F1'/
     1 '// RUN'/'// PHASE NAME-SIMUL,UNIT-R1,RETAIN-P'/'// OPTIONS UPACK
     2-R1'/'// INCLUDE NAME-SIMUL,UNIT-R1'/'// END'/'// LOAD SIMX,R1'/
     3 '// RUN')
      IF(NING)165,165,160                                         DSL04780
  160 I=7                                                         DSL04790
  163 IF(KPAR(I))161,161,164                                      DSL04800
  161 NPAR=NPAR-1                                                 DSL04810
      DO 162  J=I,NPAR                                            DSL04820
  162 KPAR(J)=KPAR(J+1)                                           DSL04830
  164 I=I+1                                                       DSL04840
      IF(I-NPAR)163,165,165                                       DSL04850
  165 NSYM=NCOM+NPAR-6                                            DSL04860
      WRITE(L,253) NINT,NSYM,KPOIT                                DSL04870
  253 FORMAT(3I4)                                                 DSL04880
      KGO=1                                                       DSL04890
      NB1=0                                                       DSL04900
      NWDS=12                                                     DSL04910
      IF(NSYM-140)46,46,155                                       DSL04920
  155 WRITE(KPT,156)NSYM                                          DSL04930
  156 FORMAT(I4,23H SYMBOLS, REDUCE TO 140)                       DSL04940
```

```
        GO TO 46                                                DSL04950
   67 IF(NSTOR)65,65,63                                         DSL04960
C...   WRITE PARAM CARD TO LOCATE STORAG VARIABLE TABLES        DSL04970
   63 DO 94  I=2,NSTOR                                          DSL04980
        J=KSTOR(I)                                              DSL04990
        KA(1)=KSYMB(J)                                          DSL05000
        KA(2)=KSYMB(J+200)                                      DSL05010
        KA(3)=KSYMB(J+400)                                      DSL05011
        KBUFR(I-1)=ICOMA                                        DSL05030
        KBUFR(I+8)=KA(1)                                        DSL05040
        KBUFR(I+17)=KA(2)                                       DSL05050
        KBUFR(I+26)=KA(3)                                       DSL05060
        KPOIN=KPOIN+LSTOR(I-1)                                  DSL05070
   94 KSTOR(I)=KPOIN                                            DSL05080
        KBUFR(1)=IBLNK                                          DSL05090
        N1=1                                                    DSL05100
        IF(NSTOR-7)69,69,64                                     DSL05110
   64 N2=6                                                      DSL05120
        GO TO 71                                                DSL05130
   69 N2=NSTOR-1                                                DSL05140
   71 WRITE(L,261) (KBUFR(I),KBUFR(I+9),KBUFR(I+18),KBUFR(I+27),DSL05150
    1 KSTOR(I+1),I=N1,N2)                                       DSL05160
  261 FORMAT(6HPARAM 6(3A2,A1,1H= I3))                          DSL05170
        IF(N2-NSTOR+1)72,65,65                                  DSL05180
   72 N1=7                                                      DSL05190
        GO TO 69                                                DSL05200
   65 N3=1                                                      DSL05210
        NDAT=NDATA                                              DSL05220
        DO 70  I=1,NDAT                                         DSL05230
        READ(12'I)LETRS(N3)
        IF(LETRS(N3)-IDOLR)73,68,73                             DSL05250
   68 N3=N3-1                                                   DSL05260
        WRITE(L,251)(LETRS(K),K=1,N3)                           DSL05270
        N3=1                                                    DSL05280
        GO TO 70                                                DSL05290
   73 N3=N3+1                                                   DSL05300
   70 CONTINUE                                                  DSL05310
  150 STOP                                                      DSL05360
        END                                                     DSL05370
 /*
```

```
// CALL FORTRN,F1
// RUN
// PRINT DEVICE-'1403,5471'
*PROCESS OBJECT(R,LIB(R1)),NOLINK
      PROGRAM TRDMP
      IMPLICIT INTEGER*2(I-N)
      INTEGER*4 M
      GLOBAL NALRM,IERR,ICONT,JPT,KEY,ISTRT,KA(6),LETRS(73),ICHAR(12),     DSL15270
     1 IRES(158),KSYMB(600),NSYMB,ITYPE,NOUT,NINP,NPAR,NINT,NDATA,NFOR,    DSL15280
     2 NSORT,ISORT,NMEM,NING,NINGL,NSTOR,NTSEQ,NREAL,MFOR,                 DSL15290
     3 KOUT(100),LFORT(100),NFORT(100),LIN(100),NIN(100),INPUT(300),       DSL15300
     4 KPAR( 60),KSTOR(10),LINT(50),INGER(10),MEMRY(15),KMEM(15),          DSL15310
     5 LSTOR(10),INGRL(10),KREAL(20),KSORT(20),KFORT(202)                  DSL15320
      M=JPT                                                                DSL15340
      WRITE(M,880) (INPUT(I), I=1,NINP)                                    DSL15350
  880 FORMAT(20I6)                                                         DSL15360
      WRITE(M,860) (KOUT(I),LFORT(I),NFORT(I),LIN(I),NIN(I),I=1,NOUT)      DSL15370
  860 FORMAT(5I6)                                                          DSL15380
      WRITE(M,880) (LINT(I),I=1,NINT),(KPAR(I),I=1,NPAR)                   DSL15390
      WRITE(M,880) (INGER(I),I=1,NING),(INGRL(I),I=1,NINGL),               DSL15400
     1 (KSORT(I),I=1,NSORT)                                                DSL15410
      WRITE(M,880) (MEMRY(I),KMEM(I),I=1,NMEM)                             DSL15420
      IF(NSTOR)890,890,885                                                 DSL15430
  885 WRITE(M,880) (KSTOR(I),LSTOR(I),I=1,NSTOR)                           DSL15440
  890 J=0                                                                  DSL15445
      DO 840  I=1,NSYMB                                                    DSL15450
      KA(1)=KSYMB(I)                                                       DSL15460
      KA(2)=KSYMB(I+200)                                                   DSL15470
      KA(3)=KSYMB(I+400)                                                   DSL15480
      J=J+1                                                                DSL15485
      KFORT(J)=KA(1)                                                       DSL15490
      KFORT(J+15)=KA(2)                                                    DSL15500
      KFORT(J+30)=KA(3)                                                    DSL15510
      IF(J-15)840,830,830                                                  DSL15520
  830 WRITE(M,850)(KFORT(K),KFORT(K+15),KFORT(K+30),K=1,15)                DSL15525
  850 FORMAT(15(2X,3A2))                                                   DSL15530
      J=0                                                                  DSL15532
  840 CONTINUE                                                             DSL15534
      IF(J)870,870,855                                                     DSL15536
  855 WRITE(M,850)(KFORT(K),KFORT(K+15),KFORT(K+30),K=1,J)                 DSL15538
  870 INVOKE SORTX
      END                                                                  DSL15550
/*
```

349

```
// CALL FORTRN,F1
// RUN
// READ DEVICE-'MFCU1,5471'
// PRINT DEVICE-'1403,5471'
*PROCESS OBJECT(R,LIB(R1)),NOLINK
      PROGRAM SIMX
      IMPLICIT INTEGER*2(I-N)
      GLOBAL NALRM,NINTG,NSYMB,KPT,KPRNT,KRANG,KFINI,KTITL,KREL,KABS,   DSL48250
     1 KMEM,IDUMY(393),DELTG,JDUMY(428),C(500)                          DSL48260
      IF(KPT-INPT(3))1,3,1                                              DSL48270
    1 KPT=INPT(3)                                                       DSL48280
      CALL INITL                                                        DSL48290
    3 CALL INTRA                                                        DSL48300
      IF(DELTG)5,5,4                                                    DSL48310
    4 INVOKE PLOTG
    5 CONTINUE
      INVOKE SIMUL
      END                                                               DSL48340
/*
```

```
// CALL FORTRN,F1
// RUN
// PRINT DEVICE-'1403,5471'
// READ DEVICE-'MFCU1,5471'
*PROCESS OBJECT(R,LIB(R1)),NOLINK
      PROGRAM SIMUL
      IMPLICIT INTEGER*2(I-N)
      GLOBAL NALRM,NINTG,NSYMB,KPT,KPRNT,KRANG,KFINI,KTITL,KREL,KABS,   DSL33280
     1 KMEM,KPOIN,IDUMY(822),C(500)                                     DSL33290
      K1=KPOIN                                                          DSL33300
      K2=KPOIN+NINTG                                                    DSL33310
      CALL MILNE(C(KMEM),C(KABS),C(KREL),C(K1),C(K2))                   DSL33320
      INVOKE SIMX
      END                                                               DSL33340
/*
```

350

```
// CALL FORTL,F1
// RUN
// PHASE NAME-OUTPX,UNIT-R1,RETAIN-P
// OPTIONS UPACK-R1
// INCLUDE NAME-OUTPX,UNIT-R1
// END
// CALL FORTL,F1
// RUN
// PHASE NAME-DSL,UNIT-R1,RETAIN-P
// OPTIONS UPACK-R1
// INCLUDE NAME-DSL,UNIT-R1
// END
// CALL FORTL,F1
// RUN
// PHASE NAME-SORTX,UNIT-R1,RETAIN-P
// OPTIONS UPACK-R1
// INCLUDE NAME-SORTX,UNIT-R1
// END
// CALL FORTL,F1
// RUN
// PHASE NAME-SIMX,UNIT-R1,RETAIN-P
// OPTIONS UPACK-R1,CORE-28K
// INCLUDE NAME-SIMX,UNIT-R1
// END
// CALL FORTL,F1
// RUN
// PHASE NAME-TRDMP,UNIT-R1,RETAIN-P
// OPTIONS UPACK-R1
// INCLUDE NAME-TRDMP,UNIT-R1
// END
```

```
// END
// LOAD DSL,R1
// FILE UNIT-R1,PACK-R1R1R1,NAME-FT00012,LABEL-FILE12,RETAIN-T,TRACKS-20
// FILE UNIT-R1,PACK-R1R1R1,NAME-FT00011,LABEL-FILE11,RETAIN-T,TRACKS-20
// RUN
TITLE   DYSAC TEST PROB. USING TRNFR                                     DSL90640
TITLE   DSL/1800      8/25/66                                            DSL90650
RENAME   TIME=DISPL, DELT=DELP                                           DSL90660
INTGER   J                                                               DSL90670
INCON    J=1                                                             DSL90680
NOSORT                                                                   DSL90690
        GO TO(3,7),J                                                     DSL90700
      3 A(5)=1.0                                                         DSL90710
        A(4)=.864/.016                                                   DSL90720
        A(3)=3.268/.016                                                  DSL90730
        A(2)=3.42/.016                                                   DSL90740
        A(1)=1./.016                                                     DSL90750
        J=2                                                              DSL90760
      7 CONTINUE                                                         DSL90770
SORT                                                                     DSL90780
PARAM   INPUT=20.4, FGAIN= 10.                                           DSL90790
STORAG   A(5), B(2)                                                      DSL90800
TABLE     B(1-2)=312.5, 375.                                             DSL90810
PARAM ERR=0.01,Y0=0.5
      YY=IMPL(Y0,ERR,FY)
      FY=0.8+SIN(YY)
PARAM Y0=0.1
      LOAD=TRNFR(1,4,B,A,ERROR,LOAD)                                     DSL90820
      ERROR=INPUT-FGAIN*LOAD                                             DSL90830
      WW=DEBUG(5,0.)                                                     DSL90840
INTEG    RKS                                                             DSL90850
RELERR   LOAD=1.E-3                                                      DSL90855
ABSERR   LOAD=1.E-4                                                      DSL90860
CONTRL DELP=.005, FINTI=6.0                                              DSL90870
PRINT   0.2,ERROR,LOAD,DELP                                              DSL90880
END                                                                     DSL90920
STOP                                                                    DSL90930
```

352

```
                    *** DSL/1800 TRANSLATOR INPUT ***
TITLE    DYSAC TEST PROB. USING TRNFR
TITLE    DSL/1800     8/25/66
RENAME   TIME=DISPL, DELT=DELP
INTGER   J
INCON    J=1
NOSORT
         GO TO(3,7),J
      3  A(5)=1.0
         A(4)=.864/.016
         A(3)=3.268/.016
         A(2)=3.42/.016
         A(1)=1./.016
         J=2
      7  CONTINUE
SORT
PARAM    INPUT=20.4, FGAIN= 10.
STORAG   A(5), B(2)
TABLE    B(1-2)=312.5, 375.
PARAM    ERR=0.01,Y0=0.5
         YY=IMPL(Y0,ERR,FY)
         FY=0.8+SIN(YY)
PARAM    Y0=0.1
         LOAD=TRNFR(1,4,B,A,ERROR,LOAD)
         ERROR=INPUT-FGAIN*LOAD
         WW=DEBUG(5,0.)
INTEG    RKS
RELERR   LOAD=1.E-3
ABSERR   LOAD=1.E-4
CONTRL   DELP=.005, FINTI=6.0
PRINT    0.2,ERROR,LOAD,DELP
END
STOP
LINT FROM DSL
    11       0       0       0       0


OUTPUT VARIABLE SEQUENCE
    1   2   3   4   5   6   7   8  12  11   9  10  13

STORAGE USED/MAXIMUM
INTGR   4/50, IN VARS  12/300, OUT VARS  13/100, PARAMS 11/60, SYMBOLS  34/200, FC
```

```
// CALL FORTRN,F1
// RUN
*PROCESS OBJECT(R,LIB(R1))
      SUBROUTINE UPDAT
      IMPLICIT INTEGER*2(I-N)
      REAL  LOAD ,INPUT,IMPL
      INTEGER *2  J
      GLOBAL NALRM,IZZZ(833) ,DISPL,DELP ,DELMI,FINTI,DELTP,DELTC,ZZ001,
     1ZZ002,ZZ003,ZZ004,YY   ,LOAD ,ERROR,WW    ,YO  ,INPUT,FGAIN,ERR  ,
     1INZZO,J
     1 ,ZZ990( 2) ,A   ( 5) ,B    ( 2)
      GO TO(3,7),J
    3 A(5)=1.0
      A(4)=.864/.016
      A(3)=3.268/.016
      A(2)=3.42/.016
      A(1)=1./.016
      J=2
    7 CONTINUE
      ERROR=INPUT-FGAIN*LOAD
      LOAD=TRNFR(1,4,B,A,ERROR,LOAD)
 9100 YY=IMPL(499,YO,ERR,FY)
      IF(NALRM) 9200, 9200, 9300
 9300 CONTINUE
      FY=0.8+SIN(YY)
      GO TO 9100
 9200 CONTINUE
      WW=DEBUG(498,5,0.)
      RETURN
      END
/*
// CALL FORTL,F1
// RUN
// PHASE NAME-SIMUL,UNIT-R1,RETAIN-P
// OPTIONS UPACK-R1
// INCLUDE NAME-SIMUL,UNIT-R1
// END
// LOAD SIMX,R1
// RUN
    4   21   29
DISPL DELP  DELMI FINTI DELTP DELTC ZZ001 ZZ002 ZZ003 ZZ004 YY     LOAD
ERROR WW    YO    INPUT FGAIN ERR   J     A     B
PARAM   A    = 22 ,B   = 27
TITLE   DYSAC TEST PROB. USING TRNFR
TITLE   DSL/1800    8/25/66
INCON   J=1
PARAM   INPUT=20.4, FGAIN= 10.
TABLE   B(1-2)=312.5, 375.
PARAM ERR=0.01,YO=0.5
PARAM YO=0.1
INTEG   RKS
RELERR  LOAD=1.E-3
ABSERR  LOAD=1.E-4
CONTRL DELP=.005, FINTI=6.0
PRINT  0.2,ERROR,LOAD,DELP
END
```

LISTINGS FOR SOURCE CHANGES
IN DSL FOR SYSTEM/7
AND CORRESPONDING COMPILE PROCEDURES

```
**************************************************************************
*                                                                        *
*      THIS APPENDIX EXPLAINS THE PROCEDURES REQUIRED FOR GENERATING      *
*      THE DSL PROGRAM ON IBM SYSTEM/7. THE FOLOWING STEPS ARE REQUIRED   *
*      IN THIS CONVERSION.                                                *
*                                                                        *
*                                                                        *
*      A. START WITH THE SYSTEM/3 DSL SOURCE LISTING AS SHOWN IN          *
*         APPENDIX B.                                                     *
*                                                                        *
*      B. REMOVE THE FORTRAN COMPILER CONTROL STATEMENTS AND REPLACE      *
*         THEM WITH SYSTEM/7 COMPILER CONTROL STATEMENTS FOR EACH         *
*         ROUTINE. TYPICAL SYSTEM/7 COMPILER CONTROL STATEMENTS ARE       *
*         SHOWN IN ONE OF THE NEW ROUTINES FOR SYSTEM/7 DSL DESCRIBED     *
*         LATER IN THIS APPENDIX.                                         *
*                                                                        *
*      C. ADD TWO NEW ROUTINES AS SHOWN LATER. THEY ARE NAMED SHIFT       *
*         AND REMSW. REMSW EXAMINES DIGITAL INPUT GROUP 2 IN MODULE 1     *
*         OF SYSTEM/7 HARDWARE, WHERE 16 CONTACT SWITCHES ARE TO BE       *
*         CONNECTED FOR DATA SWITCH PURPOSE.                              *
*                                                                        *
*      D. ADD A NEW ASSEMBLY LANGUAGE PROGRAM NAMED DSLMAIN AS SHOWN      *
*         LATER. THIS PROGRAM PROVIDES THE SUPERVISOR REQUIRED BY         *
*         SYSTEM/7 BATCH NUCLEUS DSS/7. IF YOUR DIGITAL INPUT FOR         *
*         DATA SWITCHES IS OTHER THAN MODULE 1 GRP 2 MAKE THE SOURCE      *
*         CHANGE IN DSLMAIN.                                              *
*                                                                        *
*      E. MAKE MINOR SOURCE CHANGES IN TRANS, SORT, AND OUTPT TO          *
*         A SINGLE STORAGE LOAD FOR THE TRANSLATOR PHASE. ALSO            *
*         MAKE CHANGES IN OUTPT TO PUNCH DSL OUTPUT WITH A SPECIAL        *
*         SYSTEM/7 COMMAND CONTROL PROCEDURE TO EXECUTE THE SIMULATION    *
*         PHASE.                                                          *
*                                                                        *
*      F. SET UP CONTROL LANGUAGE PROCEDURES IN THE SYSTEM/7 PROCEDURE    *
*         LIBRARY $PROCLIB TO PERMIT COMPILE, ASSEMBLE AND LINK OF DSL    *
*         MODULES AND PLACE THE LINK OBJECT IN A SPECIAL DSL LIBRARY.     *
*         THIS LIBRARY WIL  THEN BE USED IN THE TRANSLATOR AS WELL AS     *
*         THE SIMULATION PHASE.                                           *
*                                                                        *
**************************************************************************
```

```
***********************************************************************
*                                                                     *
*                                                                     *
*            FIRST WE DESCRIBE THE COMMAND CONTROL LANGUAGE FOR ASSEMBLING   *
*            THE DSLMAIN MODULE FOLOWED BY THE SOURCE OF DSLMAIN.      *
*                                                                     *
*            NEXT WE DESCRIBE THE COMMAND CONTROL LANGUAGE FOR COMPILING     *
*            ALL THE DSL ROUTINES. WE ALSO DESCRIBE THE COMMAND CONTROL  *  *
*            LANGUAGE FOR THE SIMULATION PORTION, AS REQUIRED BY DSL    *
*            TRANSLATOR OUTPUT FROM SUBROUTINE OUTPT.                  *
*                                                                     *
*            THE LAST PART OF THIS APPENDIX SHOWS THE DSL MODULE CHANGES    *
*            IN THE DSL SOURCE. ONLY PART OF THE SOURCE IS SHOWN TO AVOID   *
*            REDUNDANCY.                                               *
*                                                                     *
***********************************************************************
//                                                                    *
***********************************************************************
*                                                                     *
*      PROCEDURE TO ASSEMBLE AND LINK DSLMAIN ASSEMBLER PROGRAM        *
*      THIS PROGRAM PROVIDES THE INTERFACE TO DSS/7 SUPERVISOR.        *
*      THE PROCEDURE ASSUMES THAT YOU HAVE A DISK LABELLED DSLS79,AND  *
*      THAT YOU HAVE A DATA SET LABELLED DSLLIB ON THAT DISK.          *
*      ALL DSL MODULES WILL GO ON TO THIS LIBRARY, SO IT SHOULD BE     *
*      SUFFICIENTLY LARGE, ABOUT 150 TRACKS.                          *
*      INPUT IS ASSUMED FROM 129 CARD READER, OUTPUT LISTINGS TO 7431  *
*      PRINTER. MAKE CHANGES FOR YOUR SYSTEM. CONSULT A SYSTEM/7 PROGRAMMER. *
*                                                                     *
***********************************************************************
R $DLT,$$
CALL $UDDLT
MA,,SYSF00,$WORK
//
R KEEP
R SYS001,,SYSF00,$WORK,SYS001,(NEW,KEEP,1700)
R SYS003,,SYSF00,$WORK,SYS003,(NEW,KEEP,1100)
R SYS002,,SYSF00,$WORK,SYS002,(NEW,KEEP,1700)
R ASMLIB,,SYSR00,$MACLIB
R ASMOUT,$SYPR
R ASMPCH,,SYSF00,$WORK,$PASS,(NEW,KEEP,400)
R ASMIN,$SYCD
*
R LNKOUT,$SYPR
R LNKLB1,,SYSF00,$MSP7LIB
R LNKOBJ,,SYSF00,$WORK,$PASS,(OLD,DLT)
R LNKIN,$SYCD
R LNKPCH,,DSLS79,DSLLIB
PARM $ASM,NOESD,NORLD,NOXREF
*
CALL $ASM
CALL $LE
*
*
*************************************************************************
*
*
*      THE FOLLOWING IS THE SOURCE FOR DSLMAIN   ASSEMBLER PROGRAM
*
*
PROC DSLAL
R ASMOUT,$SYLP
R DATALB,,SYSF00,$LINKLIB
R ASMIN,*
          PRINT NOGEN
DSLMAIN   #PROC MF=SYSTEM,OPTIONS=MAIN,WA=71,EPA=MIMENT
          EXTRN MAIN,CADCLOS2,CADCLOSE,CADERHND
```

356

```
              ENTRY  DSLMAIN
MIMENT        EQU    ¤
              LR     $XR7
              ST     SAVE7
              LXL    $XR5,@BPCOM
              USING  #BPCOMD,$XR5
              L      CLOSADR
              ST     #BPPCRA
              BALL   MAIN,$XR7
CLSEUNIT      EQU    ¤
              LXL    $XR5,FUNTB
              L      0,$XR5
              CR     $ACC
              ST     UNITCNT
              AI     1,$XR5
              STX    SAVE5,$XR5
CLSELOOP      EQU    ¤
              BALL   CADCLOS2,$XR7
              L      SAVE5
              AI     5,$ACC
              ST     SAVE5
              STR    $XR5
              AS     UNITCNT
              B      CLSELOOP
              L      SAVE7
              BA
FUNTB         DC     A(CADFUNTB)
CLOSADR       DC     A(CLSEUNIT)
UNITCNT       DS     F
SAVE5         DS     F
SAVE7         DS     F
¤             FILE DEFINITION MACROS
              #FUN   UNIT=1,AM=KB
              #FUN   UNIT=2,AM=PT
              #FUN   UNIT=3,AM=SAM,CB=UNO3
              #FUN   UNIT=4,AM=SAM,CB=UNO4
              #FUN   UNIT=5,AM=SAM,CB=UNO5
              #FUN   UNIT=6,AM=SAM,CB=UNO6
              #FUN   UNIT=7,AM=SAM,CB=UNO7
              #FUN   END
UNO3          #FBUF  128,TYPE=RANDOM
UNO4          #FBUF  128,TYPE=RANDOM
UNO5          #FBUF  128,TYPE=CARD,LRECL=256
UNO6          #FBUF  128,TYPE=PRINT,LRECL=256
UNO7          #FBUF  128,TYPE=CARD,LRECL=256
¤
¤      CHANGE THE MODULE (MA) AND ROUP NO IF YOU HAVE A DIFFERENT
¤      MODULE AND GROUP FOR DATA SWITCHES.
¤
DI            #DDIG  MA=1,GROUP=2
              ENTRY  DI
              $SIO
              $CC
¤      ENTRY POINTS TO $OPR UNUSED NUCLEAR ENTRY POINTS
$OPR          BL     @OPR
              ENTRY  $OPR
@OPRG         EQU    ¤
              ENTRY  @OPRG
@NUCOM        EQU    ¤
              ENTRY  @NUCOM
$SVERR        EQU    ¤
              ENTRY  $SVERR
@DTRNS        EQU    ¤
              ENTRY  @DTRNS
              BALL   CADERHND,$LNKR
```

```
              DC      X'0412'
              $EBAS
              $VOLN
              COPY    #BPCOMD
              END
/*
         ACTION MAP,NOAUTO
         PHASE DSLMAIN,*
          INCLUDE
          ENTRY DSLMAIN
/*
/*

//                                                                              *
********************************************************************************
*                                                                              *
*      $CPFCL - SPECIAL FORTRAN COMPILE AND LINK PROCEDURE FOR DSL ON S/7       *
*                                                                              *
*      USES CARD READER 129 FOR INPUT AND 7431 FOR OUTPUT                       *
*      ASSUMES A DATA SET NAMED DSLLIB ON DISK LABELLED DSLS79 AVAILABLE        *
*      ASSUMES THAT THE FORTRAN COMPILER IS IN $LINKLIB ON SYSF00.              *
*      OUTPUT FROM LINK EDITOR ALSO GOES TO DSLLIB.                             *
*                                                                              *
********************************************************************************
*                                                                              *
R DLT,$$
CALL $UDDLT
MA,,SYSF00,$WORK
//
*
R KEEP
R FTIN,$SYCD
R FTPR,$SYPR
R FTPU,,SYSF00,$WORK,$TEMP,(NEW,KEEP,600)
R FTWK,,SYSF00,$WORK,SYSUT1,(NEW,KEEP,1200)
*
R LNKIN,$SYCD
R LNKOUT,$SYPR
R LNKPCH,,DSLS79,DSLLIB
R LNKOBJ,,SYSF00,$WORK,$TEMP
R LNKLB1,,SYSF00,$MSP7LIB
R SYS001,,SYSF00,$WORK,SYSUT1,(OLD,DLT)
R SYS002,,SYSF00,$WORK,SYSUT2,(NEW,DLT,100)
*
CALL $FORT
*
CALL $LE
*
*      END OF SPECIAL $CPFCL
*
```

358

```
      ACTION NOMAP
      PHASE X,ROOT
      INCLUDE $FWA
      PHASE $LINKSET
      INCLUDE DSLMAIN
      INCLUDE
      ENTRY DSLMAIN
      //
      *************************************************************************
      *      $CPMCLFG - SPECIAL  FORTRAN COMPILE, LINK, FORMAT AND GO          *
      *      PROCEDURE FOR THE DSL SIMULATOR PORTION.                          *
      *      ASSUMES YOU HAVE ALL FORTRAN LIBRARY SUBROUTINES EITHER           *
      *      IN $MSP7LIB ON SYSF00 OR IN DSLLIB ON DSLS79.                     *
      *      ALSO ASSUMES ALL YOUR DSL SUBROUTINE OBJECT IS ON DSLLIB.         *
      *                                                                        *
      *************************************************************************
      R $DLT,$$
      CALL $UDDLT
      MA,,SYSF00,$WORK
      //
      *
      R KEEP
      R FTIN,$SYCD
      R FTPR,$SYPR
      R FTPU,,SYSF00,$WORK,$TEMP,(NEW,KEEP,600)
      R FTWK,,SYSF00,$WORK,SYSUT1,(NEW,KEEP,1200)
      *
      R LNKIN,,SYSF00,$PROCLIB,$CPMCLFG
      R LNKOUT,$SYPR
      R LNKPCH,,SYSF00,$WORK
      R LNKOBJ,,SYSF00,$WORK,$TEMP
      R LNKLB1,,SYSF00,$MSP7LIB
      P LNKLB2,,DSLS79,DSLLIB
      R SYS001,,SYSF00,$WORK,SYSUT1,(OLD,DLT)
      R SYS002,,SYSF00,$WORK,SYSUT2,(NEW,DLT,100)
      R SYSIN,,SYSF00,$WORK
      R SYSOUT,,SYSF00,$WORK
      *    USER TO PROVIDE REFERS FOR UNITS 3 AND 4 IF REQUIRED
      R @00005,$SYCD
      R @00006,$SYPR
      R @00001,$SYKB
      *
      R JOBLIB,,SYSF00,$WORK
      CALL $FORT
      *
      CALL $LE
      *
      CALL $UDFMT,FM,$LINKSET,A,,M,,DL
      *
      CALL PHASE$$$
      R $DLT,$$
      *
      *      END OF $CPMCLFG
```

```
      PROC $CPFCL
      R FTIN,$SYCD
      R LNKPCH,,DSLS79,DSLLIB
      PARM $FORT,NOCMPAT
      XEQ
            FUNCTION IABS1(I)
            IMPLICIT INTEGER*2(I-N)
            IABS1=IABS(I)
            RETURN
            END
      /*
               ACTION NOAUTO,NOMAP
               PHASE IABS1,*
               INCLUDE
               ENTRY
      /*
      PROC $CPFCL
      R FTIN,$SYCD
      R LNKPCH,,DSLS79,DSLLIB
      PARM $FORT,NOCMPAT
      XEQ
            SUBROUTINE REMSW(N,M)
            INTEGER*2 SPECS(2) /1,0/,DIGLOC,ERCODE
            M=2
            CALL DIW (1,SPECS,DIGLOC,ERCODE)
            IF (ERCODE.NE.1) GO TO 99
               M1 = DIGLOC
               M2 =ISHFT(M1,N)
               IF (M2.LT.0) M=1
               RETURN
         99 WRITE(1,98)
         98 FORMAT(' DIGITAL INPUT ERROR IN DATASWITCH')
            RETURN
               END
      /*
               ACTION NOAUTO,NOMAP
               PHASE REMSW,*
               INCLUDE
               ENTRY
      /*
      PROC $CPFCL
      R FTIN,$SYCD
      R LNKPCH,,DSLS79,DSLLIB
      PARM $FORT,NOCMPAT
      XEQ
            SUBROUTINE  SHIFT(N1,N2,N3)
            IMPLICIT INTEGER*2(I-N)
            IF(N3.LT.1) GO TO 1
            IF(N3.GT.1) GO TO 3
            N=ISHFT(N1,8)
            N2=ISHFT(N,-8)
            N1=ISHFT(N1,-8)
            RETURN
          1 N1=ISHFT(N1,8)
            N3=ISHFT(N2,-8)
            N3=ISHFT(N3,8)
            N1=IOR(N1,N3)
```

360

```
          RETURN
        3 N1=ISHFT(N1,-8)
          N1=ISHFT(N1,8)
          N3=ISHFT(N2,8)
          N3=ISHFT(N3,-8)
          N1=IOR(N1,N3)
          RETURN
          END
/*
          ACTION NOAUTO,NOMAP
          PHASE SHIFT,*
          INCLUDE
          ENTRY
/*
PROC $CPFCL
R FTIN,$SYCD
R LNKPCH,,DSLS79,DSLLIB
PARM $FORT,NOCMPAT
XEQ
          SUBROUTINE PACK
          IMPLICIT INTEGER*2(I-N)
          DIMENSION KX(6)
          GLOBAL IDM(6),KA(6),LETRS(72),IDMY(1808),KFORT(2)
          M=0
          DO 1 I=1,6
        1 KX(I)=KA(I)
          DO 2 I=1,5,2
        2 CALL SHIFT(KX(I),KX(I+1),M)
          KA(1)=KX(1)
          KA(2)=KX(3)
          KA(3)=KX(5)
          RETURN
          END
/*
          ACTION NOAUTO,NOMAP
          PHASE PACK,*
          INCLUDE
          ENTRY
/*
PROC $CPFCL
R FTIN,$SYCD
R LNKPCH,,DSLS79,DSLLIB
PARM $FORT,NOCMPAT
XEQ
          SUBROUTINE START
          IMPLICIT INTEGER*2(I-N)
          DIMENSION JDATA(105),JDAT2(53),JDAT1(66),JCON1(10),JCNST(12)
          DIMENSION KDATA(4)
          GLOBAL IDM(6),KA(6),LT(73),ICHAR(12),IRES(158),KSYMB(600),KX(962),
         1 KMEM(15),KDMY(269)
          DATA KDATA/'CA','LL','D ',ZFF00/
          DATA JDATA/'PA','RA','M ','CO','NS','T ','IN','CO','N ','AF','GE',
         1 'N ','NL','FG','EN','TA','BL','E ','PR','OC','ED','ME','MO','RY'
         2,'ST','OR','AG','IN','TG','ER','RE','NA','ME','IN','TG','RL','PR',
         3'IN','T ','CO','NT','RL','RE','LE','RR','AB','SE','RR','RA','NG',
         4'E ','FI','NI','SH','EN','D ',' ','ST','OP',' ','SO','RT',' ',
         5'NO','SO','RT','EN','DP','RO','TI','TL','E ','SC','OP','E ','GR',
         6'AP','H ','LA','BE','L ','SC','AL','E ','CO','NT','IN','IN','TE',
```

361

```
      7'G ','RE','SE','T ','CH','AR','T ','TY','PE',' ','DU','MP',' ',
      8'/L','OC','AL'*
       DATA JDAT2/'$$',' ',' ',' ',' ','IF','(N','AL','RM',') ','92','00',
     1           ',',' ','92','00',',',' ','93','00','$$','93','00',' ','CO',
     2           'NT','IN','UE','$$',' ',' ',' ',' ','GO',' T','O ','91',
     3           '00','$$','92','00',' ','CO','NT','IN','UE','$$','ZZ',
     4           '99','0 ','RE','AL','IN','TE','GE','R ',' 1'/
       DATA JDAT1/'TI','ME',' ',' ','DE','LT',' ',' ','DE','LM','I ','FI','NT',
     1'I ',' ','DE','LT','P ','DE','LT','C ','IN','TG','R ','RE','AL','P ',
     2'MO','DI','N ','TR','NF','R ','CM','PX','P ','LE','DL','G ','DE',
     3'BU','G ','DE','RI','V ','HS','TR','S ','IM','PU','L ','PU','LS',
     4'E ','RS','T ',' ','ZH','OL','D ','IM','PL',' ','IM','PL','C ',
     5'DE','LA','Y '/
       DATA JCON1/ 1,6,2,3,2,1,1,2,3,0/,JCNST/107,77,108,93,76,123,78,80,
     1           96,92,75,97/
C     INITIALIZE IRES **
      J=1
      DO 400 I=1,35
      IRES(I)=JDATA(J)
      IRES(I+35)=JDATA(J+1)
      IRES(I+70)=JDATA(J+2)
  400 J=J+3
      DO 401 I=1,53
  401 IRES(I+105)=JDAT2(I)
C     INITIALIZE KSYMB. NOTICE DIMENSION OF KSYMB INCREASED TO 600.
      J=1
      DO 402 I=1,22
      KSYMB(I     )=JDAT1(J)
      KSYMB(I+200)=JDAT1(J+1)
      KSYMB(I+400)=JDAT1(J+2)
  402 J=J+3
C     INITIALIZE ICHAR AND KMEM
      DO 403  I=1,12
  403 ICHAR(I)=JCNST(I)
      DO 404 I=1,10
  404 KMEM(I)=JCON1(I)
      DO 500 I=1,4
  500 KA(I)=KDATA(I)
      RETURN
      END
/*
       ACTION NOAUTO,NOMAP
       PHASE START,*
       INCLUDE
       ENTRY
/*
PROC $CPFCL
R FTIN,$SYCD
R LNKPCH,,DSLS79,DSLLIB
PARM $FORT,NOCMPAT
XEQ
      SUBROUTINE BUILD
      IMPLICIT INTEGER*2(I-N)
      GLOBAL IDM(222),IWRD(3),KA(6),LETRS(72)
      M=0
      CALL SHIFT(KA(1),KA(2),M)
      IWRD(1)=KA(1)
      CALL SHIFT(KA(3),KA(4),M)
```

362

```
         KA(2)=KA(3)
         IWRD(2)=KA(3)
         CALL SHIFT(KA(5),KA(6),M)
         IWRD(3)=KA(5)
         KA(3)=KA(5)
         RETURN
         END
/*
         ACTION NOAUTO,NOMAP
         PHASE BUILD,*
         INCLUDE
         ENTRY
/*
PROC $CPFCL
R FTIN,$SYCD
R LNKPCH,,DSLS79,DSLLIB
PARM $FORT,NOCMPAT
XEQ
         SUBROUTINE SETUP
         IMPLICIT INTEGER*2(I-N)
         INTEGER*2 JDATA(93),JDAT1(8)
         GLOBAL IDM(303),IRES(101)
         DATA JDATA/'PA','RA','M ','IN','CO','N ','CO','NS','T ','PR','IN',
        1'T ','SC','OP','E ','GR','AP','H ','RA','NG','E ','FI','NI','SH',
        2'RE','LE','RR','AB','SE','RR','LA','BE','L ','CH','AR','T ','TI',
        3'TL','E ','CO','NT','RL','CO','NT','IN','IN','TE','G ','AF','GE',
        4'N ','NL','FG','EN','ST','OP','  ','EN','D ','  ','SC','AL','E ',
        5'TA','BL','E ','RE','SE','T ','TY','PE','  ','EN','**','  ','**',
        6'**','**','MI','LN','E ','RK','S ','   ','RK','SF','X ','AD','AM',
        7'S ','TR','AP','Z '/
         DATA JDAT1/'DS','L/','18','00','OV','ER','LA','Y '/
         DATA IEOF/ZC415*
         J=1
         DO 2 I=1,31
         IRES(I)=JDATA(J)
         IRES(I+31)=JDATA(J+1)
         IRES(I+62)=JDATA(J+2)
        2 J=J+3
         DO 3 I=1,8
        3 IRES(I+93)=JDAT1(I)
         IRES(55)=IEOF
         RETURN
         END
/*
         ACTION NOAUTO,NOMAP
         PHASE SETUP,*
         INCLUDE
         ENTRY
/*
PROC $CPFCL
R FTIN,$SYCD
R LNKPCH,,DSLS79,DSLLIB
PARM $FORT,NOCMPAT
XEQ
         SUBROUTINE UNPAK
         IMPLICIT INTEGER*2(I-N)
         RETURN
         END
```

```
/*
        ACTION NOAUTO,NOMAP
        PHASE UNPAK,*
        INCLUDE
        ENTRY
/*
PROC $CPFCL
R FTIN,$SYCD
R LNKPCH,,DSLS79,DSLLIB
PARM $FORT,NOCMPAT
XEQ
        SUBROUTINE OPEN
        IMPLICIT INTEGER*2(I-N)
        GLOBAL ID(6),KA(6),LETRS(72),IDUMY(1808),KFORT(2)
        DIMENSION LX(72)
        J=1
        DO 100 I=1,36
        M2=1
        M=LETRS(I)
        CALL SHIFT(M,M1,M2)
        LX(J)=M
        LX(J+1)=M1
  100 J=J+2
        DO 101 I=1,72
  101 LETRS(I)=LX(I)
        RETURN
        END
/*
        ACTION NOAUTO,NOMAP
        PHASE OPEN,*
        INCLUDE
        ENTRY
/*
PROC $CPFCL
R FTIN,$SYCD
R LNKPCH,,DSLS79,DSLLIB
PARM $FORT,NOCMPAT
XEQ
        SUBROUTINE SPLIT
        IMPLICIT INTEGER*2(I-N)
        DIMENSION LX(80)
        GLOBAL IDM(225),KA(6),LETRS(72)
        M2=1
        J=1
        DO 100 I=1,36
        M=LETRS(I)
        CALL SHIFT(M,M1,M2)
        LX(J)=M
        LX(J+1)=M1
  100 J=J+2
        DO 101 I=1,72
  101 LETRS(I)=LX(I)
        RETURN
        END
/*
        ACTION NOAUTO,NOMAP
        PHASE SPLIT,*
        INCLUDE
```

364

```
          ENTRY
/*
PROC $CPFCL
R FTIN,$SYCD
R LNKPCH,,DSLS79,DSLLIB
PARM $FORT,NOCMPAT
XEQ
      SUBROUTINE BNDEC
      IMPLICIT INTEGER*2(I-N)
      GLOBAL ID(6),KA(6),LETRS(72),IDUMY(1808),KFORT(128)
      NB1=KA(6)/100
      NB2=KA(6)-NB1*100
      NB3=NB2/10
      NB4=NB2-NB3*10
      M=1
      IF(KA(5))5,9,6
    5 K=-KA(5)+1
      KA(1)=77
      GO TO 7
    9 KA(1)=64
      K=1
      GO TO 7
    6 K=KA(5)+1
      KA(4)=107
      M=0
    7 KA(M+1)=NB1+240
      KA(M+2)=NB3+240
      KA(M+3)=NB4+240
      CALL PACK
      KFORT(K)=KA(1)
      KFORT(K+1)=KA(2)
      RETURN
      END
/*
      ACTION NOAJTO,NOMAP
      PHASE BNDEC,*
      INCLUDE
      ENTRY
/*
/*
```

```
************************************************************************
*                                                                      *
*        THIS PORTION DESCRIBES THE SOURCE REPLACEMENT FOR SORT,        *
*         OUTPT AND TRDMP. ONLY CHANGES REQUIRED ARE SHOWN AND NOT      *
*         THE ENTIRE SOURCE OF ALL THE MODULES. NOTE THESE THREE BECOME *
*         SUBROUTINES AND NOT MAIN PROGRAM OVERLAYS AS IS THE CASE FOR  *
*         1130, 1800 AND SYSTEM/3.                                      *
*                                                                      *
************************************************************************
PROC $CPFCL
R FTIN,$SYCD
R LNKPCH,,DSLS79,DSLLIB
PARM $FORT,NOCMPAT
XEQ
      SUBROUTINE SORTX
      IMPLICIT INTEGER*2(I-N)
      GLOBAL NALRM,IERR,ICONT,JPT,KEY,ISTRT,KA(6),LETRS(73),ICHAR(12),   DSL07550
     1 IRES(158),KSYMB(600),NSYMB,ITYPE,NOUT,NINP,NPAR,NINT,NDATA,NFOR,  DSL07560
     2 NSORT,ISORT,NMEM,NING,NINGL,NSTOR,NTSEQ,NREAL,MFOR,               DSL07570
     3 KOUT(100),LFORT(100),NFORT(100),LIN(100),NIN(100),INPUT(300),     DSL07580
     4 KPAR( 60),KSTOR(10),LINT(50),INGER(10),MEMRY(15),KMEM(15),        DSL07590
     5 LSTOR(10),INGRL(10),KREAL(20),KSORT(20),KFORT(202),               DSL07600
     6        KSEQ(120),KSEQ2(30),LIN2(30),NIN2(30),KSEQ3(10)            DSL07602
      KPT=JPT                                                            DSL07604
      JINT=1                                                            DSL07606
      J2=0                                                              DSL07608
      KGO=1                                                             DSL07610
      NSOR1=1                                                           DSL07612
      NOUT1=1                                                           DSL07614
      MFOR=MFOR-1                                                       DSL07616
C     .......
C     .......
C     .........
C     THE REST OF THE SOURCE IS THE SAME AS THE DSL 1800 EXCEPT
C      THE LAST FEW STATEMENTS WHICH FOLLOW.....
C     .....
C     .....
  130 NTSEQ=J2                                                          DSL09750
      WRITE(KPT,205) (KSEQ(I),I=1,J2)                                   DSL09760
  205 FORMAT(25HOOUTPUT VARIABLE SEQUENCE /(20I4))                      DSL09770
  400 WRITE(KPT,900) NINT,NINP,NOUT,NPAR,NSYMB,MFOR,NDATA               DSL09772
  900 FORMAT(21HOSTORAGE USED/MAXIMUM /6H INTGR I4,12H/50, IN VARS I4,   DSL09773
     1 14H/300, OUT VARS I4,12H/100, PARAMS I3,12H/60, SYMBOLS I4,       DSL09774
     2 14H/200, FORT WDS I5,15H/5000, DATA WDS I4,5H/1000)              DSL09775
      IF(IERR)150,151,150                                              DSL09777
  150 WRITE(KPT,1111)
 1111 FORMAT(' NO OUTPUT FROM DSL- HARD ERROR FOUND'/)
      RETURN
  151 ISTRT=INPT(2)                                                     DSL09779
C     *****   THIS STATEMENT REMOVED IN DSL FOR S/7                      DSLS7***
      RETURN
      END                                                               DSL09790
 /*
      ACTION NOAUTO,NOMAP
      PHASE SORTX,*
      INCLUDE
      ENTRY
```

```
/*
PROC $CPFCL
R  FTIN,$SYCD
R  LNKPCH,,DSLS79,DSLLIB
PARM $FORT,NOCMPAT
XEQ
      SUBROUTINE OUTPX
      IMPLICIT INTEGER*2(I-N)
      GLOBAL NALRM,IERR,ICONT,JPT,KEY,ISTRT,KA(6),LETRS(73),ICHAR(12),   DSL01910
     1 IRES(158),KSYMB(600),NSYMB,ITYPE,NOUT,NINP,NPAR,NINT,NDATA,NFOR,  DSL01920
     2 NSORT,ISORT,NMEM,NING,NINGL,NSTOR,NTSEQ,NREAL,MFOR,               DSL01930
     3 KOUT(100),LFORT(100),NFORT(100),LIN(100),NIN(100),INPUT(300),     DSL01940
     4 KPAR( 60),KSTOR(10),LINT(50),INGER(10),MEMRY(15),KMEM(15),        DSL01950
     5 LSTOR(10),INGRL(10),KREAL(20),KSORT(20)  ,                        DSL01960
     6       KFORT(128),KSPEC(36),LOCAL(36),NLOC,NSPEC,KSEQ(120),        DSL01970
     7 KOMN(110),KBUFR(36)                                               DSL01980
      COMMON/FILE/NFILE1                                                 *****S7*
      EQUIVALENCE (IRES(106),IDOLR),(IRES(107),IBLNK),(IRES(158),IONE)   DSL01990
      DATA IAST2/'*2'/
      KPT=JPT                                                            DSL02000
      L=ISTRT                                                            DSL02005
 1111 CONTINUE
      IF(NSORT)101,101,103                                              DSL02020
  101 NTSEQ=NOUT                                                        DSL02030
      DO 102  I=1,NOUT                                                  DSL02040
  102 KSEQ(I)=I                                                        DSL02050
C...  TEST IF ALL PARAMETERS INPUT                                     DSL02060
      DO 104  I=1,NINP                                                 DSL02070
      KVAR=INPUT(I)                                                    DSL02080
      DO 105  J=1,NOUT                                                 DSL02090
      IF(KOUT(J)-KVAR)105,104,105                                     DSL02100
  105 CONTINUE                                                        DSL02110
C...  TEST IF PARAMETER OR STORAG VARIABLE                            DSL02120
      DO 106  J=1,NPAR                                                 DSL02130
      IF(KPAR(J)-KVAR)106,104,106                                     DSL02140
  106 CONTINUE                                                        DSL02150
      IF(NSTOR)117,117,108                                            DSL02160
  108 DO 109  J=1,NSTOR                                                DSL02170
      IF(KSTOR(J)-KVAR)109,104,109                                    DSL02180
  109 CONTINUE                                                        DSL02190
  117 IF(KVAR)104,104,118                                             DSL02200
  118 KA(1)=KSYMB(KVAR)                                               DSL02210
      KA(2)=KSYMB(KVAR+200)                                           DSL02220
      KA(3)=KSYMB(KVAR+400)                                           DSL02221
      WRITE(KPT,200) KA(1),KA(2),KA(3)                                DSL02240
  200 FORMAT(25H PARAM NOT INPUT, SET=0   3A2)                        DSL02250
      IF(NPAR-60)107,300,300                                          DSL02260
  300 WRITE(KPT,210) NPAR,NCOM                                        DSL02270
  210 FORMAT(33H CHECK MAX VS. PARAM, COMMON VARS 2I4)                DSL02280
      GO TO 150                                                       DSL02290
  107 NPAR=NPAR+1                                                     DSL02300
      KPAR(NPAR)=KVAR                                                 DSL02310
  104 CONTINUE                                                        DSL02320
C...  FORM COMMON VARIABLE LIST (ALSO SYMBOL TABLE)                   DSL02330
C...  SIX SYSTEMS VARS, INTEGRALS, OUTPUT VARS, PARAMS, STORAG VARS   DSL02340
  103 KA(4)=240
      KA(5)=240                                                       DSL02370
      KA(6)=64
```

```
          DO 5   I=1,6                                            DSL02380
        5 KOMN(I)=I                                               DSL02390
          IF(NINT)15,15,6                                         DSL02400
        6 NKS=1                                                   DSL02402
          M=1                                                     DSL02404
      222 NEXT=KSEQ(NKS)                                          DSL02406
C...   USE OUTPUT SEQUENCE TO SET UP INTEGRATORS IN COMMON        DSL02408
          DO 10   I=1,NINT                                        DSL02410
          J=LINT(I)                                               DSL02420
          IF(J)203,7,223
      223 IF(J-NEXT)10,8,10                                       DSL02432
      203 IF(J+NEXT)10,7,10                                       DSL02434
        7 KA(1)=233                                               DSL02440
          KA(2)=233
          KA(3)=240                                               DSL02450
          IF(KA(5)-249)3,2,2                                      DSL02460
        2 KA(4)=KA(4)+1                                           DSL02470
          KA(5)=240                                               DSL02480
          GO TO 4                                                 DSL02490
        3 KA(5)=KA(5)+1                                           DSL02500
        4 CALL PACK                                               DSL02510
          NSYMB=NSYMB+1                                           DSL02520
          KSYMB(NSYMB)=KA(1)                                      DSL02530
          KSYMB(NSYMB+200)=KA(2)                                  DSL02540
          KSYMB(NSYMB+400)=KA(3)                                  DSL02541
          KOMN(M+6)=NSYMB                                         DSL02550
          GO TO 204                                               DSL02560
        8 KOMN(M+6)=KOUT(J)                                       DSL02570
          KOUT(J)=-KOUT(J)                                        DSL02580
      204 IF(M-NINT)206,15,15                                     DSL02581
      206 M=M+1                                                   DSL02582
          I=I+1                                                   DSL02583
          IF(I-NINT)224,224,205                                   DSL02584
      224 J=LINT(I)                                               DSL02587
          IF(J)205,7,205                                          DSL02588
       10 CONTINUE                                                DSL02590
      205 NKS=NKS+1                                               DSL02592
          IF(NKS-NTSEQ)222,222,15                                 DSL02594
       15 NCOM=NINT+6                                             DSL02600
          IF(NING)115,115,110                                     DSL02610
C...   ADD DUMMY WORDS IN COMMON FOR INTEGERS TO MATCH SYMBOL TABLE  DSL02620
      110 KOUNT=IRES(151)                                         DSL02630
          DO 112   I=1,NING                                       DSL02640
          DO 112   J=7,NPAR                                       DSL02650
          IF(INGER(I)-KPAR(J))112,111,112                         DSL02660
      111 KPAR(J)=KPAR(NPAR)                                      DSL02670
          NPAR=NPAR-1                                             DSL02680
      112 CONTINUE                                                DSL02690
          DO 113   I=1,NING                                       DSL02700
          NPAR=NPAR+2                                             DSL02710
          KPAR(NPAR-1)=-1                                         DSL02720
      113 KPAR(NPAR)=INGER(I)                                     DSL02730
      115 DO 17   I=1,NOUT                                        DSL02740
          KVAR=KOUT(I)                                            DSL02750
          IF(KVAR)17,17,16                                        DSL02760
C...   SKIP IF ALSO A PARAMETER                                   DSL02762
       16 DO 13   K=7,NPAR                                        DSL02764
          IF(KVAR-KPAR(K))13,17,13                                DSL02766
```

368

```
      13 CONTINUE                                               DSL02768
         IF(KVAR-500)116,116,17                                 DSL02770
     116 DO 12  K=1,NCOM                                        DSL02780
         IF(KVAR-KOMN(K))12,17,12                               DSL02790
      12 CONTINUE                                               DSL02800
         IF(NSTOR)215,215,214                                   DSL02833
     214 DO 211  K=2,NSTOR                                      DSL02834
         IF(KVAR-KSTOR(K))211,17,211                            DSL02835
     211 CONTINUE                                               DSL02836
     215 NCOM=NCOM+1                                            DSL02840
         KOMN(NCOM)=KVAR                                        DSL02850
         IF(NCOM-110)17,17,300                                  DSL02860
      17 CONTINUE                                               DSL02870
     264 IF(KPAR(NPAR))263,263,262                              DSL02880
     263 NPAR=NPAR-1                                            DSL02882
         IF(NPAR)262,262,264                                    DSL02884
C...  OUTPUT CONTROL CARDS FOR UPDATE SUBROUTINE                DSL02888
     262 WRITE(L,250)                                           DSL02890
     250 FORMAT('PROC $CPMCLFG  '/'PARM $FORT,NOCMPAT '/ 'XEQ '/ DSL*S/7*
        1 '       SUBROUTINE UPDAT'/'       IMPLICIT INTEGER*2(I-N) ') DSL*S/7*
C...  OUTPUT COMMON VARIABLE LIST                               DSL02920
         KGO=0                                                  DSL02930
         NB1=6                                                  DSL02940
         KGO1=0                                                 DSL02950
         NWDS=10                                                DSL02960
         KOUN=0                                                 DSL02965
         KBUFR(1)=IBLNK                                         DSL02970
         KBUFR(2)=IBLNK                                         DSL02980
         IF(NREAL)51,51,9                                       DSL02990
C...  REMOVE INTEGERS FROM REAL VARIABLE TABLE                  DSL03000
       9 IF(NING)11,11,14                                       DSL03010
      14 DO 29  I=1,NING                                        DSL03020
         DO 27  J=1,NREAL                                       DSL03030
         IF(INGER(I)-KREAL(J))27,20,27                          DSL03040
      20 KREAL(J)=KREAL(NREAL)                                  DSL03050
         NREAL=NREAL-1                                          DSL03060
         IF(NREAL)51,51,29                                      DSL03070
      27 CONTINUE                                               DSL03080
      29 CONTINUE                                               DSL03090
      11 KBUFR(4)=IRES(152)                                     DSL03100
         KBUFR(5)=IRES(153)                                     DSL03110
         KBUFR(6)=IBLNK                                         DSL03120
         KSW=1                                                  DSL03130
         N3=NREAL                                               DSL03140
      18 N1=1                                                   DSL03150
         KBUFR(3)=IBLNK                                         DSL03160
      19 NBUF=NB1                                               DSL03170
         N2=N1+NWDS-1                                           DSL03180
      26 IF(N2-N3)22,22,21                                      DSL03190
      21 N4=N3-N1+1                                             DSL03200
         N2=N3                                                  DSL03210
      22 DO 30  I=N1,N2                                         DSL03220
         IF(KSW-3)23,24,80                                      DSL03230
      80 IF(KSW-7)81,25,25                                      DSL03240
      23 J=KREAL(I)                                             DSL03250
         GO TO 28                                               DSL03260
      24 J=INGER(I)                                             DSL03270
         GO TO 28                                               DSL03280
```

```
   81 J=KPAR(I)                                             DSL03290
      IF(J)128,128,28                                       DSL03300
  128 NBUF=NBUF+3                                           DSL03310
      KBUFR(NBUF-2)=IRES(3)                                 DSL03320
      KBUFR(NBUF-1)=IRES(149)                               DSL03330
      KBUFR(NBUF)=KOUNT                                     DSL03340
      KOUNT=KOUNT+256                                       DSL03350
      GO TO 30                                              DSL03360
   25 J=KOMN(I)                                             DSL03370
   28 KA(1)=KSYMB(J)                                        DSL03380
      KA(2)=KSYMB(J+200)                                    DSL03390
      KA(3)=KSYMB(J+400)                                    DSL03391
      NBUF=NBUF+3                                           DSL03410
      KBUFR(NBUF-2)=KA(1)                                   DSL03420
      KBUFR(NBUF-1)=KA(2)                                   DSL03430
      KBUFR(NBUF)=KA(3)                                     DSL03440
   30 CONTINUE                                              DSL03450
      IF(N2-N3)31,35,35                                     DSL03460
   31 KEND=36                                               DSL03470
   32 IF(KGO)33,33,83                                       DSL03480
   33 K1=NB1+3                                              DSL03490
      IF(K1-KEND)75,75,40                                   DSL03500
   75 DO 34  K=K1,KEND,3                                    DSL03510
   34 KBUFR(K)=KBUFR(K)+43                                  DSL03520
      GO TO 40                                              DSL03530
   35 KEND=NBUF-3                                           DSL03540
      KSW=KSW+1                                             DSL03550
      IF(KSW-7)32,36,36                                     DSL03560
   36 IF(NPAR-6)38,38,37                                    DSL03570
   37 IF(NBUF-36)39,31,31                                   DSL03580
   38 KSW=6                                                 DSL03590
      GO TO 32                                              DSL03600
   39 N2=NWDS-N4+6                                          DSL03610
      KSW=5                                                 DSL03620
      N3=NPAR                                               DSL03630
      N1=7                                                  DSL03640
      GO TO 26                                              DSL03650
   40 IF(KGO1)59,52,54                                      DSL03660
   54 WRITE(L,254) (KBUFR(I),I=16,NBUF)                     DSL03670
  254 FORMAT(30H        GLOBAL NALRM,IZZZ(833) , 21A2)
      KGO1=0                                                DSL03690
      KOUN=7                                                DSL03700
      GO TO 53                                              DSL03710
   59 WRITE(L,255)                                          DSL03720
  255 FORMAT(12H        COMMON)                             DSL03730
      KGO1=0                                                DSL03740
      KOUN=0                                                DSL03750
      GO TO 57                                              DSL03760
   52 IF(KOUN-33)57,56,56                                   DSL03770
   56 IF(KEND-36)83,84,84                                   DSL03775
   84 KBUFR(KEND)=KBUFR(KEND)-43                            DSL03780
      KGO1=-1                                               DSL03790
   57 KOUN=KOUN+11                                          DSL03800
   83 WRITE(L,251) (KBUFR(I),I=1,NBUF)                      DSL03810
  251 FORMAT(36A2)                                          DSL03820
   53 GO TO(41,51,41,42,41,44,41,43),KSW                   DSL03830
   41 N1=N2+1                                               DSL03840
      IF(KGO)49,49,19                                       DSL03850
```

```
            KPOIT=KPOIT+LSTOR(1)                                    DSL04240
            KBUFR(1)=IRES(149)                                      DSL04250
            KBUFR(13)=IRES(150)                                     DSL04260
            KBUFR(25)=IRES(151)                                     DSL04270
            DO 98  I=2,NSTOR                                        DSL04280
            KPOIT=KPOIT+LSTOR(I)                                    DSL04290
            J=KSTOR(I)                                              DSL04300
            KA(1)=KSYMB(J)                                          DSL04310
            KA(2)=KSYMB(J+200)                                      DSL04320
            KA(3)=KSYMB(J+400)                                      DSL04321
            KBUFR(I)=KA(1)                                          DSL04340
            KBUFR(I+12)=KA(2)                                       DSL04350
         98 KBUFR(I+24)=KA(3)                                       DSL04360
            N1=1                                                    DSL04370
            ICOMA=IBLNK+43                                          DSL04380
            IF(NSTOR-5)91,91,90                                     DSL04390
         90 N2=5                                                    DSL04400
            GO TO 92                                                DSL04410
         91 N2=NSTOR                                                DSL04420
         92 WRITE(L,260) (ICOMA,KBUFR(I),KBUFR(I+12),KBUFR(I+24),   DSL04430
           1 LSTOR(I),I=N1,N2)                                      DSL04440
        260 FORMAT(6H    1 5(   3A2,A1,1H( I3,1H)))                 DSL04450
            IF(N2-NSTOR)93,66,66                                    DSL04460
         93 N1=6                                                    DSL04470
            GO TO 91                                                DSL04480
      C...  WRITE FORTRAN STATEMENTS                                DSL04490
         66 DO 60  I=1,NTSEQ                                        DSL04500
            J=KSEQ(I)                                               DSL04510
            N1=LFORT(J)                                             DSL04520
            N2=NFORT(J)                                             DSL04530
            N3=IABS1(N2)
            IF(N2)74,60,74                                          DSL04550
         74 READ(3'N1)(KFORT(M),M=1,N3)                            *****S7*
            IF(N2)47,60,58                                          DSL04570
         47 N4=1                                                    DSL04580
            DO 50  M=1,N3                                           DSL04590
            IF(KFORT(M)-IDOLR)50,48,50                              DSL04600
         48 N5=M-1                                                  DSL04610
            WRITE(L,251) (KFORT(K),K=N4,N5)                         DSL04620
            N4=M+1                                                  DSL04630
         50 CONTINUE                                                DSL04640
            GO TO 60                                                DSL04650
         58 WRITE(L,251) (KFORT(K),K=1,N3)                          DSL04660
         60 CONTINUE                                                DSL04670
      C...  END OF UPDAT SUBROUTINE                                 DSL04680
            WRITE(L,252)                                            DSL04690
      C...  WRITE SYMBOL TABLE                                      DSL04770
        252 FORMAT('    RETURN'/'    END'/'/*')                     DSL*S/7*
            IF(NING)165,165,160                                     DSL04780
        160 I=7                                                     DSL04790
        163 IF(KPAR(I))161,161,164                                  DSL04800
        161 NPAR=NPAR-1                                             DSL04810
            DO 162  J=I,NPAR                                        DSL04820
        162 KPAR(J)=KPAR(J+1)                                       DSL04830
        164 I=I+1                                                   DSL04840
            IF(I-NPAR)163,165,165                                   DSL04850
        165 NSYM=NCOM+NPAR-6                                        DSL04860
            WRITE(L,253) NINT,NSYM,KPOIT                            DSL04870
```

```
      49 KBUFR(3)=IONE                                                    DSL03860
         NB1=3                                                           DSL03870
         NWDS=11                                                         DSL03880
         GO TO 19                                                        DSL03890
C...     WRITE DIMENSION STMNT FOR ALL STORAG + INTGRL VARS              DSL03892
      42 IF(NSTOR)134,134,130                                            DSL03893
     130 IF(NINGL)134,134,244                                            DSL03895
     244 J=2                                                             DSL03896
     249 DO 245   I=1,NINGL                                              DSL03897
         IF(KSTOR(J)-INGRL(I))245,246,245                               DSL03898
     246 NSTOR=NSTOR-1                                                   DSL03900
         IJ=INGRL(I)                                                    DSL03902
         KA(1)=KSYMB(IJ)                                                DSL03903
         KA(2)=KSYMB(IJ+200)                                            DSL03904
         KA(3)=KSYMB(IJ+400)
         WRITE(L,270) KA(1),KA(2),KA(3)                                 DSL03906
     270 FORMAT(16H        DIMENSION 2A2,A1,3H(1) )                      DSL03907
         DO 248   J2=J,NSTOR                                            DSL03910
         KSTOR(J2)=KSTOR(J2+1)                                          DSL03911
     248 LSTOR(J2)=LSTOR(J2+1)                                          DSL03912
         GO TO 247                                                      DSL03913
     245 CONTINUE                                                       DSL03914
         J=J+1                                                          DSL03915
     247 IF(J-NSTOR)249,249,134                                         DSL03916
     134 IF(NSPEC)141,141,140                                           DSL03918
     140 WRITE(L,251)(KSPEC(I),I=1,36)                                  DSL03919
     141 KGO1=1                                                         DSL03920
         NB1=15                                                         DSL03930
         NWDS=7                                                         DSL03940
      46 KSW=7                                                          DSL03950
         N3=NCOM                                                        DSL03960
         GO TO 18                                                       DSL03970
      43 KSW=5                                                          DSL03980
         N3=NPAR                                                        DSL03990
         N1=7                                                           DSL04000
         GO TO 19                                                       DSL04010
      51 IF(NING)42,42,55                                               DSL04020
      55 NB1=9                                                          DSL04030
         NWDS=9                                                         DSL04040
         KSW=3                                                          DSL04050
         N3=NING                                                        DSL04060
         KBUFR(4)=IRES(154)                                            DSL04070
         KBUFR(5)=IRES(155)                                            DSL04080
         KBUFR(6)=IRES(156)                                            DSL04090
         KBUFR(7)=IRES(157)                                            DSL04100
         KBUFR(8)=IAST2
         KBUFR(9)=IBLNK                                                 DSL04120
         GO TO 18                                                       DSL04130
      44 IF(KGO)45,45,67                                                DSL04140
      45 KPOIN=NCOM+NPAR-NING-5                                         DSL04150
         KPOIT=KPOIN                                                    DSL04160
C...     MOVE STORAG VARIABLES TO PARAMETER TABLE                       DSL04170
         IF(NSTOR)66,66,61                                              DSL04180
C...     WRITE STORAG VARIABLES   (ADD TO COMMON)                       DSL04190
      61 DO 62   J=2,NSTOR                                              DSL04200
         NPAR=NPAR+1                                                    DSL04210
      62 KPAR(NPAR)=KSTOR(J)                                            DSL04220
         LSTOR(1)=NSTOR-1                                               DSL04230
```

37'

```
    253 FORMAT(3I4)                                                      DSL04880
        KGO=1                                                           DSL04890
        NB1=0                                                           DSL04900
        NWDS=12                                                         DSL04910
        IF(NSYM-140)46,46,155                                           DSL04920
    155 WRITE(KPT,156)NSYM                                              DSL04930
    156 FORMAT(I4,23H SYMBOLS, REDUCE TO 140)                           DSL04940
        GO TO 46                                                        DSL04950
     67 IF(NSTOR)65,65,63                                               DSL04960
C... WRITE PARAM CARD TO LOCATE STORAG VARIABLE TABLES                  DSL04970
     63 DO 94  I=2,NSTOR                                                DSL04980
        J=KSTOR(I)                                                      DSL04990
        KA(1)=KSYMB(J)                                                  DSL05000
        KA(2)=KSYMB(J+200)                                              DSL05010
        KA(3)=KSYMB(J+400)                                              DSL05011
        KBUFR(I-1)=ICOMA                                                DSL05030
        KBUFR(I+8)=KA(1)                                                DSL05040
        KBUFR(I+17)=KA(2)                                               DSL05050
        KBUFR(I+26)=KA(3)                                               DSL05060
        KPOIN=KPOIN+LSTOR(I-1)                                          DSL05070
     94 KSTOR(I)=KPOIN                                                  DSL05080
        KBUFR(1)=IBLNK                                                  DSL05090
        N1=1                                                            DSL05100
        IF(NSTOR-7)69,69,64                                             DSL05110
     64 N2=6                                                            DSL05120
        GO TO 71                                                        DSL05130
     69 N2=NSTOR-1                                                      DSL05140
     71 WRITE(L,261) (KBUFR(I),KBUFR(I+9),KBUFR(I+18),KBUFR(I+27),      DSL05150
      1 KSTOR(I+1),I=N1,N2)                                             DSL05160
    261 FORMAT(6HPARAM 6(3A2,A1,1H= I3))                                DSL05170
        IF(N2-NSTOR+1)72,65,65                                          DSL05180
     72 N1=7                                                            DSL05190
        GO TO 69                                                        DSL05200
     65 N3=1                                                            DSL05210
        NDAT=NDATA                                                      DSL05220
        DO 70  I=1,NDAT                                                 DSL05230
        READ(4'I)LETRS(N3)                                             *****S7*
        IF(LETRS(N3)-IDOLR)73,68,73                                     DSL05250
     68 N3=N3-1                                                         DSL05260
        WRITE(L,251)(LETRS(K),K=1,N3)                                   DSL05270
        N3=1                                                            DSL05280
        GO TO 70                                                        DSL05290
     73 N3=N3+1                                                         DSL05300
     70 CONTINUE                                                        DSL05310
    150 RETURN
        END                                                            DSL05370
/*
        ACTION NOAUTO,NOMAP
        PHASE OUTPX,*
        INCLUDE
        ENTRY
/*
PROC $CPFCL
R FTIN,$SYCD
R LNKPCH,,DSLS79,DSLLIB
PARM $FORT,NOCMPAT
XEQ
        SUBROUTINE TRDMP
```

```
      IMPLICIT INTEGER*2(I-N)
      GLOBAL NALRM,IERR,ICONT,JPT,KEY,ISTRT,KA(6),LETRS(73),ICHAR(12),    DSL15270
     1 IRES(158),KSYMB(600),NSYMB,ITYPE,NOUT,NINP,NPAR,NINT,NDATA,NFOR,   DSL15280
     2 NSORT,ISORT,NMEM,NING,NINGL,NSTOR,NTSEQ,NREAL,MFOR,                DSL15290
     3 KOUT(100),LFORT(100),NFORT(100),LIN(100),NIN(100),INPUT(300),      DSL15300
     4 KPAR( 60),KSTOR(10),LINT(50),INGER(10),MEMRY(15),KMEM(15),         DSL15310
     5 LSTOR(10),INGRL(10),KREAL(20),KSORT(20),KFORT(202)                 DSL15320
      M=JPT                                                               DSL15340
      WRITE(M,880) (INPUT(I), I=1,NINP)                                   DSL15350
  880 FORMAT(20I6)                                                        DSL15360
      WRITE(M,860) (KOUT(I),LFORT(I),NFORT(I),LIN(I),NIN(I),I=1,NOUT)     DSL15370
  860 FORMAT(5I6)                                                         DSL15380
      WRITE(M,880) (LINT(I),I=1,NINT),(KPAR(I),I=1,NPAR)                  DSL15390
      WRITE(M,880) (INGER(I),I=1,NING),(INGRL(I),I=1,NINGL),              DSL15400
     1 (KSORT(I),I=1,NSORT)                                               DSL15410
      WRITE(M,880) (MEMRY(I),KMEM(I),I=1,NMEM)                            DSL15420
      IF(NSTOR)890,890,885                                                DSL15430
  885 WRITE(M,880) (KSTOR(I),LSTOR(I),I=1,NSTOR)                          DSL15440
  890 J=0                                                                 DSL15445
      DO 840  I=1,NSYMB                                                   DSL15450
      KA(1)=KSYMB(I)                                                      DSL15460
      KA(2)=KSYMB(I+200)                                                  DSL15470
      KA(3)=KSYMB(I+400)                                                  DSL15480
      J=J+1                                                               DSL15485
      KFORT(J)=KA(1)                                                      DSL15490
      KFORT(J+15)=KA(2)                                                   DSL15500
      KFORT(J+30)=KA(3)                                                   DSL15510
      IF(J-15)840,830,830                                                 DSL15520
  830 WRITE(M,850)(KFORT(K),KFORT(K+15),KFORT(K+30),K=1,15)               DSL15525
  850 FORMAT(15(2X,3A2))                                                  DSL15530
      J=0                                                                 DSL15532
  840 CONTINUE                                                            DSL15534
      IF(J)870,870,855                                                    DSL15536
  855 WRITE(M,850)(KFORT(K),KFORT(K+15),KFORT(K+30),K=1,J)                DSL15538
  870 CONTINUE                                                            DSLS7***
      RETURN
      END                                                                 DSL15550
/*
      ACTION NOAUTO,NOMAP
      PHASE TRDMP,*
      INCLUDE
      ENTRY
/*
/*
```

374

```
****************************************************************************
*                                                                          *
*        THIS PORTION DESCRIBES CHANGES IN TRANS, THE TRANSLATOR.          *
*        NOTE THAT TRANS NOW BECOMES A MAIN PROGRAM. IT CALLS             *
*        THE OTHER OVERLAYS AS SUBROUTINES. THEY ARE NAMELY SORTX, OUTPX, *
*        AND TRDMP. THE REQUIRED CHANGES IN THOSE THREE WERE SHOWN EARLIER.*
*        NOTE THAT THE FINAL LINK EDITOR CONTROL CARDS FOLLOW THE SOURCE, SO*
*        THAT A FULL TRANSLATOR STORAGE LOAD CAN BE CREATED FOR EXECUTION. *
*                                                                          *
****************************************************************************
PROC $CPFCL
R FTIN,$SYCD
R LNKPCH,,DSLS79,DSLLIB
PARM $FORT,NOCMPAT
XEQ
      PROGRAM MAIN
      IMPLICIT INTEGER*2(I-N)
      DEFINE FILE  3(6000,1,U,MFOR),4(2000,1,U,NDATA)              DSL11587
      GLOBAL NALRM,IERR,ICONT,JPT,KEY,ISTRT,KA(6),LETRS(73),ICHAR(12),  DSL11589
     1 IRES(158),KSYMB(600),NSYMB,ITYPE,NOUT,NINP,NPAR,NINT,NDATA,NFOR, DSL11590
     2 NSORT,ISORT,NMEM,NING,NINGL,NSTOR,NTSEQ,NREAL,MFOR,         DSL11591
     3 KOUT(100),LFORT(100),NFORT(100),LIN(100),NIN(100),INPUT(300), DSL11592
     4 KPAR( 60),KSTOR(10),LINT(50),INGER(10),MEMRY(15),KMEM(15),  DSL11593
     5 LSTOR(10),INGRL(10),KREAL(20),KSORT(20),                    DSL11594
     6     KFORT(128),KSPEC(36),LOCAL(36),NLOC,NSPEC,KEYB,KARD,ISC(5) DSL11595
      COMMON/FILE/NFILE1                                          *****S7*
      EXTERNAL TRDMP,SORTX,OUTPX
      EQUIVALENCE (IRES(106),IDOLR),(IRES(107),IBLNK),(KA(4),LOC),  DSL11596
     1 (KA(6),NMBR),(IRES(158),IONE)                               DSL11597
      DATA MASK3/' ,'/
      DATA MASK2/'  '/
      MM2=2
      KEND=0                                                       DSL11598
      IPRO=0                                                       DSL11600
      NIMP=0                                                       DSL11601
      IMP1=600                                                     DSL11602
      KDUMP=0                                                      DSL11604
      MPL=0                                                        DSL11606
C...  INITIALIZE DATA AREAS, STORE SYMBOL TABLES IN COMMON         DSL11609
      CALL TRAN1                                                   DSL11610
      CALL START                                                   DSL11611
      MAXN1=INPT(4)
      MAXN=MAXN1
      ICA=KA(1)                                                    DSL11613
      LL=KA(2)                                                     DSL11614
      ID=KA(3)                                                     DSL11615
      MASK=KA(4)                                                   DSL11616
C...  START OF LOOP FOR EACH CARD,  READ AND PRINT INPUT          DSL11620
    4 ICONT=0                                                      DSL11622
      ITYPE=0                                                      DSL11624
      MEMR=0                                                       DSL11626
      JINTG=0                                                      DSL11627
    5 CALL READR                                                   DSL11628
      ISTRT=7                                                      DSL11630
      IF(ICONT)63,2,172                                            DSL11640
    2 IF(NFOR-1)15,15,12                                           DSL11642
   12 N2=0                                                         DSL11643
```

```
         IF(MPL)320,320,310                                        DSL11644
    310  MPL=0                                                      DSL11645
C...   MULTIPLE OUTPUT BLOCK, INSERT OUTPUTS IN CALL STATEMENT      DSL11646
C...   TEST FOR LEADING BLANK OR ALPHA IN PAIR, OR RIGHT PAREN      DSL11647
         IF(KFORT(NFOR-1))323,323,322                              DSL11649
C...   INSERT BLANK-COMMA(16491), RIGHT PAREN-BLANK(23872), + COMMA(107) DSL11651
    322  KFORT(NFOR-1)=16491                                        DSL11652
         GO TO 324                                                  DSL11653
    323 CALL SHIFT(KFORT(NFOR-1),MASK3,MM2)
    324 DO 325 NF= 94,NF8                                           DSL11657
         KFORT(NFOR)=KFORT(NF)                                      DSL11658
    325 NFOR=NFOR+1                                                 DSL11659
         KFORT(NFOR)=23872                                          DSL11660
         GO TO 450                                                  DSL11661
    320 NFOR=NFOR-1                                                 DSL11662
    450 WRITE(3'MFOR)(KFORT(K),K=1,NFOR)                           *****S7*
         NFOR=1                                                     DSL11680
         IF(MFOR-5000)15,15,100                                     DSL11685
     15 IF(LETRS(1)-IBLNK)6,70,6                                    DSL11690
C...   DETERMINE DATA STATEMENT TYPE   (1-35)                       DSL11700
C...   PARAM,CONST,INCON,AFGEN,NLFGEN,TABLE,PROCED,MEMORY,STORAG,INTGER, DSL11710
C...   RENAME,INTGRL,PRINT,CONTRL,RELERR,ABSERR,RANGE,FINISH,END,STOP,   DSL11720
C...   SORT,NOSORT,ENDPRO,TITLE,SCOPE,GRAPH,LABLE,SCALE,CONTIN,INTEG,    DSL11730
C...   RESET,CHART,TYPE,DUMP,*LOCAL                                 DSL11740
      6 DO 7  I=1,35                                                DSL11750
         IF(LETRS(1)-IRES(I))7,8,7                                  DSL11760
      8 IF(LETRS(2)-IRES(I+35))7,9,7                                DSL11770
      9 IF(LETRS(3)-IRES(I+70))7,10,7                               DSL11780
      7 CONTINUE                                                    DSL11790
C...   TEST FOR D IN COL 1                                          DSL11800
         IF(LETRS(1)-ID)70,28,70                                    DSL11805
     10 ITYPE=I                                                     DSL11810
     63 IF(ITYPE-7)14,21,64                                         DSL11820
     64 IF(ITYPE-12)21,21,66                                        DSL11830
     66 IF(ITYPE-21)14,51,67                                        DSL11840
     67 IF(ITYPE-23)52,53,25                                        DSL11850
     25 IF(ITYPE-34)14,36,26                                        DSL11860
     36 KDUMP=1                                                     DSL11864
         GO TO 4                                                    DSL11866
     26 NLOC=1                                                      DSL11870
         DO 27  I=1,36                                              DSL11880
     27 LOCAL(I)=LETRS(I)                                           DSL11890
         GO TO 4                                                    DSL11900
     28 DO 29  I=2,36                                               DSL11910
     29 KSPEC(I)=LETRS(I)                                           DSL11920
         KSPEC(1)=IBLNK                                             DSL11930
         NSPEC=1                                                    DSL11940
         GO TO 4                                                    DSL11950
C...   STORE DATA CARD AS VARIABLE WORD RECORD WITH $$ LAST WORD    DSL11960
     14 IEND=36                                                     DSL11970
     16 IF(LETRS(IEND)-IBLNK)18,17,18                               DSL11980
     17 IEND=IEND-1                                                 DSL11990
         GO TO 16                                                   DSL12000
     18 IEND=IEND+1                                                 DSL12010
         LETRS(IEND)=IDOLR                                          DSL12020
         DO 1002 K=1,IEND
   1002 WRITE(4'NDATA)LETRS(K)                                      DSL12030
         IF(NDATA-1000)20,20,100                                    DSL12040
```

376

```
       20 IF(KEND)61,11,24                                              DSL12050
C...      NOTE - ADD 30 TO KEY NAME SEQUENCE NO. TO GET FORTRAN STATEMENT NODSL12060
       11 IF(ITYPE-19)21,49,24                                          DSL12070
       24 IF(ITYPE-20)4,50,4                                            DSL12080
       21 CALL OPEN                                                     DSL12090
          IF(ITYPE-12)22,42,43                                          DSL12100
       22 GO TO(31,31,31,34,34,43,37,38,39,40,41),ITYPE                 DSL12110
       31 CALL SCAN                                                     DSL12120
          L=1                                                           DSL12130
          GO TO(61,61,62,61,61,4,48),KEY                               DSL12140
       61 NALRM=6                                                       DSL12150
          GO TO 99                                                      DSL12160
       62 DO 122  M=1,NPAR                                              DSL12170
          IF(LOC-KPAR(M))122,123,122                                    DSL12180
      122 CONTINUE                                                      DSL12190
          IF(NPAR-60)108,100,100                                        DSL12200
      108 NPAR=NPAR+1                                                    DSL12210
          KPAR(NPAR)=LOC                                                DSL12220
      123 IF(L-1)61,31,43                                               DSL12230
       34 IF(ICONT)43,65,61                                             DSL12240
       65 CALL SCAN                                                     DSL12250
          L=2                                                           DSL12260
          IF(KEY-3)65,62,61                                             DSL12270
C...      PROCEDURAL BLOCK                                              DSL12280
       37 IPRO=1                                                        DSL12290
          NFOR1=MFOR                                                    DSL12300
          GO TO 68                                                      DSL12310
C...      MEMORY BLOCK SPECIFICATION                                    DSL12320
       38 CALL SCAN                                                     DSL12330
          IF(KEY-4)61,150,149                                           DSL12340
      149 IF(KEY-5)61,61,4                                              DSL12350
      150 IF(NMEM-15)151,100,100                                        DSL12360
      151 NMEM=NMEM+1                                                    DSL12370
          MEMRY(NMEM)=LOC                                               DSL12380
          CALL NUMRC                                                    DSL12390
          KMEM(NMEM)=NMBR                                               DSL12400
          GO TO 38                                                      DSL12410
C...      STORAG BLOCK SPECIFICATION                                    DSL12420
       39 CALL SCAN                                                     DSL12430
          IF(KEY-4)61,160,159                                           DSL12440
      159 IF(KEY-5)61,61,4                                              DSL12450
      160 IF(NSTOR)162,162,163                                          DSL12460
      162 NSTOR=2                                                        DSL12470
          GO TO 164                                                     DSL12480
      163 IF(NSTOR-10)161,100,100                                       DSL12490
      161 NSTOR=NSTOR+1                                                  DSL12500
      164 KSTOR(NSTOR)=LOC                                              DSL12510
          CALL NUMRC                                                    DSL12520
          LSTOR(NSTOR)=NMBR                                            DSL12530
          GO TO 39                                                      DSL12540
       40 CALL SCAN                                                     DSL12550
          IF(KEY-2)4,110,4                                              DSL12560
      110 IF(NING-10)111,100,100                                        DSL12570
      111 NING=NING+1                                                    DSL12580
          INGER(NING)=LOC                                               DSL12590
          GO TO 40                                                      DSL12600
C...      RENAME                                                        DSL12610
       41 CALL SCAN                                                     DSL12620
```

```
      IF(KEY-3)61,112,61                          DSL12630
  112 IF(LOC-6)113,113,61                         DSL12640
  113 I=LOC                                        DSL12650
      CALL SCAN                                    DSL12660
      KSYMB(I)=KSYMB(NSYMB)                        DSL12670
      KSYMB(I+200)=KSYMB(NSYMB+200)                DSL12680
      KSYMB(I+400)=KSYMB(NSYMB+400)
      NSYMB=NSYMB-1                                DSL12690
      IF(ISTRT-72)41,4,4                           DSL12700
C...  INTEGRAL OUTPUT VARIABLE SPECIFICATION       DSL12710
   42 CALL SCAN                                    DSL12720
      IF(KEY-2)61,170,61                           DSL12730
  170 IF(NINGL-10)171,100,100                      DSL12740
  171 NINGL=NINGL+1                                DSL12750
      INGRL(NINGL)=LOC                             DSL12760
      IF(ISTRT-72)72,4,4                           DSL12770
C...  TEST FOR CONTINUATION OF DATA CARD           DSL12780
   43 KPT=0                                        DSL12790
      I=IEND+IEND-2                                DSL12800
   44 IF(LETRS(I)-64)4,45,46                       DSL12810
   45 I=I-1                                        DSL12820
      GO TO 44                                     DSL12830
   46 IF(LETRS(I)-75)4,47,4                        DSL12840
   47 KPT=KPT+1                                    DSL12850
      IF(KPT-3)45,48,48                            DSL12860
   48 ICONT=-1                                     DSL12870
      GO TO 5                                      DSL12880
C...  END AND STOP CARDS                           DSL12890
   49 KEND=1                                       DSL12900
      GO TO 4                                      DSL12910
   51 IF(ISORT)264,264,4                           DSL12950
  264 ISORT=1                                      DSL12960
      IF(NOUT-1)4,4,175                            DSL12970
  175 NSORT=NSORT+2                                DSL12980
      KSORT(NSORT-1)=NOUT                          DSL12990
      GO TO 4                                      DSL13000
   52 IF(ISORT)265,4,265                           DSL13005
  265 ISORT=0                                      DSL13010
      IF(NOUT-1)176,176,177                        DSL13020
  176 NSORT=0                                      DSL13030
      GO TO 4                                      DSL13040
  177 KSORT(NSORT)=NOUT-1                          DSL13050
      GO TO 4                                      DSL13060
   53 IPRO=0                                       DSL13070
      NFORT(NOUT1)=NFOR1-MFOR                      DSL13080
      LFORT(NOUT1)=NFOR1                           DSL13090
      LIN(NOUT1)=NINP1                             DSL13100
      NIN(NOUT1)=NINP-NINP1+1                      DSL13110
      GO TO 4                                      DSL13120
C...  START SCAN OF DSL STRUCTURE STATEMENT        DSL13130
C...  CONTINUATION OF STRUCTURE STATEMENT          DSL13140
  172 NB1=NMBR/2                                   DSL13150
      NFOR=NFORA+NB1+MEMR+1                        DSL13160
      MEMR=0                                       DSL13164
      KFORT(NFOR-1)=IDOLR                          DSL13170
      IF(NMBR-NB1-NB1)94,174,70                    DSL13180
  174 KFORT(NFOR-2)=KFORT(NFOR-2)-11               DSL13190
   70 IEND=36                                      DSL13200
```

```
 71 IF(LETRS(IEND)-IBLNK)73,72,73                                    DSL13210
 72 IEND=IEND-1                                                       DSL13220
    GO TO 71                                                          DSL13230
 73 NFORA=NFOR                                                        DSL13240
    DO 74  K=1,IEND                                                   DSL13250
    KFORT(NFOR)=LETRS(K)                                              DSL13260
 74 NFOR=NFOR+1                                                       DSL13270
    IF(NFOR- 200)116,116,100                                          DSL13280
116 CALL OPEN                                                         DSL13290
C... TEST FOR COMMENT (ASTERISK COL. 1), REFERENCE NUMBER, CONTINUATION DSL13300
    IF(LETRS(1)-92)92,91,92                                           DSL13310
 91 NFOR=NFOR-IEND                                                    DSL13320
    GO TO 5                                                           DSL13330
 92 DO 95  K=1,5                                                      DSL13340
    IF(LETRS(K)-64)94,95,93                                           DSL13350
 93 IF(LETRS(K)-240)94,95,95                                          DSL13360
 95 CONTINUE                                                          DSL13370
    IF(ICONT)94,210,211                                               DSL13380
210 IF(LETRS(6)-64)94,96,94                                           DSL13390
 94 NALRM=9                                                           DSL13400
    GO TO 99                                                          DSL13410
211 KFORT(NFORA+2)=IONE                                               DSL13420
 96 IF(IPRO)120,121,121                                               DSL13430
120 KFORT(NFOR)=IDOLR                                                 DSL13440
    NFOR=NFOR+1                                                       DSL13450
    GO TO 4                                                           DSL13460
121 IF(ICONT)94,101,102                                               DSL13470
102 KFORT(NFOR)=IDOLR                                                 DSL13480
    NFOR=NFOR+1                                                       DSL13490
    NFORT(NOUT1)=NFOR1-NFOR                                           DSL13500
    IF(L-2)77,94,82                                                   DSL13510
C... FIND ALL OUTPUT VARIABLES                                        DSL13520
101 LFORT(NOUT)=MFOR                                                  DSL13530
    NFORT(NOUT)=IEND                                                  DSL13540
    NFOR1=NFORA                                                       DSL13550
 68 NOUT1=NOUT                                                        DSL13560
    L=1                                                               DSL13570
 77 CALL SCAN                                                         DSL13580
C... BRANCH ON NUMERIC,OPERATOR,=,(,GT 5 CHAR.,EOR,CONTINUE           DSL13585
    GO TO(94,79,78,263,130,94,103),KEY                                DSL13590
C... TEST IF END OF CARD BEFORE COMMA OR EQUALS                       DSL13600
 79 IF(ISTRT-72)350,190,190                                           DSL13610
350 MPL=1                                                             DSL13614
    GO TO 355                                                         DSL13616
263 L=3                                                               DSL13618
    GO TO 355                                                         DSL13619
190 IF(NSYMB-LOC)94,191,94                                            DSL13620
191 NSYMB=NSYMB-1                                                     DSL13630
130 IF(ISORT)94,131,94                                                DSL13640
131 KOUT(NOUT)=0                                                      DSL13642
    NOUT=NOUT+1                                                       DSL13643
    GO TO 4                                                           DSL13644
 78 L=2                                                               DSL13645
355 KSTR=1                                                            DSL13647
C... TEST IF OUTPUT IS STORAG VAR                                     DSL13649
    IF(NSTOR)353,353,351                                              DSL13650
351 DO 352 I=2,NSTOR                                                  DSL13651
    IF(LOC-KSTOR(I))352,354,352                                       DSL13652
```

```
      354 MPL=1                                                          DSL13653
          KSTR=LSTOR(I)                                                  DSL13654
          GO TO 353                                                      DSL13655
      352 CONTINUE                                                       DSL13656
C...  TEST IF OUTPUT VARIABLE IS AN INTEGRAL                             DSL13660
      353 IF(NINGL)196,196,192                                           DSL13661
      192 DO 193   I=1,NINGL                                             DSL13662
          IF(INGRL(I)-LOC)193,194,193                                   DSL13663
      194 IF(NINT- 50)195,100,100                                        DSL13664
      195 LINT(NINT+1)=NOUT                                              DSL13665
          NINT=NINT+KSTR                                                 DSL13666
          JINTG=1                                                        DSL13667
          GO TO 196                                                      DSL13668
      193 CONTINUE                                                       DSL13669
      196 GO TO(76,356,262),L                                           DSL13672
      356 IF(MPL)76,76,272                                               DSL13674
      262 MPL=0                                                          DSL13678
          GO TO 225                                                      DSL13680
      272 NF1=(ISTRT-1)/2                                                DSL13685
          NF8=NF1+90                                                     DSL13686
          IF(ISTRT-NF1-NF1-2)276,275,100                                DSL13687
C...  ADD BLANK-ZERO (16384)                                             DSL13689
      275 KFORT(NF1+1)=LETRS(ISTRT)+16384                                DSL13690
          GO TO 277                                                      DSL13691
      276 KFORT(NF1)=LETRS(ISTRT-2)
          MM0=0
          CALL SHIFT(KFORT(NF1),MASK2,MM0)
      277 DO 273   NF=4,NF1                                              DSL13698
      273 KFORT(NF+ 90)=KFORT(NF)                                        DSL13699
          IF(NF1-5)271,279,278                                          DSL13700
      271 NALRM=8                                                        DSL13701
          GO TO 99                                                       DSL13702
      278 NF3=5                                                          DSL13703
          DO 274   NF=NF1,NFOR                                           DSL13704
          KFORT(NF3)=KFORT(NF)                                           DSL13705
      274 NF3=NF3+1                                                      DSL13706
          NF3=NF1-5                                                      DSL13707
          MEMR=MEMR-NF3                                                  DSL13709
          NFOR=NFOR-NF3                                                  DSL13710
C...  INSERT CA-LL IN 4 AND 5                                            DSL13711
      279 KFORT(4)=ICA                                                   DSL13712
          KFORT(5)=LL                                                    DSL13713
          NFORT(NOUT1)=NFORT(NOUT1)+3                                    DSL13714
       76 IF(NOUT-100)109,109,100                                       DSL13715
      109 IF(NIMP)94,225,220                                            DSL13718
      220 IF(LOC-IMP3)225,221,225                                       DSL13720
      221 LOC=IMP2                                                       DSL13730
          NIMP=-2                                                        DSL13740
      225 KOUT(NOUT)=LOC                                                 DSL13750
          NOUT=NOUT+1                                                    DSL13850
          IF(L-1)94,77,81                                               DSL13860
C...  FIND ALL INPUT VARIABLES                                           DSL13870
       81 NINP1=NINP+1                                                   DSL13880
       82 CALL SCAN                                                      DSL13890
          L=3                                                            DSL13900
C...  BRANCH ON NUMERIC,OPERATOR,=,(,GT 5 CHAR.,EOR,CONTINUE            DSL13905
          GO TO(94,84,94,85,94,89,103),KEY                              DSL13910
      103 ICONT=1                                                        DSL13920
```

380

```
            GO TO 5                                              DSL13930
     84 IF(NINP1-NINP)138,138,106                                DSL13960
    138 DO 104  M=NINP1,NINP                                     DSL13970
            IF(INPUT(M)-LOC)104,82,104                           DSL13980
    104 CONTINUE                                                 DSL13990
            IF(NINP-300)106,100,100                              DSL14000
    106 IF(NIMP)228,230,228                                      DSL14010
    228 IF(LOC-IMP2)230,229,230                                  DSL14020
    229 LOC=IMP1                                                 DSL14030
    230 NINP=NINP+1                                              DSL14040
            INPUT(NINP)=LOC                                      DSL14050
            GO TO 82                                             DSL14060
C...    TEST IF INTEGRAL BLOCK (INTGR,REALP,MODIN,TRNFR,CMPXP,LEDLG) DSL14070
     85 IF(LOC-7)82,141,148                                      DSL14090
    148 IF(LOC-12)141,145,180                                    DSL14100
    141 IF(JINTG)266,266,82                                      DSL14103
    266 IF(NINT)143,143,147                                      DSL14105
    147 IF(LINT(NINT)-NOUT1)240,142,143                          DSL14110
    240 IF(LINT(NINT)+NOUT1)143,244,143                          DSL14120
    143 ICH=ISTRT-6                                              DSL14130
    246 ICH=ICH-1                                                DSL14140
            IF(LETRS(ICH)-64)103,246,248                         DSL14150
    248 IF(LETRS(ICH)-126)249,247,246                            DSL14155
    249 IF(LETRS(ICH)-123)241,247,241                            DSL14160
    241 NIN(NOUT1)=NINP-NINP1+1                                  DSL14170
            LIN(NOUT1)=NINP1                                     DSL14180
            NOUT1=NOUT                                           DSL14190
            NOUT=NOUT+1                                          DSL14200
            NINP1=NINP+1                                         DSL14210
            GO TO 245                                            DSL14220
    244 LINT(NINT)=0                                             DSL14230
    245 LINT(NINT+1)=-NOUT1                                      DSL14240
            GO TO 107                                            DSL14250
    142 LINT(NINT)=0                                             DSL14260
    247 LINT(NINT+1)=NOUT1                                       DSL14270
    107 NINT=NINT+1                                              DSL14280
            IF(LOC-10)146,144,145                                DSL14290
    144 CALL NUMRC                                               DSL14300
            CALL NUMRC                                           DSL14310
            NINT=NINT+NMBR-1                                     DSL14320
    146 IF(NINT-50)82,82,100                                     DSL14330
    145 NINT=NINT+1                                              DSL14340
            LINT(NINT)=NOUT1+1                                   DSL14350
            GO TO 146                                            DSL14360
C...    TEST IF SYSTEM MEMORY BLOCK, DELAY BLOCK,                DSL14370
C...    OR IMPLICIT FUNCTION (IMPL, IMPLC)                       DSL14380
    180 DO 181  I=1,NMEM                                         DSL14390
            IF(MEMRY(I)-LOC)181,183,181                          DSL14400
    181 CONTINUE                                                 DSL14410
            GO TO 82                                             DSL14420
    183 IF(NFORT(NOUT1))255,256,256                              DSL14422
    255 NFORT(NOUT1)=NFORT(NOUT1)-2                              DSL14424
            GO TO 257                                            DSL14426
    256 NFORT(NOUT1)=NFORT(NOUT1)+2                              DSL14430
    257 NF1=(ISTRT-1)/2                                          DSL14440
            NF2=NFORA+NF1+MEMR-1                                 DSL14450
            MEMR=MEMR+2                                          DSL14455
            NF3=NFOR-1                                           DSL14460
```

```
          NFOR=NFOR+2                                          DSL14470
   184 KFORT(NF3+2)=KFORT(NF3)                                 DSL14480
          NF3=NF3-1                                            DSL14490
          IF(NF3-NF2)185,185,184                               DSL14500
   185 IF(ISTRT-NF1-NF1-2)179,178,100                          DSL14510
   178 KA(5)=-NF2                                              DSL14520
C...   TEST IF PAREN FROM 026 OR 029                           DSL14530
          IF(KFORT(NF2+3)-20000)281,280,280                    DSL14540
   280 KFORT(NF2+3)=KFORT(NF2+3)-256                           DSL14550
          GO TO 188                                            DSL14560
   281 KFORT(NF2+3)=KFORT(NF2+3)+7680                          DSL14570
          GO TO 188                                            DSL14580
   179 KA(5)=NF2                                               DSL14590
C...   TEST IF DELAY BLOCK                                     DSL14598
   188 IF(I-10)187,186,187                                     DSL14600
   186 CALL NUMRC                                              DSL14610
          KMEM(I)=NMBR+NMBR+5                                  DSL14620
   187 MAXN=MAXN-KMEM(I)                                       DSL14630
          NMBR=MAXN+1                                          DSL14640
          CALL BNDEC                                           DSL14650
C...   TEST IF IMPLICIT FUNCTION BLOCK                         DSL14658
          IF(I-8)82,88,189                                     DSL14660
   189 IF(I-9)82,88,82                                         DSL14670
    88 KFORT(NFORA)=IRES(138)                                  DSL14680
          KFORT(NFORA+1)=IRES(139)                             DSL14690
          DO 250  KR=106,131                                   DSL14700
          KFORT(NFOR)=IRES(KR)                                 DSL14710
   250 NFOR=NFOR+1                                             DSL14720
          NFORT(NOUT1)=-NFORT(NOUT1)-26                        DSL14730
          IMP1=IMP1+1                                          DSL14740
          IMP2=KOUT(NOUT1)                                     DSL14750
          KOUT(NOUT1)=IMP1                                     DSL14760
          NIMP=-1                                              DSL14770
          GO TO 82                                             DSL14780
    89 IF(NIMP)267,270,270                                     DSL14790
   267 IF(NIMP+2)94,268,269                                    DSL14800
   268 NIMP=0                                                  DSL14810
          IF(NFORT(NOUT1))223,224,224                          DSL14820
   223 NFOR=NFOR-1                                             DSL14830
          NFORT(NOUT1)=NFORT(NOUT1)-17                         DSL14840
          GO TO 226                                            DSL14850
   224 NFORT(NOUT1)=-NFORT(NOUT1)-18                           DSL14860
   226 DO 222  KR=131,148                                      DSL14870
          KFORT(NFOR)=IRES(KR)                                 DSL14880
   222 NFOR=NFOR+1                                             DSL14890
          IRES(125)=IRES(125)+1                                DSL14900
          IRES(139)=IRES(139)+1                                DSL14910
          IRES(142)=IRES(142)+1                                DSL14920
          IRES(116)=IRES(142)                                  DSL14930
          IRES(119)=IRES(142)                                  DSL14940
          IRES(122)=IRES(125)                                  DSL14950
          GO TO 270                                            DSL14960
   269 NIMP=1                                                  DSL14970
          IMP3=INPUT(NINP)                                     DSL14980
          INPUT(NINP)=0                                        DSL14990
   270 IF(IPRO)4,97,98                                         DSL15000
    97 NIN(NOUT1)=NINP-NINP1+1                                 DSL15010
          LIN(NOUT1)=NINP1                                     DSL15020
```

382

```
         GO TO 4                                              DSL15030
  98 IPRO=-1                                                  DSL15040
         GO TO 4                                              DSL15050
 100 NALRM=10                                                 DSL15060
         NSORT=0                                              DSL15070
         IERR=1                                               DSL15080
         CALL ERR1                                            DSL15090
         GO TO 50                                             DSL15100
  99 CALL   ERR1                                              DSL15110
         GO TO 4                                              DSL15120
  50 IF(KDUMP)55,56,55                                        DSL15130
  55 CALL TRDMP
  56 CONTINUE
         CALL SORTX
         CALL OUTPX
1000 EXIT
         END                                                  DSL15170
/*
         ACTION NOAUTO,NOMAP
         PHASE TRANS,*
         INCLUDE
         ENTRY
/*
C $UDDLT,MA,,SYSF00,$WORK
R KEEP
R LNKOUT,$SYPR
R LNKIN,$SYCD
R LNKPCH,,DSLS79,DSLLIB
R LNKOBJ,,DSLS79,DSLLIB
R LNKLB1,,DSLS79,DSLLIB
*
*      YOUR FORTRAN SUBROUTINE LIBRARY SHOULD BE IN $FORTSUB ON S7FORT
*      CHANGE THE FOLLOWING CARD IF DIFFERENT
*
R LNKLB2,,S7FORT,$FORTSUB
R SYS001,,SYSF00,$WORK,SYSUT1,(NEW,DLT,1200)
R SYS002,,SYSF00,$WORK,SYSUT2,(NEW,DLT,120)
C $LE
         ACTION AUTO,MAP
         PHASE X,ROOT
         INCLUDE $FWA
         PHASE TRANSX,*
         INCLUDE DSLMAIN
     INCLUDE TRANS
         INCLUDE SORTX
         INCLUDE OUTPX
         INCLUDE TRDMP
         ENTRY DSLMAIN
/*
R KEEP
R SYSIN,,DSLS79,DSLLIB,TRANSX
R SYSOUT,,DSLS79,DSLLIB
C $UDFMT,FM,TRANSX,A,DSL,M,,DL
/*
```

```
*************************************************************************
*                                                                       *
*       THE FOLLOWING IS THE SET OF CONTROL CARDS REQUIRED FOR EXECUTION OF DSL *
*       NOTE THAT IF YOU HAVE DIFFERENT DEVICES FOR INPUT/OUTPUT YOU CAN  *
*       ALSO CHANGE THEM PRIOR TO CALLING DSL. REPLACE $SYCD AND $SYPR   *
*       APPROPRIATE DEVICES WHICH YOU HAVE.                              *
*                                                                       *
*************************************************************************
R KEEP
R @00001,$SYKB
R @00002,$SYKB
R @00003,,DSLS79,DSLLIB,NEW3
R @00004,,DSLS79,DSLLIB,NEW4
R @00005,$SYCD
R @00006,$SYPR
R JOBLIB,,DSLS79,DSLLIB
CALL DSL
TITLE DSL TEST ON SYSTEM/7
PARAM A=0.,B=1.,C=2.
INTEG MILNE
      D=INTGR(A,E)
      E=B**2-SIN(C**.3)
      F=D+C
PARAM DELT=.1
PARAM FINTI=10.
PRINT 1.,A,B,C,D,E,F
END
STOP
```

384

APPENDIX

D REFERENCES

1. Acrivos, A., M. J. Shah, and E. E. Petersen, *AICHE Journal, 6,* 1960, pp. 312–317.

2. Brennan, R. D., and H. Sano, "PACTOLUS—A Digital Simulator Program for the IBM 1620," *1964 Fall Joint Computer Conference, AFIPS Conference Proceedings, 26,* Oct. 1964.

3. Byrne, E. R., "JANIS—A Program for a General Purpose Digital Computer to Perform Analog Type Simulations," conference paper, ACM National Conference, Denver, Aug. 1963.

4. Chambers, R. P., "Random-Number Generation," *IEEE Spectrum,* Feb. 1967.

5. Dost, M. H., "Simulation Languages, Ideal Analysis Tools for the Control Engineer," *IFAC Symposium,* Budapest, Sept. 6, 1971.

6. Dost, M. H., and R. R. Barker, *Simulation,* Nov. 1967, pp. 237–247.

7. Gaskill, R. A., J. W. Harris, and A. L. McKnight, "DAS—A Digital Analog Simulator," *1963 Spring Joint Computer Conference, AFIPS Conference Proceedings, 23,* 1963.

8. Harnet, R. T., and J. F. Sansom, "MIDAS Programming Guide," *Report No. SEG-TDR-64-1,* Analog Computation Division, Systems Engineering Group, Research and Technology Division, Air Force Systems Command, Wright-Patterson Air Force Base, Dayton, Ohio, Jan. 1964.

9. Harnett, R. T., R. J. Sansom, and L. M. Warshawsky, "MIDAS...An Analog Approach to Digital Computation," *Simulation, 3,* No. 3, Sept. 1964.

10. Hurley, J. R., and J. J. Skiles, "DYSAC—A Digitally Simulated Analog Computer," *1963 Spring Joint Computer Conference, AFIPS Conference Proceedings, 23,* 1963.

11. *IBM 1130 Continuous System Modeling Program, Program Description and Operations Manual,* SH20-0905, IBM Program Information Department, Hawthorne, N.Y.

12. *IBM 1130 Distributed System Programs, General Information Manual,* GH20-0800, IBM Distribution Center, Mechanicsburg, Pa. 17055.

13. *IBM 1130 FORTRAN Language,* Form C26-5933, IBM Distribution Center, Mechanicsburg, Pa. 17055.

14. *IBM 1130/1800 Plotter Subroutines,* Form C26-3755, IBM Distribution Center, Mechanicsburg, Pa. 17055.

15. *IBM Application Description Manual, 360 Continuous Systems Modeling Program* (program nos. 360A-CX-16X and 5734-X59), Form H20-0240, IBM Distribution Center, Mechanicsburg, Pa. 17055.

16. *IBM Application Description Manual, 1130 CSMP,* Form H20-0209, IBM Distribution Center, Mechanicsburg, Pa. 17055.

17. *IBM DSL/1800, Program for the Simulation of Continuous Process Dynamics,* Program no. 1800-15.1-001, IBM Program Information Department, Hawthorne, N. Y.

18. *IBM System/3 Disk FORTRAN IV Reference Manual,* SC28-6874, IBM Distribution Center, Mechanicsburg, Pa. 17055.

19. *IBM System/3 Model 10 Disk System, Operator's Guide GC21-7508,* IBM Distribution Center, Mechanicsburg, Pa. 17055.

20. *IBM System/7 FORTRAN IV Language,* GC28-6876, IBM Distribution Center, Mechanicsburg, Pa. 17055.

21. *IBM System/7: Combining FORTRAN IV and MSP/7: A Programmer's Reference Guide,* SC34-0025, IBM Distribution Center, Mechanicsburg, Pa. 17055.

22. Lapidus, L., *Digital Computation for Chemical Engineers,* McGraw-Hill, New York, 1962.

23. Levine, L., "A New Digital Computer for Solving Differential Equations," *Simulation, 4,* No. 3, Apr. 1965 (DES-1).

24. Milne, W. E., and R. R. Reynolds, "Fifth-Order Methods for the Numerical Solution of Ordinary Differential Equations," *ACM Journal,* Jan. 1962.

25. O'Brien, J. P., "Note on Decomposition in First Order of Multi-Order Linear Differential Equations with Constant Coefficients," *Computer Journal, 2,* No. 3, July 1959, p. 144.

26. Rideout, V. C., and L. Tavernini, "MADBLOCK—A Program for Digital Simulation of a Hybrid Computer," *Simulation, 4,* No. 1, Jan. 1965.

27. Samsom, F. J., H. E. Peterson, and L. M. Warshawsky, "MIMIC—A Digital Simulator Program," *SESCA Internal Memo 62,* System Engineering Group, Wright-Patterson Air Force Base, Dayton, Ohio, 1965.

28. Schubert, R., "Single Precision Floating Point Runge-Kutta Integration," *IBM SHARE Library #BJ11B*, Aerospace Corp., San Bernadino Operation.

29. Selfridge, R. G., "Coding a General-Purpose Digital Computer To Operate as a Differential Analyzer," *Proceedings, 1955 Western Joint Computer Conference (IRE)*, 1955.

30. Shah, M. J., *Industrial and Engineering Chemistry, 59*, No. 1, 1967, pp. 72–83.

31. Shah, M. J., *Industrial and Engineering Chemistry, 59*, No. 4, 1967, pp. 70–85.

32. Stein, M. L., and J. Rose, "Changing from Analog to Digital Programming by Digital Techniques," *Journal of the Association for Computing Machinery*, Jan. 1960.

33. Stein, M. L., J. Rose, and D. B. Parker, "A Compiler with an Analog-Oriented Input Language," *Proceedings of the 1959 Western Joint Computer Conference*, pp. 92–102.

34. Stover, R. F., and H. A. Knudtson, "H-800 PARTNER—Proof of Analog Results Through a Numerically Equivalent Routine," *Doc. No. U-ED 15002*, Aerospace Division, Honeywell Corp., 1962.

35. Strauss, J. C., "Committee Report," Ad Hoc Committee on Simulation Software, Western Simulation Council, Nov. 1965.

36. Strauss, J. C., and W. L. Gilbert, "SCADS: A Programming System for the Simulation of Combined Analog Digital Systems," Carnegie Institute of Technology, March 1964.

37. Strauss, J. C., et al. of SCi Simulation Software Committee, "The SCi Continuous System Simulation Language (CSSL)," *Simulation, 9*, No. 6, Dec. 1967, pp. 281–303.

38. Syn, W. M., and R. N. Linebarger, "DSL/90—A Digital Simulation Program for Continuous System Modeling," Spring Joint Computer Conference Proceedings, Boston, Apr. 1966.

39. Syn, W. M., and D. G. Wyman, "DSL/90 Digital Simulation Language User's Guide," *IBM Technical Report No. TR 02,355*, IBM, San Jose, Calif. (also *SHARE No. IW DSL 3358*).

40. Warten, W. R., and M. E. Fowler, "A Numerical Integration Technique for Ordinary Differential Equations with Widely Separated Eigenvalues," *IBM Journal, 11*, No. 9, Sept. 1967.

41. Wegstein, J. H., "Accelerating Convergence of Iterative Processes," National Bureau of Standards Report, Washington, D.C.

INDEX